R. Bargagli

Antarctic Ecosystems

Environmental Contamination, Climate Change, and Human Impact

With 50 Figures and 18 Tables

Springer

Professor Dr. Roberto Bargagli
University of Siena
Department of Environmental Sciences
Via P.A. Mattioli, 4
53100 Siena
Italy

email: bargagli@unisi.it

Cover illustration: A view of an iceberg in the Ross Sea from an ice-free area on Prior Island (northern Victoria Land, Antarctica)

ISSN 0070-8356
ISBN 978-3-540-74005-6 Springer-Verlag Berlin Heidelberg New York

Library of Congress Control Number: 2007932172

Springer-Verlag is a part of Springer Science+Business Media
springer.com

© Springer-Verlag Berlin Heidelberg 2005, 2008

Editor: Dr. Dieter Czeschlik, Heidelberg, Germany
Desk editor: Dr. Andrea Schlitzberger, Heidelberg, Germany

To Guido and Pietro,
the grandchildren born when I was writing this book,
wishing them a world where man will be reconciled with the environment

Preface

The picture of Antarctica as the remotest continent and the symbol of the last great wilderness and pristine environment on Earth has changed considerably in the last two decades. Environmental problems such as the recurring appearance of the "ozone hole" and the break-up of Antarctic Peninsula ice shelves have shown that Antarctica is inextricably linked to global atmospheric, oceanographic and climatic processes, and is therefore exposed to the impact of human activities in the rest of the world. Possible effects of global warming on the stability of ice sheets and the consequent rise in sea level are stimulating interest in this continent. There is also an increased awareness that near-pristine Antarctic ice, marine sediments and biota are archives of climatic and evolutionary history, providing essential data for a global baseline against which to monitor global changes. Although Antarctica is perceived as the last unspoiled region of the Earth and as a symbol of global conservation, its environment is not pristine, especially near scientific stations or areas affected by accidental oil spills. The application of the Protocol on Environmental Protection to the Antarctic Treaty is helping to significantly reduce the impact of human activities in Antarctica and the Southern Ocean. However, most persistent contaminants in the Antarctic environment originate from anthropogenic sources in the Southern Hemisphere and the rest of the world, rather than from local sources.

The Scientific Committee on Antarctic Research (SCAR) favours the implementation of coordinated national and international Antarctic research programmes on global-scale processes. However, the rapid development of analytical techniques and remote-sensing technology requires continuous updating of results which are dispersed in many journals. This book presents an overview of available data on environmental contaminants in Antarctica in order to evaluate whether current levels represent (on a local scale) a threat to the sustainability of ecosystems and whether climate change and probable future increases in atmospheric contaminant inputs from other continents of the Southern Hemisphere may compromise the scientific value of the near-pristine Antarctic environment. The idea for this overview and the discussion of possible interactions between climate change and spatio-temporal patterns of contaminant deposition in Antarctica arose during the preparation of two reviews of trace metal distribution in Antarctic ecosystems (Bargagli 2000, 2001; Review of Environmental Contamination and Toxicology 166:129–173; 171:53–110). I myself was surprised by the vast amount of literature data on concentrations of persistent contaminants in a number of environmental matrices, by the adoption of unreliable sampling and analytical procedures, and by widespread misconceptions about the functioning of Antarctic ecosystems and/or the sensitivity of Antarctic organisms to heavy metals. Although it has been known since the 1990s that concentrations of Cd, Hg and other potentially toxic elements in some species of Antarctic organisms are naturally higher than in related species from polluted environments of the Northern Hemisphere, some researchers still use Antarctic data on Cd and other elements as global reference values.

In contrast to multi-author books, which often examine a few issues in great depth, this book aims to provide an overview of the Antarctic environment, the functioning of terrestrial and aquatic ecosystems, and the occurrence and cycling of persistent contaminants. Attempts were made to interpret possible interactions between predicted climate change and pathways of persistent atmospheric contaminants from anthropogenic sources in the Southern Hemisphere, and to provide suggestions for large-scale, long-term environmental monitoring of different regions of Antarctica and the Southern Ocean. It is very difficult to give an overview of interactions between climatic and environmental factors, and of the possible impact of climate changes and human activities in Antarctica and the rest of the world on Antarctic ecosystems. In recent years there has been an extraordinary development of models on future trends in air temperature, atmospheric precipitation, sea-ice cover and deposition patterns of a growing number of persistent contaminants in polar regions. The publication of this book in *Ecological Studies* continues the series initiated by Beyer and Bölter (eds) with *Geoecology of Antarctic Ice-Free Coastal Landscapes*, and is indicative of the general, growing interest in Antarctic ecosystems. Although this book reflects the state of the art at the time of writing, many new data will likely become available in the near future.

A real effort has been made to pass on the wisdom gained from personal research experience in the Ross Sea and Victoria Land environment, probably the regions most cited in this book. The book spans diverse topics such as climatology, meteorology, glaciology, atmospheric chemistry, oceanography, pedology, hydrology, environmental biogeochemistry, and marine and terrestrial biology and ecology. Some areas of research and interest probably received only limited attention. However, an extensive bibliography is included at the end of the book for readers wishing to pursue issues of interest in greater depth. In addition to any deficiencies, there may be some errors and misconceptions – I would be grateful to colleagues for constructive criticism.

The book was organised in relatively self-contained sections but, due to the interdisciplinary nature of the subject and to interrelations between topics, it was necessary to cross-reference chapters, figures and tables; some repetitions could not be avoided. The first three chapters are devoted to general aspects of the environment in Antarctica and the Southern Ocean – climate trends, glacial systems, and the structure and functioning of terrestrial, freshwater, and marine ecosystems. Early responses of Antarctic ecosystems to climate change and the possible effects of enhanced UV-B radiation on phytoplankton and primary productivity were emphasised. Chapter 4 deals with persistent atmospheric contaminants and their sources in Antarctica and elsewhere in the Southern Hemisphere. Chapter 5 discusses the deposition of atmospheric contaminants and their incorporation into ice, reports available data on chemicals in snow, ice, soils, lakes, and cryptogamic organisms, and suggests some approaches for monitoring atmospheric contaminants around scientific stations. Chapter 6 addresses the chemical composition and biogeochemical cycle of trace elements in Southern Ocean seawater, and discusses local environmental pollution in marine coastal ecosystems near scientific stations, disused whaling stations or in areas affected by accidental oil spills, and effects of pollutants on benthic communities. Chapter 7 gives a comprehensive account of the accumulation of trace metals and persistent organic compounds in Antarctic marine organisms, the transfer of contaminants in pelagic and neritic food chains, and discusses the potential role of the most widespread species of organisms as biomonitors of contaminants and environmental changes. The last chapter discusses the potential impact of human activities in the Southern Hemisphere on the Antarctic climate and environment, and suggests the development of long-term circum-Antarctic monitoring networks. The book concludes with a discussion on the possible role of Antarctic research in implementing the protection of the global environment.

Overall, I would be delighted if this book can serve as a reference for researchers interested in local or large-scale monitoring surveys, management of the Antarctic environment, in climate change and its effects on polar regions. The description of Antarctic ecosystems, their structure and functioning may serve terrestrial and marine ecologists, and conservationists. As the problem of environmental contamination in Antarctic ecosystems is intermeshed with global processes, the book may also be useful to students of atmosphere physics and chemistry, oceanography, and glaciology.

I am grateful to Fabrizio Monaci for preparing all the illustrations, Arabella Palladino for the revision of the English, Otto Lange for constructive review of the manuscript, several colleagues for suggestions, and to co-workers in the Department of Environmental Sciences and the National Antarctic Museum

Felice Ippolito at the University of Siena for their help in collecting and archiving the literature. I am particularly grateful to my wife and entire family for their understanding and patience which enabled me to devote many weekends and the summer holidays to writing this book.

Siena, September 2004   *Roberto Bargagli*

Contents

List of Acronyms and Abbreviations

ACC	Antarctic Circumpolar Current
AChE	Acethylcholinesterase
AEON	Antarctic Environmental Officers Network
ALH84001	Mars meteorite found on December 1984 at Allan Hills, Victoria Land
AMAP	Arctic Monitoring and Assessment Programme
AMIEREZ	Antarctic Marine Ice Ecosystem Research at the Ice Edge Zone
ANARE	Australian National Antarctic Research Expedition
ANL	Argonne National Laboratory
APF	Antarctic Polar Front
ASMA	Antarctic Special Managed Area
ASPA	Antarctic Specially Protected Area
ATCPs	Antarctic Treaty Consultative Parties
ATS	Antarctic Treaty System
BAS	British Antarctic Survey
BIOMASS	Biological Investigations of Antarctic Systems and Stocks
BIOTAS	Biological Investigations of Terrestrial Antarctic System
BROD	Benzyloxyresurfin-O-deethylase
CCAMLR	Commission for Conservation of Antarctic Marine Living Resources
CEC	Cation Exchange Capacity
CEMP	CCAMLR Ecosystem Monitoring Programme
CFCs	Chlorofluorocarbons
CHLS	the sum of cis-chlordane, $trans$-chlordane, cis-nanochlor, $trans$-nanochlor
COMNAP	Council of Managers of National Antarctic Programme
CRAMRA	Convention on the Regulation of Antarctic Mineral Resources Activities
CS-EASIZ	Coast and Shelf-Ecology of the Antarctic Sea-Ice Zone
DDT	Dichlorodiphenyltrichloroethane
DFs	Dibenzofurans

DMHg	Dimethylmercury
DMS	Dimethylsulphide
DMSP	Dimethylsulphonioproprionate
DPASV	Differential Pulse Anodic Stripping Voltammetry
ENSO	El Niño–Southern Oscillation Phenomenon
EPA	Environmental Protection Agency
EPICA	European Project for Ice Coring in Antarctica
EROD	Ethoxyresurfin-O-deethylase activity
GCTE	Global Change and Terrestrial Ecosystems
GLOBEC	Global Ocean Ecosystem Dynamics Research
GLOCHANT	Global Change and the Antarctic
GWPs	Global Warming Potentials
HCB	Hexachlorobenzene
HCHs	Hexachlorocyclohexanes
HNLC	High Nutrient Low Chlorophyll (waters)
ICP-MS	Inductively Coupled Plasma-Mass Spectrometry
IGAC	International Global Atmospheric Chemistry
IGBP	International Geosphere–Biosphere Program
IGY	International Geophysical Year
IPCC	Intergovernmental Panel on Climate Change
ITASE	International Trans-Antarctic Scientific Expedition
IUCN	International Union for the Conservation of Nature
IWC	International Whaling Commission
JGOFS	Joint Global Ocean Flux Study
LEAFS	Laser Excited Atomic Fluorescence Spectrometry
LTER	Long-Term Ecological Research
MAA	Mycosporine-Like Amino Acids
MeHg	Methylmercury
NAS	National Academy of Science
NASA	National Aeronautical and Space Administration
NGO	Non-Governmental Organization
NMHCs	Non-Methane Hydrocarbons
NOAA	National Oceanic and Atmospheric Administration
NSF	National Science Foundation
PAGES	Past Global Environmental Changes
PAHs	Polycyclic Aromatic Hydrocarbons
PASC	Polar Atmospheric Snow Chemistry
PAR	Photosynthetic Active Radiation
PCBs	Polychlorinated Biphenyls
PCDDs	Polychlorinated Dibenzo-p-dioxins
PCDFs	Polychlorinated Dibenzofurans
PCFs	Perfluorcarbons
PCTs	Polychlorinated Terpenyls
PNRA	Italian National Antarctic Research Programme

POPs	Persistent Organic Pollutants
RiSCC	Regional Sensitivity to Climate Change
ROS	Reactive Oxygen Species
SCAR	Scientific Committee on Antarctic Research
SDS	Sodium Dodecyl Sulphate
SRES	Special Report on Emission Scenarios (commissioned by the IPCC)
UCDW	Upper Circumpolar Deep Water
UKMO	United Kingdom Meteorological Office
UNEP	United Nations Environmental Programme
USAP	United States Antarctic Program
UV	Ultraviolet
UV-B	Ultraviolet B radiation
VAI	Volcanic Aerosol Index
VOCs	Volatile Organic Compounds
WG-PACA	Working Group on Physics and Chemistry of the Atmosphere
WHO	World Health Organization
WMO	World Meteorological Organization

1 Antarctica: Geomorphology and Climate Trends

1.1 Introduction

Several regions of the Earth, such as Lake Baikal and the Himalayas, are usually described through several superlatives, but none can compete with Antarctica: the remotest, coldest, windiest, highest continent, with the biggest and thickest ice sheet. The ice moves towards the sea and calves the world's largest icebergs in the Southern Ocean, which has the deepest continental shelf, the largest wind-driven oceanic current, the highest number of endemic species and the largest seasonal variation in ice cover. Antarctica is a unique continent: it contains almost 80 % of the world's freshwater, yet it is the largest cold desert on Earth. Although it receives much more solar radiation in the summer than anywhere else in the world, it is the coldest place on Earth. In contrast with other continents, Antarctica is not located in plates with constructive and destructive margins; during the last 100 Ma, it has thus occupied a quite stable position with respect to the South Pole. Climatic changes in Antarctica during this period are therefore mainly due to global changes.

Through a better understanding of these unique features, it will become obvious to the reader why several research activities cannot be performed at more convenient locations and why this remote, cold and forbidding place, where field research is very difficult and expensive, has become a continent for science. The ice sheet, which deposited over thousands or millions of years, preserves a record of changes of atmospheric composition and climate, and the collection of meteorites in the ice ablation areas provides clues about the history of the solar system. The elevation of the continent, its dry, cold, clean atmosphere, and geomagnetic latitude allow unique astronomical and astrophysical observation and investigation of Earth's magnetosphere and ionosphere.

In spite of its remoteness, Antarctica is linked to lower latitudes through the circulation of the atmosphere and oceans. The large equator-to-pole temperature difference drives the poleward transport of heat and determines the general circulation of the atmosphere, making Antarctica the main heat sink of the Southern Hemisphere. The continuous low-level drainage of the continental surface by katabatic winds is compensated by the inflow of relatively warm air masses which converge and subside in the troposphere over Antarctica. The flow of cold air to the ocean in the shallow boundary layer, coupled with the tropospheric and stratospheric circulation, gives Antarctica a major role in the global climate system and makes it a sink for persistent atmospheric pollutants.

Global climate models predict that the greatest changes will occur at high latitudes. Feedback mechanisms might easily magnify relatively small changes in sea-ice extent and ice-sheet balance, and these changes are likely to be of global importance. Only Antarctica can provide essential data for better understanding these processes and the response of ecosystems to climatic and environmental change. Without Antarctic data, global models would not be able to accurately predict climate change and the impact of persistent airborne pollutants.

This chapter outlines the continental features (morphology, geology, climate) and involvement of Antarctica in global climate processes, with particular emphasis on its important role in establishing global baselines against which to monitor climate change and the impact of human activity. Given its potential contribution to the global increase in sea levels, the stability of Antarctic ice is of general concern and interest. The chapter also reviews available data on climatic variability and change in Antarctica, and estimates of how the Antarctic climate may respond to increasing concentrations of greenhouse gases.

In the last two decades, following the discovery of the recurring formation of the ozone "hole" in Antarctica, the possible effects of global warming on the stability of ice sheets, increasing sea levels and global environmental change, many books have been published on the climate, geography, geology, glaciology, environment and resources of Antarctica. This chapter will only briefly review these topics, with particular emphasis on climatic and atmospheric processes affecting the transport and deposition of persistent environmental pollutants. The reader wishing to further pursue issues of interest in greater depth can refer to specific books and papers quoted in the bibliography.

1.2 Physical Characteristics

The word Antarctic originates from the Greek name of the polar constellation (*arktos*, the bear) and indicates the region which lies opposite to it (anti-Arctic or Antarctic). The terms Antarctic and Antarctica are often used interchangeably, but it seems more proper to use the first to denote the region (i.e. the area of the Earth south of 60° S, which includes the continent, isolated islands and a large part of the Southern Ocean) and the second for the continent itself (Hansom and Gordon 1998). In spite of the theoretical hypothesis of ancient Greeks and Romans (the Latin geographer Pomponio Mela envisaged a southern continent, *Antipodi*, inhabited by the *Antictoni*), Antarctica was omitted in many geographical maps until the 18th and 19th centuries, when the seas around the "*Terra Australis nondum cognita*" became of interest for the sealing and whaling industries. Although parts of the coast and interior began to appear in some maps at this time, the cartography of the Antarctic region was only completed in the last century (e.g. Sugden 1982; Simpson-Housley 1992; Chaturvedi 1996).

Except for the northern part of the Antarctic Peninsula (Fig. 1), the continent lies entirely within the Antarctic Circle (i.e. the parallel at 66° 33′S, corresponding to the angle between the Earth's rotation axis and the plane of its orbit round the Sun). The continent, along with its islands and ice shelves, covers about 13.66×10^6 km^2, representing about 10% of the world land surface and 30% of that in the Southern Hemisphere. Excluding ice

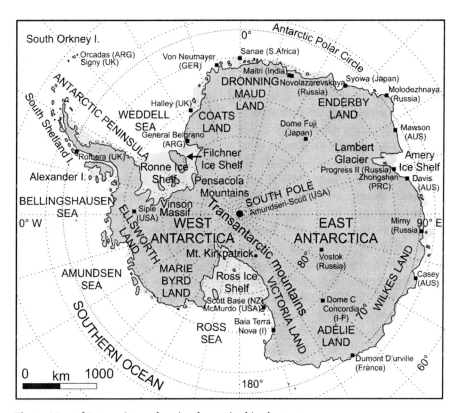

Fig. 1. Map of Antarctica and main places cited in the text

shelves, the land surface of the continent and islands covers 12.09×10^6 km², with the continent constituting more than 98% of this area. The continental landmass under the ice sheet consists of two main units (East and West Antarctica) separated by the Transantarctic Mountains, which extend across the continent from the Ross Sea to the Weddell Sea. East Antarctica, or Greater Antarctica, comprises the Transantarctic Mountains and the large area extending from these mountains to the Indian Ocean (more than 10×10^6 km²); it has an approximately circular, symmetric shape, with the coastline following the 62° S line of latitude, except at the Lambert Glacier–Amery Ice Shelf indentation (Fig. 1). The Transantarctic Mountains are a large mountain range, stretching for 3,500 km from Victoria Land to the Pensacola Mountains; Mt. Kirkpatrick (4,528 m) is the highest peak. In general, East Antarctica is characterised by narrow coastal strips and steep slopes rising sharply to the high Antarctic plateau. This huge mass of ice (about 28.5×10^6 km³; more than 80% of the world's freshwater) has an average elevation of about 2,300 m and, although its surface appears rather flat, in some zones the ice sheet may rise to altitudes of more than 4,000 m (4,776 m in Adélie Land, at 69° 54'S, 135° 12'E). The bedrock lies mostly close to or below sea level, with a maximum depression of –2,555 m in the subglacial Bentley Basin (81° S, 110° W).

West Antarctica, or Lesser Antarctica (surface area with islands and ice shelves of about 3.42×10^6 km²), lies mostly to the west of longitude 165 to 315° E and comprises Marie Byrd Land, Ellsworth Land and the Antarctic Peninsula, a bedrock archipelago straggling 1,200 km northwards. Although the average elevation of West Antarctica is about 850 m and its ice sheet is generally lower than that of its eastern neighbour, several summits lie above 3,000 m, including the Vinson Massif (4,897 m), the highest peak in Antarctica. The Antarctic Peninsula constitutes a narrow north–south mountain barrier with an average width of 70 km and a mean height of 1,500 m. This barrier affects atmospheric circulation and contributes to determine markedly different climatic conditions between the west coast, facing the Bellingshausen Sea, and the east coast in the Weddell Sea.

From the interior, where it builds up, the ice flows down in ice streams at speeds of about 500 m year^{-1}. Ice streams at the edge of the continent may form large floating ice shelves, such as the Ross and Ronne ice shelves (about 0.5×10^6 km² each; Fig. 1), or ice tongues, which break up into widespread tabular icebergs throughout the Southern Ocean. The ice shelves make up more than 40% of the continental coastline, with grounded ice walls, ice streams and outlet glaciers forming the remaining 60% of coastline. Only a small percentage (about 5%) of the coastline is covered by rocky cliffs or beaches (Drewry 1983). Thus, a large portion of the continental shelf is covered by ice shelves, and has undergone extensive glacial erosion, especially during the Pliocene–Pleistocene glacial advances (J.B. Anderson 1991). The removal of bedrock and the huge weight of the ice sheet have helped make the Antarctic continental shelf much deeper (about 500 m) and wider (mean 200 km) than other shelves. At the margin of the shelf, the continental slope falls rather steeply to the ocean basin at depths of 3,000–6,000 m. Mountainous submarine ridges connected to the break-up of the Gondwanaland supercontinent and the tectonic development of Antarctica rise from the ocean floor, encircling the continent.

1.3 Geology and Mineral Resources

1.3.1 Geology

Besides the huge mass of ice, several geological peculiarities, such as negligible seismic activity, the local concentration of meteorites on ice ablation surfaces, and subduction processes which led to the formation of West Antarctica through the aggregation of microcontinents (Ricci et al. 2001), distinguish Antarctica from other continents. In contrast with the other continents which are located in plates with constructive and destructive margins, Antarctica is completely surrounded by the sea and is located in a continuously expanding lithospheric plate. During the last 100 Ma, the continent has therefore occupied a quite stable position with respect to the South Pole. This is an important peculiarity, because it means that climate changes in Antarctica mainly reflect global changes.

In spite of difficulties due to the lack of rock outcrops (there are about 331,000 km^2 of ice-free areas, corresponding to less than 3 % of the continental area), the geology and evolution of Antarctica is becoming quite well known (e.g. Splettstoesser and Dreschoff 1990; LeMasurier and Thomson 1990; Tingey 1991; Thomson et al. 1991; Stump 1995; Ricci 1997; J.B. Anderson 1999). The broad structure of the continent is related to the amalgamation (about 500–550 Ma ago) and break-up (about 150 Ma ago) of Gondwanaland and earlier continents. Antarctica was once the central keystone of Gondwanaland, which also included South America, Africa, Madagascar, Arabia, Ceylon, India, Australia and New Zealand. After the fragmentation of Gondwana, the continental blocks dispersed and Antarctica drifted towards polar latitudes. Due to its central position in the supercontinent, the main geological structures, especially Antarctic orogenic belts such as the Transantarctic Mountains, are an extension of similar structural units in South America, Africa and Australia (Ricci 1991). Much palaeontological and palaeoenvironmental evidence indicates that the Gondwana continental blocks have a common history. The fossil fern *Glossopteris* and herbivorous reptile *Lystrosaurus* are among the organisms which lived on these continents until 180 Ma ago (Crame 1989; Gee 1989; Olivero et al. 1991; Crame 1992). The period of continent drifting from Antarctica is well documented by the age of marine sediments in peri-Antarctic oceanic basins (Ricci et al. 2001).

Three main episodes (Storey 1995) have been identified in the break-up of the Gondwana supercontinent (Fig. 2). Initial rifting led to the formation of a seaway between West (South America and Africa) and East Gondwanaland (Antarctica, Australia, India and New Zealand); in a second stage (about 120–130 Ma ago), South America and Antarctica separated from the African-Indian plate and finally, Australia and New Zealand separated from Antarctica (about 60 Ma ago). In the meantime Antarctica had drifted to the polar position; the separation of the Antarctic Peninsula from South America (about 25–30 Ma ago), with the opening of the Drake Passage, was the final event in the break-up of Gondwanaland. This quite recent separation is testified by remains of the southern beech *Nothofagus* and even of marsupials in Seymour Island (Antarctic Peninsula; Francis 1991). The formation of the Drake Passage led to the isolation of the Antarctic continent, with the establishment

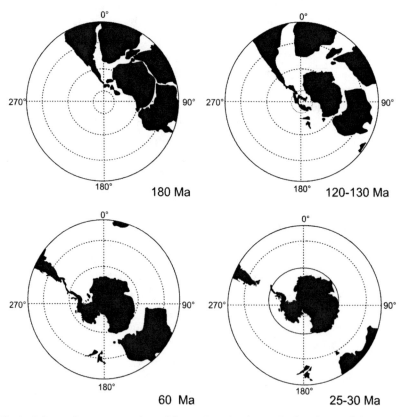

Fig. 2. Schematic representation of the main episodes in the break-up of the Gondwana supercontinent

of westerly circumpolar oceanic circulation and the development of the Antarctic ice cap.

The general tectonic framework of Antarctica is characterised by the stable ancient shield of Precambrian igneous and metamorphic rocks in East Antarctica. This shield consists of several stable cratons, separated by younger mobile belts and flanked on the Pacific side by younger orogenic rocks (Tingey 1991). Rock outcrops in coastal areas of East Antarctica are mostly metamorphosed rocks and subordinate igneous and sedimentary rocks. Analogous Gondwanian sequences, including Devonian to Triassic sedimentary rocks (Beacon Supergroup) and the Jurassic continental Ferrar Supergroup, outcrop in India and in the continents of the Southern Hemisphere. The Beacon sediments contain sandstones, shales and conglomerates with coal-bearing Permian strata, and are characterised by plant, fish and palynomorph assemblages correlating with those of eastern Australia and other parts of Gondwanaland (Truswell 1991). Likewise, the Ferrar Supergroup dolerites provide evidence of a link with similar rocks found in Tasmania, Australia and the Karoo dolerites of South Africa. The Napier Complex in Enderby Land comprises very old rocks (about 3,900 Ma) with mineral associations indicating very high temperatures (about 1,000 ° C; Ricci et al. 2001). In the Transantarctic Mountains, subordinate or sporadic metamorphic rocks of igneous and sedimentary origin form the basement or are incorporated into thick pelagic sequences (mainly Precambrian and Palaeozoic turbidites). These rocks were deformed, metamorphosed and intruded by plutonites in the Cambro-Ordovician, during the Ross Orogeny.

West Antarctica consists of several microplates sharing a common history with South America. This history can be referred to the Phanerozoic (<600 Ma), with the oldest igneous and metamorphic basement rocks underlying sedimentary and volcanic sequences of probable Palaeozoic and Mesozoic age. Two orogenic belts, both sub-parallel to the Pacific margin, can be traced on the basis of radiometric measurements: the Ellsworth Orogen and Andean Orogen. Proterozoic gneisses and schists, and some igneous and metamorphic rocks occur in Ellsworth Land and Marie Byrd Land. Volcanic and plutonic rocks make up most of the Antarctic Peninsula and the southernmost Andes of South America. Volcanism dating from the Middle Tertiary has continued until recently, and volcanic activity has historically occurred in the South Sandwich Islands and very recently in Deception Island and Bridgeman Island, at the northern end of the Antarctic Peninsula. Volcanic activity elsewhere in Antarctica is associated with the major rifting of West Antarctica during the Cenozoic (LeMasurier and Rex 1991). The rift system is bounded on its poleward side by the Pacific flank of the Transantarctic Mountains, and in West Antarctica by a rift running from the western Ross Sea to Ellsworth Land. On the other side of the rift, the Transantarctic Mountains are the product of rapid uplift which began about 50 Ma ago. Volcanism extends from the Balleny Islands to as far south as Mt. Early, at the head of the Scott Glacier. The

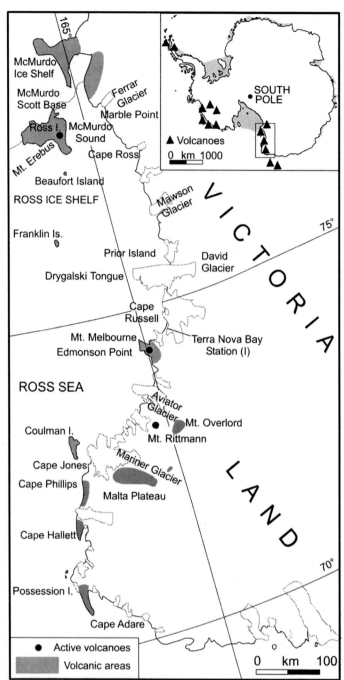

Fig. 3. Schematic map of the main volcanic areas in the McMurdo Sound and northern Victoria Land

Terror Rift in the western Ross Sea extends between the Mount Erebus and Mount Melbourne active volcanoes and volcanic rocks, collectively known as the McMurdo Volcanics (an association of trachytes, olivine basalts, and phonolites; Harrington 1958), occur in the McMurdo Sound region and northern Victoria Land (Fig. 3).

1.3.2 Geochemical Anomalies and Mineral Resources

On the basis of the above-reported arrangement of tectonic plates, it has been hypothesised that zones of Antarctica which were once juxtaposed with the mineral deposits of other continents may contain similar mineral and/or hydrocarbon resources (de Wit 1985). However, most claims of potential mineral wealth are mere speculations because they are based on superficial continental comparisons without adequate knowledge of geological environments and processes (Hansom and Gordon 1998). Comparisons between geological formations and terranes in Antarctica and the rest of the world should consider rocks of the same age, referring to the same continental refit. This is a difficult task because Antarctica was formed by the amalgamation of fragments from supercontinents older than Gondwanaland, and the juxtaposed parts are now widely dispersed in the Southern and Northern Hemispheres. The continental refit is irrelevant not only for Antarctic terranes older than 180–500 Ma (i.e. the age of Gondwanaland) but also in zones, such as the Antarctic Peninsula, which formed more recently. Like other continents, Antarctica has areas with geochemical anomalies and mineral deposits (metals, non-metals, oil, gas, coal) which could be exploited, but their location is largely unknown. Moreover, the Antarctic Treaty System is a unique international agreement, preventing (at least for the near future) the exploitation of mineral deposits.

Several oil companies carried out exploration during the 1970s (Auburn 1982) and claimed the mineral wealth of Antarctica. In 1981, within the framework of the Antarctic Treaty System, a proposal was made to regulate the exploitation of minerals. The Convention on the Regulation of Antarctic Mineral Resource Activities (CRAMRA) was negotiated in 12 sessions plus numerous informal consultations during a six-year period, before being finalised (ATCP 1988). During the negotiations, the possible environmental impact of mining activities, and the sinking and fuel spill of the *Bahia Paraiso* near Anvers Island raised public concern. Nongovernmental organisations contributed to the decision of some countries to stall CRAMRA and support a ban of mining activities in Antarctica. Although the Convention was never ratified, it provided an important basis and stimuli for negotiating the comprehensive protection of the Antarctic environment (Blay 1992; Francioni 1993). In October 1991 the Consultative Parties adopted the Protocol on Environmental Protection to the Antarctic Treaty (the Madrid Protocol; ATCP

1992) which places an indefinite ban on all mineral activities. This ban may be reviewed after 50 years, or before, if there is consensus. In the future, states seeking Consultative Party status will be required to ratify the Madrid Protocol.

The assessment of metal concentrations in biotic and abiotic components of Antarctic ecosystems is one of the aims of this book, and it therefore seems appropriate to mention areas of the continent affected by geochemical anomalies which may enhance natural concentrations of some elements in freshwaters, soils and organisms (Fig. 4). In general, it is believed that the gold, nickel, uranium, copper and iron deposits of Western Australia and those of silver, lead, copper and zinc in central Australia continue in East Antarctica (Wilkes Land and Terre Adélie respectively; Willan et al. 1990). A 400-m-thick exposure of a banded-iron rock formation (iron content about 32 %) has been found on Mt. Ruker (Prince Charles Mountains, between Wilkes Land and Enderby Land) in East Antarctica. Aeromagnetic surveys have shown that the iron anomaly extends westwards for about 180 km (Rowley et al. 1991). Abundant erratic rocks of banded iron occur at Vestfold Hills (Ravich et al. 1982), and other iron-bearing rocks occur in Enderby Land (Neumann Nunatak)

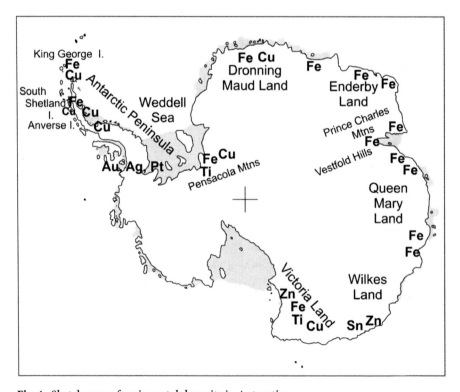

Fig. 4. Sketch map of main metal deposits in Antarctica

and Dronning Maud Land (sometimes associated with low copper and lead contents; Rowley et al. 1983).

The Ferrar Supergroup in the Transantarctic Mountains comprises gabbroic intrusions of the Dufek Massif (Pensacola Mountains), which show iron-titanium oxides and copper and iron sulphides in the exposed top layers (Ford 1990). The presence of minerals of the platinum group in the unexposed middle portion of these rocks has also been speculated. Layered igneous intrusions like those in the Dufek Massif have also been reported in Victoria Land (Hamilton 1964).

The Antarctic Peninsula has often been considered one of Antarctica's most important zones for mineral resources because its geologic and tectonic setting is similar to that of the South American Andes, which bear some of the world's largest copper, antimony, tin, molybdenum, silver, lead, iron, tungsten, zinc and gold deposits. Copper mineralisations are known in the north-western portion of the peninsula and in surrounding islands such as the King George, South Shetland and Anvers islands (Fig. 4); however, the potential of these resources is unknown (Pride et al. 1990).

Thick beds of coal with high ash and low sulphur contents are known in the Permian sandstones of the Beacon Supergroup in the Prince Charles Mountains, Victoria Land, the Weddell Sea coast, and especially in the Transantarctic Mountains (Coates et al. 1990; Rowley et al. 1991).

1.4 The Antarctic Climate and its Role in the Global Climate System

1.4.1 The Antarctic Climate

The geographic position, astronomical factors (e.g. the Earth's maximum distance from the Sun during the austral winter), ice cover (over the continent and Southern Ocean) and average altitude are the main factors affecting the climate of Antarctica. The continent is the coldest place on Earth because of the long polar night and the low inclination of solar rays during the summer, the permanent ice cover which prevents the adsorption of most incident heat, and the average elevation (about 2,300 m) which determines the lack of the lower part of the troposphere (i.e. the major repository for heat in the rest of the world). Because it is the coldest place on Earth, Antarctica plays a very important role in the global climate system as the greatest heat sink of the Southern Hemisphere.

Temperature gradients from the equator to the pole, and atmospheric and oceanic dynamics in the Southern Hemisphere are driven by temperatures and atmospheric pressures in Antarctica, the extent of sea ice, and water temperatures in the Southern Ocean (Yuan et al. 1996). Unfortunately, the limited number of climate stations and the scarce temporal significance of many records do not allow an exhaustive identification of recent trends in the Antarctic climate. However, some general features, such as the high atmospheric pressure at the more southerly latitudes, have been described since early Antarctic expeditions (e.g. Bellinghshausen circum-Antarctic navigation in 1819–1821). During the winter of 1898 de Gerlache's ship, the *Belgica*, was trapped in sea ice at 72° S and 90° W; during this period, the scientists Arctowski and Dobrowolski carried out observations on clouds, frost and other meteorological parameters. At the beginning of the 20th century, during the Antarctic expeditions of Shackleton, Scott, Amundsen, Filchner and Mawson, some stations were established and a range of meteorological data were collected. Most of these expeditions experienced and described the katabatic winds, one of the most striking phenomena in many coastal areas of Antarctica.

The collection of data on the Antarctic climate began with the establishment of the first research stations in the 1950s, such as Mawson in 1954, McMurdo, Dumont d'Urville, Halley Bay and Mirny in 1956, Davis, Wilkes and South Pole in 1957, and Vostok in 1958, and the development of research activities within the framework of the International Geophysical Year of 1957–1958. Although most Antarctic stations were far apart and often located at coastal sites, the first weather forecasts were issued for the Antarctic Peninsula, the peri-Antarctic islands and the area of operation of the whaling fleet (King and Turner 1997). In spite of the scarce number of records, the relatively uniform climatic conditions on the Antarctic plateau enabled the International Weather Central Forecast Office established at Little America V (Bay of Whales) in 1957–1958, then at McMurdo and finally at the Australian Weather Bureau in Melbourne, to develop synoptic surface charts for the Antarctic and the Southern Ocean. However, acceptable forecasting of atmospheric conditions in this part of the world has only been possible since the 1970s, with the introduction of operational polar orbiting satellites, powerful computers and global numerical models. During the last two decades, the establishment of suitable operational communication systems and increasing international cooperation between scientists and operational meteorologists have significantly contributed to a better understanding of the Antarctic climate and its role as a heat sink in the Southern Hemisphere.

Low solar radiation at the top of the polar atmosphere and the albedo over snow and ice surfaces are the main cooling factors. Reflection by seasonal sea ice in the Southern Ocean is more exposed to the effects of global climate variations. A small increase in temperature could reduce the ice cover, amplifying the initial temperature increase. This ice–albedo feedback mechanism is probably one of the main factors determining variations in the Antarctic climate and, consequently, in the global climate system. In contrast to the Northern Hemisphere, which has large continental areas with seasonal snow cover, Antarctica is permanently covered by snow and ice; the response of sea ice in the Southern Ocean to increases in temperature due to the effects of albedo feedback cannot be as rapid as that of Arctic soils and rocks. Contrary to Arctic pack ice, that in the Southern Ocean melts during summer, with considerable year-to-year variations in its extent. Sea ice not only increases the albedo of the ocean surface but also reduces heat fluxes between the air and sea, and water evaporation during the winter (Parkinson 1992). Thus, sea ice may affect atmospheric circulation and precipitation at great distances from the immediate area. The formation of sea ice at the end of summer involves the rejection of dense, brine-rich waters, determining the formation of cold deep waters. Other dense bottom waters are formed by melting at the base of ice shelves. These cold, saline waters influence global oceanic circulation because they flow northwards along the ocean bottoms and can be detected well into the Northern Hemisphere.

Changes in the spatial extent, thickness, persistence and annual distribution of sea ice have a major influence on the climate of Antarctica, consequently affecting global atmospheric and oceanic circulation. However, 25 years of satellite monitoring of the extent and variability of sea ice in the Southern Ocean has not revealed overall trends (Zwally 1991), and attempts to link changes in sea-ice extent to climate change have not yielded conclusive results (Hanna 1996; Murphy and King 1997; Jacka and Budd 1998).

Melting of the Antarctic ice sheets, which would increase global sea levels, is one of the major concerns of global climate change. Maintenance of the ice sheets is due to the balance between snowfall over the continent and the rate of ice discharge in the Southern Ocean. This balance strictly depends on temperature, the extent of sea ice, and water evaporation in the Southern Ocean, which affect the transport of heat, air masses and precipitation from mid-latitudes. It is very difficult to measure changes in the ice sheet over an area of 14×10^6 km^2; however, changes of a few percent in the total volume of ice would affect the volume of the world's oceans. It has been estimated, for instance, that the West Antarctica ice sheet contains enough ice to raise global sea level by about 6 m (IPCC 2001). Most of this ice sheet is grounded below sea level, and the grounded portion has been the most dynamic part of Antarctic ice sheets in the recent geological past. It is therefore receiving particular attention from scientists and, although its potential collapse in response to climate change is still debated, it is now widely agreed that major losses of grounded ice in this region are very unlikely during the 20th century. According to Titus and Narayanan (1996), there is only a 5% probability that it will raise global sea levels more than 0.16 m by the year 2100. Nonetheless, on a longer timescale, changes in ice dynamics could result in significant increases in the outflow of ice into the ice shelves and a retreat of the grounding line (Huybrechts and de Wolde 1999). New satellite altimeters provide near-global, homogeneous coverage of the world's oceans and ice sheets, and will become important tools for better understanding the dynamics of Antarctic ice sheets and changes in global sea levels (Cazenave et al. 1998; Nerem 1999).

1.4.2 Solar Radiation

Many aspects of the Antarctic climate, such as katabatic winds and the stability of the atmosphere, are affected by the amount of solar radiation. This radiation is a function of latitude, the elevation of receiving surfaces, the time of year, the amount of clouds, water vapour and aerosol concentrations in the atmosphere. On the Antarctic plateau, radiation travels through a short atmospheric path; there are usually few clouds and very low levels of aerosol and water vapour (about ten times less moisture than in temperate air; Barry and Chorley 1992). Thus, up to 90 % of the radiation arriving at the top of the atmosphere may reach ice and snow surfaces. As a result, during the austral summer, the coldest place in the world receives more solar radiation than anywhere else on Earth (Zillman 1967). During the dark polar winters, the plateau receives less solar radiation than anywhere else on Earth. Nevertheless, the total annual irradiation of the plateau may be higher than that of Antarctic coastal areas. Faraday Station, for instance, is located north of the Antarctic Circle (65.3° S, 63.3° W) and has no days of complete darkness in the middle of winter, but the average incoming annual radiation (3,156 MJ m^{-2}) is about 32 % lower than that at Vostok Station (78.5° S, 106.9° E; King and Turner 1997). This is due to intense irradiation during the continuous summer daylight, the elevation, the clean atmosphere and, above all, the limited cloud cover at Vostok Station. Clouds have a dominant effect on the planetary albedo, reflecting about 20 % of the solar radiation reaching the Earth, compared to the 6 % reflectance of gas molecules and aerosols (Loeb 2002). Cloud cover in Antarctica varies strongly: the average cloudiness is usually high over the ocean and in coastal areas of the Antarctic Peninsula and West Antarctica, and lower along the East Antarctic coastline, with minimum values in the interior.

Absorbed solar radiation (i.e. the difference between intercepted and reflected radiation) is the principal source of energy which drives the climate system. About one third of the radiation reaches the Earth as direct radiation and the remainder as a diffuse (scattered) component. The very clear atmosphere on the Antarctic plateau determines high ratios between direct and diffuse radiation. However, the amount of absorbed radiation is very low because snow and ice surfaces reflect 80–90 % of the incoming solar radiation (Carroll 1982; Warren 1982; Wendler et al. 1988); low levels of atmospheric moisture and aerosol allow long-wave radiation to escape from the atmosphere without yielding heat. In coastal ablation zones, persistent katabatic winds originate snow-free areas of blue ice which have lower albedos with respect to snow-covered areas on the plateau (Fig. 5). In the few areas of Antarctica where the ground is free of snow in the summer, a large amount of heat is absorbed; this may promote the warming of the lowest layers of the atmosphere, shallow convection and the formation of cumulus clouds. Clouds reduce the amount of absorbed radiation and decrease the loss of terrestrial radiation to space by absorbing and then re-emitting part of this radiation back down to the sur-

Fig. 5. Typical albedos (%) of different natural surfaces compared with the whole Earth average value

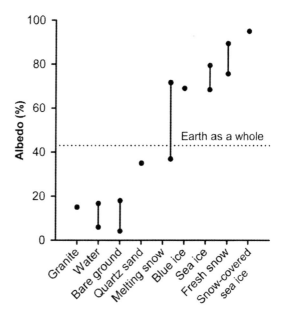

face. The net effect of possible changes in cloudiness over Antarctica is not easy to predict because it depends on cloud characteristics such as cloud height and water content (Arking 1991). In general, low stratiform clouds are believed to have a strong cooling effect, while high cirrus clouds probably have a warming effect.

Only about 2 % of the total surface of Antarctica is free of ice, but in several regions global warming could lead to the melting and replacement of bright, highly reflective surfaces by darker underlying surfaces. The consequent increase in the absorption of solar radiation would enhance warming ("the ice–albedo feedback"), and for this reason many models predict greater warming at high latitudes in response to increasing concentrations of greenhouse gases.

The sum of the effective short-wave and long-wave components (i.e. the net radiation, or radiation balance) indicates the extent to which a surface is receiving energy. In the Southern Hemisphere the largest net differences in flux occur in regions of layered clouds at mid-latitudes. At low latitudes, albedo and greenhouse effects almost balance each other, whereas in Antarctica the balance is negative, especially during the winter when there is scarce or no incoming short-wave radiation from the Sun. Owing to relatively higher surface temperatures, net radiation losses are greater over coastal ablation zones and blue ice than in the interior of the continent. Variations of surface radiation balance in the same or different Antarctic areas are mainly due to cloud cover, which reflects some of the incident solar radiation and increases the downward component of long-wave radiation. Little radiation is adsorbed over the Antarctic ice sheets and sea ice, and the cloud cover increases the net radiative warming of surfaces. Based on 20-year measurements, the net radiation for the Faraday and Halley (75.5° S, 26.6° W) stations was only about 10 % of the total global solar radiation, and the greater variability of cloud cover at Faraday Station determined the higher inter-annual variability of radiation at this station (King and Turner 1997). According to these authors, the average annual net long-wave cooling of Antarctic surfaces exceeds net solar heating by 10–20 W m^{-2}.

A positive average annual radiation balance occurs in Antarctic areas free of ice; however, their role in the radiation balance of the continent is almost negligible because of their small surface area compared to that of Antarctica. Thus, in the global energy budget, the radiational losses of Antarctica play a very important role in determining the poleward flow of warm air masses from tropical regions.

Although for decades the duration of bright sunshine in Antarctic stations has been the most widespread radiation measurement for climatological purposes, the springtime depletion of stratospheric ozone over Antarctica discovered in the 1980s (Farman et al. 1985) prompted interest in ground-level measurements of ultraviolet (UV) radiation, which is expected to adversely affect the polar biosphere. Due to the wavelength dependence of the ozone absorption cross section, a decrease in the total ozone column determines a greater increase in UV-B radiation (280 to 315 nm) than in UV-A (315 to 400 nm) and PAR (Photosynthetically Active Radiation; 400 to 700 nm) radiation. UV-A radiation plays an important role in photo-repair mechanisms and is essentially unaffected (for wavelengths>330 nm) by ozone depletion. Hence, while stratospheric ozone depletion leads to a dramatic increase in damaging UV-B radiation, the energy necessary for photo-repair processes does not change (Roy et al. 1994) and primary productivity may decrease (Smith et al. 1992).

In 1998 the US National Science Foundation established a network of scanning UV-B spectroradiometers comprising three Antarctic sites: McMurdo Station (77.51° S, 166.40° E), the South Pole (90° S, 0° E) and Palmer Station (64.46° S, 64.03° W). Data from these sites and others in the Northern Hemisphere show that the maximum monthly and yearly erythemal doses (sunburning irradiance) occur in San Diego (California), but that the maximum weekly doses occur at the South Pole and the maximum daily and hourly doses at Palmer Station (Booth et al. 1994). There is evidence that at comparable latitudes, locations in the Southern Hemisphere receive more UV radiation than those in the Northern Hemisphere (about 15 %, Bodeker 1997), and that erythemal UV radiation at southern latitudes can exceed that in Europe by more than 50 % (Seckmeyer et al. 1995). According to the latter authors, summertime monthly measurements of daily erythemal doses show a very small latitudinal gradient in the Southern Hemisphere when compared to the large gradient in the Northern Hemisphere.

Surface UV irradiance is affected by the solar zenith angle (which determines the optical path length), cloud cover and particularly surface geometry, ozone (both the vertical distribution and integrated column amount), surface albedo, aerosol concentrations in the atmosphere, the Earth–Sun distance (which modulates extraterrestrial solar irradiance) and altitude (model calculations indicate that UV-B radiation increases by about 5 % each kilometre of altitude; Bodeker 1997). The minimum Antarctic total-column ozone level is usually reached in the first week of October which, in combination with the small solar zenith angle, produces high UV-B values (Hermann et al. 1995).

1.4.3 Temperature

The above-reported radiation balance, the geographical location of Antarctica and its distance from the moderating influence of the ocean (especially when the Southern Ocean is covered by ice) determine very low temperatures over most of the continent. In general, the surface air temperature is related to latitude and/or altitude and declines on ice sheets by about 1 °C every 100 m of altitude (Fortuin and Oerlemans 1990). This linear relationship shows different regression coefficients in the continental interior, coastal escarpments and ice shelves. However, considering that more than 50 % of the continent lies at an elevation of more than 2,000 m and about 25 % of it is more than 3,000 m above sea level, the average annual temperature on the plateau (–50 °C) is about 30 °C lower than at coastal sites at the same latitude (Fig. 6).

The continentality of Antarctica together with the features of solar and long-wave radiation determine large seasonal temperature variations. The annual cycle of surface air temperatures is characterised by a broad summer maximum (about –30 °C on the plateau and –4 °C at coastal locations) and a minimum in July or August (–70 °C on the plateau, with a very low record of –89.5 °C in July 1983, at Vostok, and about –25 °C at the coast). A peculiar feature of the temperature regime is the rapid seasonal transition, with a short summer period (on the plateau only 6–8 weeks centred around the beginning of January). Surface temperatures in the interior decrease by about 25 °C from late January to late March, and "coreless" winter temperatures are reached in April, with minimal variations for the next 5 months. Coastal areas show a similar trend, although the "coreless" winter phenomenon occurs for a shorter period, indicatively from May to August (Fig. 7). Latitudinal variations in surface temperatures sharply increase during the austral autumn and remain large until the return of the Sun. Typical annual values on the continent do not apply to the Antarctic Peninsula, which is at a lower latitude and for most of its length is divided into two distinct zones by a 2,000-m-high mountain range. The west coast of the peninsula has a relatively mild maritime climate, and the annual average temperature (about –1.8 °C) is about 7 °C higher than

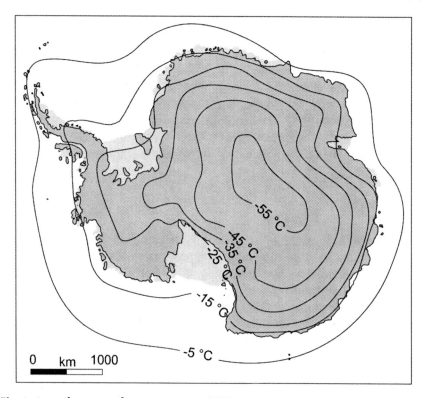

Fig. 6. Annual mean surface temperature (°C) in Antarctica

that of the east coast, at the same latitude. In fact, the latter zone is more affected by southerly winds and the greater extent of sea ice.

A remarkable feature of temperature regimes in Antarctica is the inversion of surface temperatures due to strong radiational cooling over snow and ice surfaces. The strongest inversions occur on the plateau during winter, but significant inversions may also occur in coastal regions throughout most of the year (Phillpot 1985). Inversions are usually associated with calm anticyclonic conditions occurring in winter, and their thickness ranges from 10 m to a few hundred metres. Temperatures above the inversion layer may sometimes be 20–30 °C higher than those at the surface (Schwerdtfeger 1970, 1984). The terrain slope, elevation, strong winds and influx of thick cloud shields over the continent (associated with cyclonic activity near the coast) can disrupt surface inversions (Carroll 1994). Many Antarctic blizzards are therefore accompanied by a rise in temperature which provides no relief to humans because of wind chill.

Although in recent years the situation has been improving through the installation of automatic weather stations (Stearns et al. 1993), available data

Fig. 7. Monthly mean surface temperature at stations at different latitudes and altitudes in Antarctica (for references, see text)

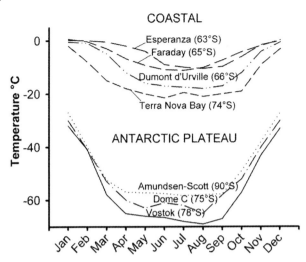

are still insufficient to allow a reliable interpretation and identification of spatio-temporal trends in Antarctic temperature regimes. In recent years, to overcome these difficulties, geographical coverage has been improved by supplementing direct measurements of air temperature with measurements of snow surface temperatures through infrared radiometric observations from polar orbiting satellites. This approach is based on the measurement of snow surface radiance and its conversion into snow surface temperatures, based on the known emissivity of snow surfaces and on corrections to compensate for atmospheric transmission. Although these measurements can only be made under cloud-free conditions and in snow-covered areas, Comiso (1994) traced maps of monthly and annual Antarctic surface temperatures using radiometric data from the Nimbus 7 satellite.

Another technique for tracing maps of surface temperatures and validating the results of remote sensing observations consists in the measurement of snow temperatures 10 m below the surface. The amplitude of the annual temperature wave is attenuated at this depth, and the measured value is close to the annual average surface temperature. During the last decade a large number of 10-m snow temperatures were recorded throughout the continent on traverses performed within the framework of ITASE (International Trans-Antarctic Scientific Expedition; Mayewski and Goodwin 1997).

1.4.4 Clouds and Precipitation

Given the low temperatures, the air in Antarctica can hold only a small amount of water vapour, even when close to saturation. At –20 °C, for instance, the water content is reduced to 4 % of that at 20 °C. Annual average values of moisture in total air columns measured through radiosondes at several coastal stations of East Antarctica are about 3.3 kg m^{-2}, whereas on the plateau the values are typically <0.5 kg m^{-2} (Connolley and King 1993). Data from radiosondes have been used to estimate the transport of water vapour across the coast of East Antarctica (Bromwich 1990). Through numerical analyses of all available meteorological observations, it has been found that about 40 % of water vapour falling as snow on Antarctica reaches the continent through West Antarctica, i.e. through the area between the Ross Ice Shelf and the Antarctic Peninsula (Bromwich et al. 1995). This sector has the largest inter-annual moisture variability in Antarctica, particularly in conjunction with the El Niño–Southern Oscillation (ENSO) phenomenon (Trenberth and Hoar 1996).

Imagery from geostationary and polar orbiting satellites shows that the area of the Southern Ocean near 60°S is the cloudiest place in the Southern Hemisphere (about 85–90 % cloud cover throughout the year). In coastal regions of Antarctica near 70° S, the total cloud cover is about 45–50 %, and a further decrease occurs inland from the coast. In contrast to oceanic and coastal areas, the amount of clouds near the South Pole shows large seasonal variations (from about 35 % in autumn and winter to nearly 55 % in spring and summer; King and Turner 1997). There are also latitudinal changes in cloud types: a large amount of stratus occurs close to 60° S, altostratus and cirrus prevail at 70° S, and cirrus is the most commonly reported type of cloud in the interior of Antarctica.

In Antarctic coastal regions, atmospheric precipitation comes mostly from thin clouds associated with major low-pressure, frontal cyclonic systems. The most important mechanism for the production of precipitation is adiabatic cooling of moist air masses as they rise up the steep slopes of the plateau. The intrusion of mild air masses beyond the coastal zone is very rare. In 3 years of surface observations at Vostok Station, Phillpot (1968) recorded 42 depressions, several of which produced precipitation. Most precipitation on the plateau comes from isolated clouds or even from apparently clear skies (e.g. "diamond dust"). Clear-sky precipitation is very common; it originates from a cloud of ice crystals, too thin to be seen either from the ground or in satellite imagery. It obviously determines a very low accumulation rate.

It is important to understand the distribution of atmospheric precipitation and snowfall over the continent in order to evaluate the mass balance of snow deposition and possible global sea-level variations, to study past climates through the analysis of ice cores, and to estimate the deposition rate of persistent airborne pollutants. As a rule, solid precipitation occurs over the continent; rain occasionally occurs during the summer in the northerly latitudes, for example, in the western Antarctic Peninsula.

Snow accumulation at a site is determined by the amount of precipitation, together with snow lost or gained through surface wind redistribution, sublimation or melting. Thus, conventional snow gauges, stakes and pit measurements cannot give accurate measurements. Furthermore, the amount of snowfall on the plateau may be less than the minimum resolution of snow gauges (Bromwich 1988). Although winds make it difficult to distinguish between falling and blowing snow, and although sublimation during the summer may determine a significant loss of snow (Stearns and Weidner 1993), snow accumulation on the continental scale is a convenient approach to estimate the time-averaged spatial distribution of annual precipitation. A large number of time-series of snow accumulation rates are collected during ITASE traverses as a proxy for precipitation. A variety of physical and chemical properties (e.g. snow density cycles, stable isotopes, electrical conductivity) can be used to estimate the annual accumulation rate. Measurements are performed over periods of one year to a decade on marker canes placed on the surface of the ice sheet, through snow radar detection of distinct snow layers, downhole gamma ray detectors or measurements of gross beta from caesium[137] (Mayewski and Goodwin 1997). Maps of the annual accumulation of snow (Giovinetto and Bentley 1985; Zwally and Giovinetto 1995) show that highest accumulation occurs on the steepest slopes slightly inland of the coast, especially in West Antarctica. In the coastal region of the south-eastern Bellingshausen Sea, which is at the end of a depression track and in an area of frequent cyclogenesis, more than 800 mm (water equivalent) of snow accumulate each year. A rather high annual precipitation occurs in the western Antarctic Peninsula and in Marie Byrd Land. On the contrary, in the coastal belts of East Antarctica the annual water-equivalent precipitation is about 200–300 mm, with very low values in the Ronne and Ross ice shelves (probably in relation to the lack of topographic lifting of mild air masses).

Precipitation is inversely correlated with altitude and decreases by an order of magnitude from the coast to the interior. A large proportion of the plateau receives less than 50 mm of snow each year (i.e. a value comparable with that of the Sahara). As desert climates are characterised by less than 250 mm of annual precipitation, Antarctica is the largest cold desert in the world.

Negative snow-accumulation values occur in net ablation areas such as Dronning Maud Land and other zones of East Antarctica (Giovinetto and Bentley 1985). There is a good overall fit between the accumulation pattern described above and that resulting from the calculated atmospheric moisture budget (Bromwich and Parish 1998). The latter approach is based on the net precipitation (which approximates the accumulation rate), calculated as the difference between precipitation and evaporation/sublimation (i.e. from the convergence of moisture transport in the overlying atmospheric volume plus the impact of changes in moisture storage; Yamazaki 1992; Bromwich et al. 1995). In general, the difference between precipitation and evaporation/sublimation shows a broad winter maximum and a summer minimum (Bromwich and Parish 1998).

1.4.5 Wind Regime

Since early exploration, one of the most impressive aspects of the Antarctic climate has been the strength and persistence of winds. Most surface winds blowing over ice sheets are katabatic winds generated by the outward and downward flow towards the coast of cold, dense air masses from the boundary layer of the interior of Antarctica. Intense radiative cooling over sloping ice surfaces produces a horizontal pressure-gradient force with a downslope direction. Katabatic winds affect only the first few hundred metres above the ground and their velocity is proportional to the steepness of the underlying terrain. The minimum velocity therefore coincides with the top of the inversion layer, and the strongest winds occur in coastal escarpments with very steep slopes, smooth ice surfaces (minimum turbulence) and a topography which channels the katabatic outflow into small stretches (Parish 1988). It has been suggested (Radok 1973) that the continental drainage flow is also accelerated by the replacement of relatively warm air by cold air moving downslope and by sublimational cooling of drifted snow.

Katabatic surface winds show several remarkable features such as their suddenness (in just a few minutes wind speeds can surge from near calm to up to 50 m s^{-1}, generating eddying walls of drifting snow and rapid variations in surface pressure, temperature and humidity; Phillpot 1985), persistence and directional constancy. Their effects are very localised, and the speed usually drops remarkably a few kilometres offshore or a few hundred metres above the ground. These features have been experienced and described by early Antarctic explorers. At Cape Denison (Adélie Land), Mawson's expedition experienced an incessant katabatic wind for nearly two years (1912–1913). In this period the mean wind speed was 19.8 m s^{-1}, with the highest monthly mean in July 1913 (24.9 m s^{-1}) and the windiest day on 16 August 1913 (36.0 m s^{-1}). Even the lowest monthly mean (February 1912, 11.7 m s^{-1}; Loewe 1972) during the 2-year period was higher than the annual mean wind speed (5–10 m s^{-1}) for most of the Antarctic coast (King and Turner 1997). As shown in Fig. 8, streamlines of cold, negatively buoyant air become concentrated into restricted pathways in the confluence zone upslope. Owing to the local topography, Adélie Land is an area of confluence of cold air streams which feed intense, persistent katabatic winds downstream, at Cape Denison and Dumont d'Urville (Wendler et al. 1988). The Nansen Ice Sheet (Terra Nova Bay, Victoria Land) is another windy coastal zone in East Antarctica (Bromwich et al. 1993). In February 1912, six men of Scott's northern party were stranded on an island in this zone; they experienced 7 months of strong and exasperating winds which continuously cleared the sea ice, producing a coastal polynya. At the end of the winter the polynya forced the six men to climb up the Drygalski Ice Tongue in a successful attempt to rejoin the survivors of Scott's expedition at Cape Evans. The place was understandably named Inexpressible Island.

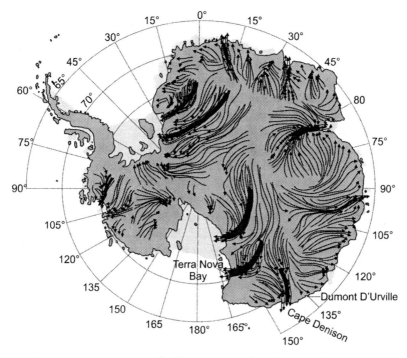

Fig. 8. Schematic representation of cold air streams (katabatic winds) over Antarctica

The speed and direction of katabatic winds at several scientific stations have been measured in the last four decades. Since the 1980s, the US Antarctic Program (USAP) has installed automatic weather stations at remote locations of Antarctica to enhance the continental meteorological network (Stearns et al. 1993). More than 50 of these stations are currently operating in Antarctica, recording monthly means and extremes for air temperature, air pressure, wind speed, and wind direction which are reported in tables. Most annual surface wind data indicate the unidirectional nature of katabatic winds, usually oriented 20–50° to the left of the fall line (according to the Coriolis deflection of a gravity-driven airflow) and in a narrow sector (about 30°, Bromwich and Parish 1998). The outflow of cold surface air hugs the coastline of the continent as an anticyclonic vortex (King and Turner 1997).

The wind direction at a site can be deduced from the alignment of snow erosional features (sastrugi). Maps of surface streamlines over the continent have been produced by combining data recorded in stations with sastrugi observations made during traverse expeditions (Parish and Bromwich 1987). As a rule, coastal stations exhibit a wider range of directional constancy because the wind regime is affected by katabatic winds and synoptically forced winds, while plateau surface winds are largely controlled by local topography.

Atmospheric circulation in Antarctica reflects theoretical models for polar locations: during their progressive subsidence along a gradient towards the pole, the cold and dense air masses are deflected by the rotation of the Earth, forming a clockwise vortex. In winter, the steep temperature gradient between the air in the vortex and that to the north acts as a barrier. As a result, high-pressure systems are dominant in the South Pole, accompanied by the circumpolar flow of strong westerly winds (Drake 1995). The mountains in West Antarctica and the Antarctic Peninsula are a significant topographic barrier to the circumpolar airflow, and determine local increases in windiness and precipitation as the cyclones rise and cool over the mountains. In East Antarctica, not only the atmospheric circulation in winter but also the high topography and steep slopes limit the penetration of cyclones moving polewards and impinging the coastline. The sharp increase in altitude reduces the depth of the vortex column and the associated circulation of cyclones.

The winter polar vortex has several important environmental implications. On the one hand it prevents the penetration of warmer air masses carrying persistent pollutants from lower latitudes, and on the other it plays a prominent role in the destruction of stratospheric ozone (Molina and Rowland 1974; Solomon 1990). The lack of sunlight determines very low temperatures in the vortex and the formation of ice crystals in polar stratospheric clouds. Chlorine compounds adhere to these crystals, and spring sunlight converts them into reactive species (chlorine monoxide) which destroy ozone. The catalytic process has a positive feedback: the destruction of ozone decreases the absorption of solar energy, and the consequent cooling of the atmosphere promotes the formation of new ice clouds and the destruction of ozone. As solar radiative warming increases, the ice crystals evaporate and the polar vortex weakens, thus allowing the partial regeneration of ozone and its influx from lower latitudes. Whereas under natural conditions the destruction and production of ozone are thought to be broadly self-regulating, the progressive introduction of large quantities of very persistent pollutants such as halocarbons into the atmosphere is determining a statistically significant decrease (about 30 %) of historical springtime ozone concentrations in Antarctica. The "ozone hole" has been observed since the spring of 1985 up to 60° S and the maximum depletion was recorded in 1993 (Farman et al. 1985; Jones and Shanklin 1995).

At the beginning of summer, as the atmosphere warms, the vortex disappears and the atmospheric pressure and winds become more variable. Thus, during the polar spring and summer, there is an enhanced influx of air masses and persistent atmospheric pollutants from lower latitudes.

1.4.6 Atmospheric Interactions of Antarctica with Lower Latitudes

In spite of its remoteness, Antarctica is linked to the lower-latitude regions through atmospheric and oceanic circulation. The large equator-to-pole temperature difference drives the general circulation of the atmosphere through the poleward transport of heat. Katabatic winds play an important role in this circulation system because they are responsible for the continuous low-level drainage of the continental surface, which is compensated by the inflow of relatively warm air masses which converge and subside in the troposphere over Antarctica. The flow of cold air to the ocean in the shallow boundary layer affects the tropospheric and stratospheric circulation, attributing to Antarctica a major role in the Southern Hemisphere as a sink for heat and persistent airborne pollutants.

The transport of heat is mainly due to transient eddies (extra-tropical cyclones). When cyclones formed at mid-latitudes reach the Antarctic coasts, they are in a mature phase and are beginning to decay; however, they can be responsible for wind and precipitation in coastal regions. In general, upper-air long waves in the Southern Hemisphere are weaker than those in the northern one, but they have a large amplitude and are responsible for major precipitation events and rapid increases in surface temperatures (King and Turner 1997). Sinclair (1981), for instance, investigated an event (25–29 December 1978) with intrusion of warm and humid air from the Atlantic and Indian ocean sectors of Antarctica, which produced unusually high temperatures (+9.6 °C at McMurdo, –15.7 °C at Vostok and –13.6 °C at the South Pole).

The middle to upper tropospheric air masses converge over Antarctica and broadly subside approximately south of 70° S (Bromwich and Parish 1998). The katabatic outflow moves near-surface air masses over the Southern Ocean north of the continent, and these masses rise (between 65 and 55° S) in areas with cyclonic activity. The circulation between Antarctica and sub-polar latitudes is responsible for a seasonal cycle of mass loading onto the continent during the spring, and net transport away from it in autumn. This mass flux across the Antarctic coastline is due to the solar insolation cycle, which is responsible for very large changes in surface atmospheric pressures up to the subtropics. According to Bromwich and Parish (1998), seasonal changes in surface pressure over Antarctica influence mass redistribution over nearly the entire Southern Hemisphere. Interactions between Antarctica and the subtropics are also affected by the El Niño–Southern Oscillation phenomenon (ENSO), which increases the transfer of heat from the ocean to the atmosphere and influences global mean surface temperatures (Sun and Trenberth 1998). The ENSO is generated by ocean–atmosphere interactions internal to the tropical Pacific and the overlying atmosphere. During El Niño events the temperature of the eastern equatorial Pacific increases, and the normally high temperature difference in the sea surface across the tropical Pacific decreases. Consequently, trade winds weaken and the sea-level pressure between Tahiti and Darwin (the Southern Oscillation) becomes anomalously negative. As the warm water extends eastwards along the equator, sea levels fall in the west and rise in the east by as much as 25 cm, and the weakened trade winds reduce the upwelling of cold water, thereby strengthening the temperature anomaly

(Neelin et al. 1998). The altered atmospheric heat patterns force large-scale waves in the atmosphere, establishing teleconnections with mid and high latitudes through the alteration of winds and changes in jet streams and storm tracks (Trenberth et al. 2001). Ever since Savage et al. (1988) observed a connection between the ENSO event and significantly lower air temperatures at the South Pole in 1983, there has been increasing evidence of possible interactions between Antarctica and the subtropics. Associations have been hypothesised between El Niño events and variations in Antarctic surface temperatures and pressures (Smith and Stearn 1993), the polar jet stream (Chen et al. 1996), low pressures in the Amundsen Sea, the extent of ice cover in the Southern Ocean, and the precipitation rate in West Antarctica (Gloersen 1995).

1.5 Global Warming and Climate Variations in Antarctica

1.5.1 Climate Variability and Changes Due to Human Activities

The climate varies naturally on all timescales as a consequence of internal (interactions within and between the atmosphere, hydrosphere, biosphere, and cryosphere) and external factors (volcanic eruptions, variations in Earth's orbit, tectonic variations in the position of continents, asteroid impacts). According to Goudie (2002), most of the variability in any climate record on timescales up to a century can often be attributed to such simple processes that it is debatable whether it should be considered a fluctuation in climate at all. Several oscillations have been indicated as possible sources of climate fluctuations on different timescales, such as the biennial variability in the Asian monsoon, the every three-to-ten years ENSO, decadal or inter-decadal variability in mid-latitudes, centennial variability in oceanic thermohaline circulation, and the Pleistocene ice-age cycles (a 22,000-year cycle over which the solstice and perihelion years move in and out of phase with each other). The shape of the Earth's orbit around the Sun varies about every 100,000 years (the so-called Milankovitch cycle) and is believed to be responsible for the periodic onset and decay of Northern Hemisphere ice sheets which have characterised the climate of the Pleistocene (Kutzbach 1992). According to the Milankovitch theory, the current interglacial period will terminate with another ice age at some point in the next couple of thousand years. However, the impact of anthropogenic activity on the composition of the atmosphere threatens to invalidate this prediction. The increase in concentrations of greenhouse gases contributes to positive radiative forcing in the earth–atmosphere system and may determine global warming (i.e. amplify climatic changes due to Milankovitch forcing; Allen 2002). Human activity will probably continue to affect the climate system for many decades until a new equilibrium is reached. At present, one of the main difficulties for scientists is the distinction between anthropogenic effects (climate change) and the intrinsic steady-state equilibrium variability of climate (climate variations), especially over timescales of a few decades. It cannot be excluded that the effects of human activity on the climate system will lead to a "breakpoint", with dramatic climatic changes occurring in just over a decade or so (Adams et al. 1999). Several abrupt climate changes, with large-scale and possibly irreversible changes in the terrestrial biosphere, took place in the past, when anthropogenic perturbations were lacking or negligible. Besides the events associated with major extinction episodes, especially that at the end of the Cretaceous (65 Ma ago), which has been the subject of great debate and controversy, there is also evidence of abrupt climate changes in the Holocene. This period has often been thought of as rather stable, but abrupt events have been dated to about 8,200 years ago (Alley et al. 1997), and between 3,900 and 3,500 years ago (Anderson et al. 1998). The development of the Sahara Desert (Claussen et al. 1999) and the collapse of civilisation in Mesopotamia and elsewhere were probably due to the onset of acute arid phases (Goudie 2002).

In recent years knowledge of the Earth's climate has grown enormously as a result of the growing use of satellite remote sensing, the availability of digital data, and the very rapid progress in technologies for their elaboration and dissemination. Available instrumental temperature records indicate that mean global values (air temperature over land, and sea surface temperature) have risen between 0.4–0.6 °C since 1900, at a rate broadly consistent with that expected from the measured rise in atmospheric concentrations of greenhouse gases (Kemp 2002). These temperature variations are close to those of the Northern Hemisphere during the Medieval Warm Period, when vineyards grew in the United Kingdom as far north as York. However, the global-scale accuracy of these estimates is affected by gaps in the spatio-temporal observational coverage and by increasing urbanisation around meteorological stations, with the development of urban heat islands. Furthermore, there has not been a steady increase in mean global temperatures, and records show a great deal of variability. Warming during the 20th century mostly occurred in two periods: 1910–1945 and 1976–2000 (IPCC 2001). Despite the increase (about 2 parts per million per year) in carbon dioxide concentrations and those of other greenhouse gases, the mean global temperature decreased in the period 1946–1975, particularly in the Northern Hemisphere. According to Kemp (2002), in this period many meteorological stations were moved from urban centres to rural or semirural airport locations, and this may have produced an apparent, artificial cooling. In any case, the 1990s was a very warm decade, and 1998 was the warmest year in the instrumental record since 1861.

Proxy data from the Northern Hemisphere indicate that the temperature rise in the 20th century was probably the largest of any century during the past 1,000 years (IPCC 2001). Thus, if in the 1980s the majority of scientists involved in the study of climate change were not completely confident that global temperature increases were beyond the range of natural variations, during the last decade, based on the measured warming trend, a better understanding of global processes, and progress in the differentiation between natural and anthropogenic influences, scientists from the Intergovernmental Panel on Climate Change (IPCC 1996) concluded that there is a discernible impact of human activity on recent changes in the climate.

Although the phrase "global climatic or environmental change" has come into wide use only in the past few decades, the issue of human alteration of the environment is not new. It was addressed, for instance, by Buffon (1707–1788) in *Histoire Naturelle*; Marsh (1801–1882) wrote *The Earth as Modified by Human Action*, and in 1866 the Finnish lichenologist Nylander (1822–1899) ascribed the scarcity of epiphytic lichens on the trunks of chestnut trees in a Parisian park to coal-burning and the poor "salubrité de l'air". Other scientists realised the possible large-scale effects of human activity. The Swedish chemist Arrhenius (1859–1927) suggested that the increase in the Earth's temperature was a consequence of the increase in atmospheric concentrations of carbon dioxide from the combustion of coal. Vernadsky (1863–1945) described the role of human activity in natural processes and in the transition from the biosphere to noosphere (i.e. the sphere of reason) in *The Biosphere and the Noosphere*. Although this author confided in the power of human consciousness, in his last paper, written during World War II, he was in doubt and added ... "if man will not use his reason and activities for self-destruction". This doubt soon spread to the scientific community and to the public in general. In fact, the post-World War II years were a turning point in the development of reliable approaches and instruments to detect and monitor the impact of human activity on the natural flux of elements and global-scale processes (Meyer 2002). Technological advancements in the collection and analysis of data made it possible to record trends in change such as that of carbon dioxide concentrations or radioactive fallout from the atmospheric testing of nuclear weapons. Data on the production of xenobiotic compounds and the discovery that persistent molecules and metals accumulate even in organisms from the remotest regions of the Earth increased public awareness of the environmental impact of human activity. The progressive acidification of many terrestrial and aquatic ecosystems in remote regions of the Northern Hemisphere, the discovery of the role of chlorofluorocarbons (CFCs) in the recurring formation of the "ozone hole", and the understanding of the possible consequences of climate change on global sea levels, biodiversity, and human health increased global environmental concern and became the subject of many discussions and debates.

According to the Assessment Report of the Working Group of the Intergovernmental Panel on Climate Change (IPCC 2001), radiative forcing (i.e. the index, expressed in W m^{-2}, of the influence of a factor on the balance between

incoming and outgoing energy in the earth–atmosphere system) due to natural factors (e.g. changes in solar output or explosive volcanic activity) does
not explain global warming in the second half of the 20th century. The best
agreement between model simulations and observations from the last
140 years is found when anthropogenic and natural forcing factors are combined. The impact of human activity is mainly due to the combustion of fossil
fuels and land-use changes, which have increased carbon dioxide concentrations (360 parts per million; ppm) to the highest levels in the past
420,000 years, and at a rate of increase (0.4 % year^{-1} over the past two decades)
without precedent in at least the past 20,000 years. Concentrations of other
well-mixed greenhouse gases, such as methane, carbon monoxide, nitrous
oxides, perfluorocarbons (PFCs), sulphur hexafluoride (SF_6) and other synthetic compounds, are increasing. From 1750 to 2000 radiative forcing due to
the increase of these gases has been estimated to be 2.43 W m^{-2} (IPCC 2001;
Fig. 9). As these gases have long lifetimes and a nearly uniform spatial distribution, a few observations coupled with an understanding of their radiative
properties suffice to yield estimates of radiative forcing and of related global
mean surface-temperature responses (Shine and Forster 1999). Whereas well-
mixed greenhouse gases cause radiative forcing everywhere on the globe,
forcing due to aerosols, tropospheric and stratospheric ozone, and other
short-lived compounds varies spatially (e.g. sulphate and carbonaceous
aerosols, and tropospheric ozone prevail in the Northern Hemisphere,

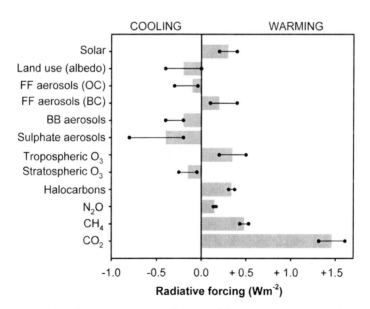

Fig. 9. Estimated global mean radiative forcing of the climate system for the year 2000,
relative to 1750 (data from IPCC 2001; *BC* fossil fuel-burning black carbon, *OC* organic
carbon, *BB* biomass burning)

whereas stratospheric ozone and biomass burning aerosol prevail in the Southern Hemisphere). Different radiative forcing mechanisms lead to differences in the partitioning of perturbations between the atmosphere and surface. Increases in temperature are thus not evenly distributed in the two hemispheres and show significant variations at regional scales. In the Northern Hemisphere, for instance, the warming rates of three regions between 50 and 70° N (i.e. in Siberia near Lake Baikal, in western Alaska near Nome, and in Canada's Prairie Province in the North West Territories) are roughly three times greater than the global average (Cuff 2002).

The impact of anthropogenic and naturally emitted substances depends on their radiative properties and on the timescale characterising their removal from the atmosphere. Global Warming Potentials (GWPs) are a measure of the relative radiative effect of a given substance compared to another, integrated over different time horizons (i.e. a simplified index allowing estimates of the potential future impact of substances on the climate system, in a relative sense). The WMO (1999) gave the most recent GWPs evaluations for 20-, 100- and 500-year time horizons. Although estimates from the different scenarios vary greatly, the study predicts that global mean temperatures and sea levels will continue to increase for hundreds of years after the stabilisation of greenhouse gas concentrations (even at present levels). This large interval of time is due to the long timescales at which deep ocean waters adjust to climate change.

Based on scientific observation and models, the physical plausibility of projections in all commonly used scenarios and expert judgement, the Third Assessment Report of the IPCC (2001) identified the changes which will probably occur over nearly all (or most) land areas during the 21st century: (1) an increase in maximum and minimum temperatures, the number of hot days, and heat indexes; (2) a decrease in the number of cold days, frost days and diurnal temperature ranges; and (3) more intense precipitation events.

Polar regions will play an important role in this scenario. Models of global climate change indicate that the largest equilibrium warming occurs in polar regions in winter and, as will be discussed in Chapter 2, there is increasing evidence of environmental changes in ecosystems situated at latitudes between 40° and 70° . Antarctica is providing important geophysical data which will increase our understanding of past climate change, the clearest link between atmospheric concentrations of greenhouse gases and surface temperatures (Lorius et al. 1985), and of physico-chemical processes leading to the formation of the ozone "hole" (Molina and Rowland 1974). Its very sensitive ecosystems will provide useful advanced warning of some of the effects of wider-scale ozone depletion and warming.

The recent break-up of Antarctic Peninsula ice shelves (Vaughan and Doake 1996) and the calving of very large icebergs (e.g. B-15 in March 2000 from the Ross Ice Shelf, and A-43 on May 2000 from the Ronne Ice Shelf) give cause for concern and are relevant to discussions on global warming. However, there are many misconceptions among the media about the extent of recent climate change in Antarctica and its possible impact on the rest of the world. The following sections will deal with variations in historic climate records from Antarctica and interactions with the global climate system.

1.5.2 Trends of Surface Air Temperature in Antarctica

The climate of polar regions usually shows much greater inter-annual or inter-decadal variability than that of lower latitudes. An understanding of the factors driving this variability and its extent are important to foresee how the Antarctic climate might change as a result of the increase in greenhouse gases and other emissions produced by human activity. Several measurements of air temperature in Antarctica were performed during early expeditions, and comparisons between mean air temperature values at the beginning of the last century and more recent records indicate an average increase of about 1 °C, mostly occurring during the last 40 years (Jones 1995). Systematic recording of temperatures in the continent began only in 1957–1958 during the International Geophysical Year, and are restricted in spatial coverage because most climate stations have been located in easily accessible coastal zones of East Antarctica and the Antarctic Peninsula. Only two stations have recorded about four decades of temperature data on the high, vast interior.

It is hard to assess the statistical significance of temperature variations in most stations due to the short time interval covered by records and the high degree of inter-annual variability. Sansom (1989), for instance, found that temperature trends at Scott Base and the Faraday, Mirny and Amundsen-Scott stations for the period 1957–1986 were not statistically significant if the year-to-year persistence was appropriately considered. It was later shown (King 1994; Stark 1994) that the warming trend at Faraday Station was highly significant in the period 1945–1990. At the same station King and Turner (1997) found a significant year-to-year persistence of surface-temperature anomalies (+0.052 °C year^{-1} from 1945 to 1993) which is much greater than the range of natural variations. An increase of about 2.5 °C in 50 years has been reported for the Antarctic Peninsula (e.g. Doake and Vaughan 1991; Stark 1994; Morris et al. 1997), indicating that some parts of this region are experiencing one of the largest warming trends in the world.

Figure 10 compares trends in mean surface temperatures during the last 45 years for three Antarctic climatic zones with those for the Southern Hemisphere and global average values. Although this is a rough comparison which takes into consideration only a few decades and refers to available mean values obtained from a different number of records, Fig. 10 shows an anomalous warming trend for the Antarctic Peninsula, while most of the continent is rather cold in the same period of time.

As predicted by global climate models, at Faraday Station the winter months gave the greatest contribution to inter-annual temperature variability.

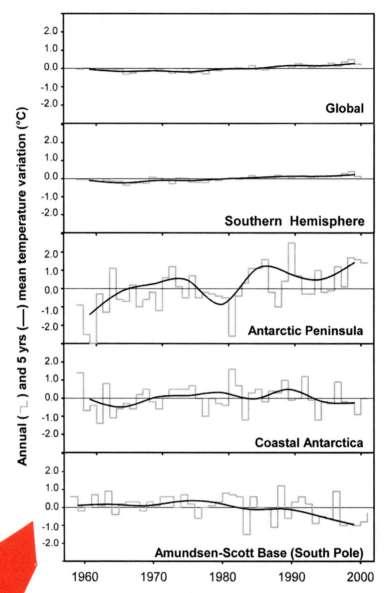

Trends of annual and 5-year mean temperature variations expressed with
...t to the 1957–2001 average value (for references, see text)

Significant warming trends have also been recorded at Marguerite Bay (about
300 km south of Faraday Station) and further south up to the latitude of about
70° S, at Alexander Island (King 1994). Like in the Northern Hemisphere, the
surface-temperature increase in the Antarctic Peninsula does not seem evenly
distributed. Records at Faraday Station show a much higher warming trend

than those at Orcadas Station, and correlate less with data from the South Shetland Islands, at the tip of the Antarctic Peninsula, than with records from the southern part of the peninsula.

Some of the longest air temperature records come from sub-Antarctic islands such as Signy Island Station (South Orkney Islands) and South Georgia. From the beginning of the last century, the mean annual air temperature increased much less in these islands (BAS 1987; Stark 1994) than at Faraday Station. Likewise, the warming trend of other sub-Antarctic islands such as Îles Kerguelen (Frenot et al. 1995), Marion Island, Macquarie Island and Heard Island (Chown and Smith 1993) is lower than that of Faraday Station.

According to King (1994), the high persistence of temperature anomalies from one year to the next at Faraday Station is probably in relation to changes in oceanic circulation and temperature. He found that the west coast of the Antarctic Peninsula is the only region of Antarctica showing a significant relationship between air temperature and sea-ice extent. A significant inverse relationship between winter air surface temperature and the extent of sea ice has also been found in the Bellingshausen Sea (King and Turner 1997). The coast of the west Antarctic Peninsula lies close to the ice edge during most of the winter, and the sea ice affects the absorption of sun radiation by seawater and the flow of heat from the sea to the atmosphere. Routine satellite records show considerable inter-annual variability in the winter extent of ice in the Bellingshausen Sea, although remote sensing records are not long enough to determine whether there is a significant relationship between the sea-ice extent and winter warming (Gloersen and Campbell 1988). Besides the sea-ice extent, other factors also probably contribute to the increase in surface temperature in the western Antarctic Peninsula. A strong northerly component of atmospheric circulation, with the advection of warmer air masses, is associated with anomalous winter temperatures (King 1994). Increased cloudiness probably also contributes to warming, because in winter the surface albedo is more important than direct insolation.

As will be discussed in Chapter 2, the progressive increase of mean annual temperatures in some western coastal areas of the Antarctic Peninsula is affecting small ice caps, low-lying glaciers, and ice shelves, the reproduction and survival of organisms, and colonisation processes. Warming is mainly due to local features and processes, and this trend probably cannot be extrapolated to the continent or sub-Antarctic islands (e.g. Ellis 1991; Zwally 1991; King and Turner 1997). A different history of air temperatures on the continent and peninsula has been reported by Jones (1990), who compared air temperature records from early expeditions in the late 19th century onwards with those of the period 1957–1986 and found a warming (from 0.3 to 3.0 °C) on the Antarctic Peninsula. However, 20th century expedition records from the McMurdo Sound region (Ross Sea) show no significant increasing or decreasing temperature trend. Likewise, King and Turner (1997) examined the longest series of mean annual surface air temperatures recorded in selected

Antarctic stations and found a very complex spatial pattern of temperature variations. Large temporal variations on the west coast of the Antarctic Peninsula were poorly correlated with temperature trends in East Antarctica. It was therefore assumed that the longer series of observations from stations in the Antarctic Peninsula and sub-Antarctic islands cannot be used to infer the history of temperature in the continent.

Data from East Antarctic stations referring to the period 1957–1991 show more subtle warming trends and, in the same period, values measured at Amundsen-Scott Station indicated a slight ongoing cooling (Fig. 10). The topography of Antarctica probably plays a very important role in determining the high variety of regional climates in the continent. The air temperature and pressure records collected by about 30 automatic weather stations installed in the 1980s at remote locations by the US Antarctic Program (USAP), to supplement measurements in research stations, show that climate regimes in coastal areas of Antarctica are very different from those in the interior (Stearns et al. 1993). Moreover, the climate changes significantly in areas of the plateau at different elevations, e.g. Dome C (above 2,500 m) vs. the Siple and Byrd stations in West Antarctica (1,000–1,500 m). The coastal climate can be even more varied depending on the slope of the terrain (i.e. the katabatic flow). Although Stearns et al. (1993) found similar monthly mean air temperatures at Dome C and Clean Air (South Pole Station), during the second half of the 20th century the instrumental records showed different trends at Vostok Station (a slight warning trend) and a slight cooling at Amundsen-Scott Station (South Pole). Owing to the uniform morphological characteristics of the polar plateau, large areas surrounding the two stations are probably characterised by these two opposite trends (King and Turner 1997). The South Pole has been cooling slightly since 1976, and Dutton et al. (1991) pointed out an increased cloud cover in January–February from 1976 to 1985, while summer insolation partially recovered in the period 1986–1989. Neff (1992) suggested that cooling could be related to the beginning of spring depletion of stratospheric ozone, which may reduce the stability of the stratosphere and allow warm, moist air to reach the plateau. Unfortunately, the processes linking changes in surface air temperature to variations in tropospheric and stratospheric circulation are particularly complex, and they are largely unknown in Antarctica because of the sparse radiosonde network.

1.5.3 Extending Spatio-Temporal Temperature Trends

Land-based instrumental measurements of air temperature in the Southern Hemisphere began in the mid-19th century, and some records have been used to compile datasets (e.g. the World Weather Records and the Global Historic Climate Network; Jones 1994). Marine records began in the same period; they are largely confined to major shipping routes from Europe to South America and Australasia, and have been collected in the Comprehensive Ocean Atmosphere Data Set (Wooddruff et al. 1987) and marine data bank of the UKMO (Parker et al. 1995). These datasets are affected by different recording practices and sites; however, after corrections and homogeneity assessment, they have been used to investigate temporal trends in the average air temperature of the Southern Hemisphere (Jones 1994; Jones and Allan 1998). In the period 1858–1996 the land-based data show a warming of 0.4 °C, common to all seasons. After the two coldest decades (the 1880s and 1890s), there was an almost steady warming, with a total increase in temperature of about 0.6 °C during the last century. This trend is corroborated by historical records of air temperature over the oceans (Parker et al. 1995), which are characterised by a much lower year-to-year variability and a clear increase in average temperatures since 1910, with a marked cooling in the mid-1940s. The temperature rise in the Southern Hemisphere shows a much more linear trend than that of the Northern Hemisphere, and all seasons show a quite similar increase in temperature. Several zones display a warming trend, and the late 1980s and early 1990s are among the warmest years on record. In general, regional characteristics of temperature and precipitation series relate to their location and connection with ENSO events (Jones and Allan 1998). However, 1992 and 1993 were exceptionally cool years in New Zealand while, in contrast to most regions of the Southern Hemisphere, Antarctica (except for the Antarctic Peninsula) experienced a rather cold phase (Comiso 1999; Fig. 10).

High-resolution proxy records (i.e. those allowing an annual resolution and spanning up to a few millennia) can be used to evaluate whether air temperature variations and changes in the period covered by instrumental records are typical or anomalous with respect to the recent past and to collect information on areas without historic records (such as most of the Antarctic plateau). Although preliminary studies on tree rings, peat bogs and coral growth have been performed in different regions of the Southern Hemisphere (e.g. Lough et al. 1996; Villalba et al. 1997), most available records come from ice core proxies. The oxygen isotope ratio (largely dependent on the temperature at which the water vapour condensed to form precipitation) recorded in ice layers may provide an indication of past climatic conditions. This isotopic thermometer can be calibrated by comparing oxygen isotope ratios in the upper layers of an ice core with temperature records from meteorological stations and temperature profiles from ice boreholes (Clow et al. 1996). The analysis of soluble and insoluble constituents (e.g. main marine anions and cations, continental dust, volcanic dust, radionuclides from nuclear fallout, persistent organic pollutants) allows the dating of cores and the identification of spatio-temporal variations in atmospheric composition and circulation.

High-elevation ice caps (about 5,500–6,000 m) in the Peruvian Andes yield the longest proxy records for low-latitude regions of the Southern Hemisphere. Isotope and snow-accumulation records in two ice cores from Quelcaya (Thompson 1996) provide evidence of the Little Age cooling and a warming during the 20th century. The ice layers show a marked seasonality of oxygen isotope variations, and alternations between dry spells and high snow-accumulation rates in two periods: from 750 to 1000, and from 1500 to 1700 A.D.

Several ice cores have been drilled in the Antarctic Peninsula and, in general, the results show a strong warming trend since the 1950s and warmer conditions in the early 19th century (Peel 1992). The cooling trend indicated from the late 19th century to 1950 seems in contrast with the rather steady warming trend detected through instrumental records. However, the isotopic signal in the snow depends on the condensation temperature at the core site and on the source region of water. In this case, the isotopic anomaly was probably due to a polynya in the Weddell Sea (Jones et al. 1993) which supplied cold water (compared with the open-ocean water from the north) for precipitation falling in the eastern Antarctic Peninsula. As water vapour from relatively cold seas can give rise to isotopically anomalous signals, changes in additional isotope ratios such as that between hydrogen and deuterium have to be checked in order to separate the effects of temperature changes in the source from those of the deposition region.

Many ice cores have been drilled in the Antarctic continent, but most are characterised by a scarce accumulation of snow and do not allow a clear identification of annual layers before the 20th century. In general, isotopic records show a complex pattern of temperature variations over the past 500 years (Mosley-Thompson 1992). Two of the probably best-dated ice cores were drilled at Law Dome (Morgan and van Ommen 1997) and Plateau Station (Mosley-Thompson 1996). However, no evidence of long-term changes in temperature was found in either place. The main finding was that autumn and winter isotope values tended to decrease (cf. cooler temperatures) in the late 18th and early 19th centuries.

1.5.4 Moisture and Precipitation Trend

Processes in the climate system determine its response to natural or anthropogenic forcings. Feedback processes are very important to estimate the sensitivity and evolution of climate, because they can amplify (positive feedback) or reduce (negative feedback) the response to an initial perturbation. The water-holding capacity of the atmosphere increases with temperature and, because water vapour is a powerful greenhouse gas, the increase in atmospheric water vapour produces a major feedback process.

Models of global weather patterns predict that, in response to a doubling of carbon dioxide concentrations (e.g. Cess et al. 1997; Hall and Manabe 1999; Held and Soden 2000), the warming produced by water vapour feedback alone would be approximately double that produced by fixed values of water vapour. Furthermore, water vapour feedback amplifies other feedback mechanisms such as cloud feedback and ice or snow albedo feedback. However, as most of the atmosphere is unsaturated with respect to water vapour, an increase in air temperature does not necessarily mean that the eventual warming will be characterised by a proportional increase in the concentration of water vapour.

In recent years the treatment of water vapour in models has dramatically improved, but factors governing its behaviour in the free troposphere (roughly above the 1–2 km thick boundary layer) remain uncertain, and there are discrepancies between the modelled and observed distribution of water vapour (IPCC 2001). Even greater uncertainties in climate projections arise from cloud feedback. In this case, the sign (positive of negative) of feedback is unknown (Watterson et al. 1999; Meleshko et al. 2000). Clouds can cool the Earth's surface by absorbing and reflecting solar radiation and by warming the surface through the absorption and emission of long-wave radiation. The balance between these two effects depends on changes in cloud height, thickness and radiative properties; in turn, these properties depend on the evolution of atmospheric water vapour and aerosols, water drops, and ice particles.

Despite uncertainties in the possible effects of water vapour and cloud feedback mechanisms, all climate models predict a general increase in mean atmospheric precipitation in the tropics and at mid and high latitudes, and a general decrease in rainfall in the subtropical belts (IPCC 2001). Antarctica is characterised by unique climatic processes, involving complex interactions and feedback loops, which may ultimately lead to a glacial–interglacial climatic transition (Simmonds 1998; Petit et al. 1999). All models predict a wetter and warmer climate over the 21st century. In the Special Report on Emissions Scenarios (SRES) commissioned by the IPCC (2000), the projections for polar regions are well beyond the range of variability of the current climate, and the increases in temperature and precipitation are among the largest for any part of the globe. With respect to the present Antarctic climate summarised by Carter et al. (2000), the IPCC-SRES climate scenarios for 2080 predict a +2 to 17 % increase in precipitation and +0.0–2.8 °C increase in surface temperature during summer, and a +5–20 % increase in precipitation and +0.5–5.0 °C increase in temperature during winter. These large ranges indicate that the future climate of Antarctica remains uncertain. In any case, the projected temperature increase will probably have little impact on the melting of the ice sheets because almost all the continent will remain well below freezing. Satellite altimetry shows that, at present, most portions of the Antarctic ice sheets are nearly in equilibrium (Wingham et al. 1998; i.e. the accumulation of snow and ice on the continent roughly corresponds to iceberg calving and basal melting of ice shelves). With the exception of glaciers in some coastal zones and the Antarctic Peninsula, this balance will probably be scarcely affected by global warming in the next century.

Unfortunately, the Antarctic Peninsula is not well resolved and local effects cannot be reproduced in the present generation of global climate models. Furthermore, local responses to small climate changes may be exaggerated by local climatic and environmental features, and it is impossible to discriminate the effects of global and regional processes. Local anomalies and documented changes in precipitation and temperature in the Antarctic Peninsula (e.g. Drewry 1991; Peel 1992; King and Turner 1997; King and Harangozo 1998; Smith et al. 1999) seem to mostly corroborate projected climate-change scenarios for 2100.

Since the beginning of the 1990s it has been suggested (Warrick and Oerlemans 1990; Simmonds 1992) that, if the emission rate of greenhouse gases does not change, atmospheric precipitation over Antarctica will increase, with a likely local thickening of ice sheets. Fortuin and Oerlemans (1990) predicted that a 1 °C uniform warming would produce higher snow accumulation and stronger evaporation in coastal areas, with a negative contribution (-0.27 mm year^{-1}) to sea-level change. Despite model uncertainties in the future climate of Antarctica, moisture and precipitation seem to have increased over the continent since the 1960s (Morgan et al. 1991). Among scientists (e.g. Ohmura et al. 1996; Smith et al. 1998; Vaughan et al. 1999), there is widespread consensus that the enhanced accumulation of snow will contribute to the lowering of sea levels for some hundred years, until the increased ice discharge and melting finally produces a rise in sea level.

The increase in precipitation over Antarctica will probably not be directly related to small variations in surface air temperatures recorded in the same period and consequent changes in the moisture-holding capacity of air. At Faraday Station in the period 1956–1993, temperature increased significantly and precipitation increased by about 20 %, but very little year-to-year correlation was found between the two parameters (King and Turner 1997). Based on the analysis of ice core data, time-series data of mean annual surface pressures over the Bellingshausen Sea, wind components and cloudiness, researchers concluded that changes in atmospheric circulation and enhanced cyclonic activity are the primary sources of the inter-annual variability of precipitation over the Antarctic Peninsula. In several sub-Antarctic islands, despite a significant warming trend, there is also a remarkable decrease in precipitation (Chown and Smith 1993; Frenot et al. 1995). These rather surprising climate changes are probably also linked to changes in the frequency and track of cyclones at latitudes between 40 and 60° S.

Antarctica is a sink for water vapour produced in the tropics by excess evaporation with respect to precipitation. If this flow were to increase, the continent would also receive increasing amounts of persistent atmospheric pollutants. However, it seems very likely that their deposition throughout the continent would show a rather complex pattern. The isotopic composition of snow at Syowa Station (coastal East Antarctica; Bromwich and Weaver 1983) indicates that the main source of water vapour was located in the Southern Ocean, north of the pack ice. The most plausible explanation for this result is that moist air masses moving southwards from the subtropics lose some of their water vapour through precipitation, which is then replaced by evaporation in the Southern Ocean. According to Bromwich (1988), evaporation between 62° and 65° S is sufficient to supply precipitation in Antarctica.

Model studies by Ciais et al. (1995) indicated a completely different source (between 20° and 40° S) for precipitation over the Antarctic plateau. At the South Pole the extension of large-amplitude long waves with warm, moist air from mid-latitudes is also supported by some precipitation events characterised by rapid increases in surface temperature (Sinclair 1981). Based on these indications and available data on clouds, circulation regimes and precipitation processes, King and Turner (1997) suggested different sources for water vapour in precipitation falling in coastal areas and the interior of Antarctica.

The most striking feature of the mean surface pressure field in the Southern Hemisphere is the circumpolar trough which encircles the continent between 60° and 70° S. The trough changes its position and intensity through the year; it is furthest south and most pronounced in the spring and autumn and moves north, weakening, in the summer and winter. This semi-annual oscillation is due to the phase difference between the seasonal cycle of surface pressure values over Antarctica and sub-Antarctic latitudes. It affects the strength of the westerlies and atmospheric precipitation from latitudes of 40° to 60° S (King and Turner 1997). Changes to circulation in the cyclogenetic area of the mid-latitudes affect the position and depth of the circumpolar trough, and consequently the climate and deposition of long-range transported pollutants in coastal Antarctica.

The ENSO phenomenon is the largest inter-annual climatic variation on Earth and, during the last decade, a number of studies have attempted to use climate models to assess the changes which might occur in ENSO as a result of climate warming. Unfortunately, most models have shown conflicting results, and it is very difficult to attribute past and recent changes in the amplitude and frequency of ENSO to external forcing factors. The ENSO pattern can apparently occur at a variety of timescales, either without any change in forcing or in response to external forcing such as an increase in carbon dioxide (Knutson and Manabe 1998; Boer et al. 2000). In the Southern Hemisphere, most recorded series of regional temperature and precipitation relate to their location with respect to the ENSO. Although several claims have been made, the phenomenon only slightly influences Antarctica, South America (south of 40° S) and south-western parts of Africa (Jones and Allan 1998). It will thus be even more difficult to single out possible in-phase and out-of-phase relationships between the relatively short data records from Antarctica and ENSO events. Furthermore, there is now considerable evidence that quasi-biennial and lower-frequency signals of the ENSO are also modulated by decadal-multidecadal fluctuations in the climate system (e.g. Allan et al. 1995; Wang and Ropelewski 1995).

The connection between the Antarctic climate and meteorological phenomena in the Southern Hemisphere and teleconnections with the rest of the global atmosphere are only beginning to be explored. The major challenge for future research will be the acquisition of more exhaustive, improved datasets, for a better understanding of the physical processes underlying the spatial and temporal variability of the Antarctic climate. This knowledge is a prerequisite for the development of more reliable global and regional models and for improving predictions.

1.6 Summary

The recent development of remote sensing techniques and technologies for automatic data collection and transmission from the most inaccessible areas of Antarctica is greatly improving our knowledge of climate in the interior of the continent. Although much remains to be understood about the geological history, advances in geophysical techniques have allowed researchers to outline the sub-ice topography of Antarctica and reconstruct its evolution, which is linked to the amalgamation and break-up of Gondwanaland and earlier continents. The stable shield of East Antarctica is formed by a complex of Precambrian cratons overlain by sedimentary rocks (Beacon Supergroup) and intruded by basaltic rocks of the Ferrar Group. West Antarctica was formed by the aggregation of several microcontinents sharing a common geological history with South America. For the purposes of this book, one of the most important features of Antarctica is its location in an expanding lithospheric plate which has been in a quite stable position with respect to the South Pole during the last 100 Ma. This means that climatic and environmental changes in this period mainly reflect global changes.

Antarctica's potential to yield mineral, coal and hydrocarbon resources such as those found in rocks of formerly contiguous continents has been the subject of much speculation in the past. Although the presence of exploitable mineral and hydrocarbon deposits is very probable, their location is largely unknown. In any case, exploitation is prohibited, at least for the near future, by a moratorium (to protect the environment) and by the severity of environmental conditions.

The dynamics of the atmosphere and oceans in the Southern Hemisphere are driven by temperatures in Antarctica, which redistribute the polar cold to lower latitudes, and replace it with warmth. A small increase in temperatures at high latitudes may be amplified by ice–albedo feedback mechanisms, with significant repercussions on the global climate system. The subsidence of cold polar air and Coriolis deflection produce an intense, clockwise polar vortex in winter, which is accompanied by high-pressure systems at the South Pole and the circumpolar flow of strong westerly winds. This vortex play an important role in the destruction of stratospheric ozone; when it disappears in the spring, there is an enhanced influx of warm, moist air masses from higher latitudes bearing persistent pollutants.

Almost all precipitation in the continent falls as snow, and East Antarctica is the largest cold desert in the world because most areas receive less than 250 mm (water equivalent) per year of snow. Although characterised by large inter-annual variability, atmospheric precipitation is much higher (up to 800 mm water equivalent) in coastal areas of West Antarctica. Under the present climatic conditions, the accumulation of snow and ice on the continent seems to roughly compensate iceberg calving and basal melting of ice shelves. Furthermore, satellite monitoring of sea-ice extent and variability in the Southern Ocean during the last 25 years has revealed no general trends. However, these conditions of relative equilibrium are expected to be significantly affected by global warming. The most recent trends in records and projections of models on global climate change indicate that the climate of Antarctica will become wetter and warmer in the 21st century.

The potential contribution of Antarctic ice sheets to the increase in global sea levels has caused general concern and interest in how the Antarctic climate may respond to increasing concentrations in greenhouse gases and other emissions produced by human activity. While some zones, such as the west coast of the Antarctic Peninsula, are already showing one of the largest warming trends in the world, East Antarctica is one of the few regions in the Southern Hemisphere experiencing a quite cold phase. Antarctic ice sheets, even those of West Antarctica (which were the most dynamic Antarctic ice sheets in the recent geological past and are grounded below sea level), are expected to scarcely contribute to global sea-level changes in the 21st century.

Although connections between the Antarctic climate, ENSO events and global atmospheric processes are only beginning to be investigated, it is now widely acknowledged that the enhanced accumulation of snow in Antarctica will contribute to a positive mass balance of Antarctic ice sheets, with a negative contribution to sea-level rise in the near future. In the longer term, the warming of waters in the Southern Hemisphere and changes to circulation will probably trigger processes which could last for millennia, long after greenhouse gas emissions have stabilised. These processes will cause basal melting of ice shelves and progressive, irreversible impacts on ice sheets, especially those of West Antarctica. The possible consequences of these long-term changes in global climate and sea levels are such that research on the Antarctic climate and its teleconnections with climate and environmental changes in the Southern Hemisphere and the rest of the world is a major priority for the 21st century.

2 Glacial, Terrestrial and Freshwater Ecosystems

2.1 Introduction

The most impressive feature of the Antarctic environment is undoubtedly the huge amount of ice in both the continent and Southern Ocean (in winter). Antarctic ice sheets lock up about 70–80 % of the Earth's freshwater, enough to raise the global sea level by some 60–70 m. Global climate models predict that, in the short term, changes in air temperature, clouds and precipitation will probably increase the accumulation of snow in Antarctica. This accumulation, together with possible changes in Southern Ocean temperatures and circulation, will have long-term effects on ice flow and ice ablation, with enhanced return of water to the ocean, through icebergs and meltwater runoff.

This chapter gives a brief account of the flow and mass balance of Antarctic ice sheets, and emphasises the role of Antarctic ice cores in research on global processes. As successive snow layers build up, those beneath are gradually compressed into solid ice, which preserves a unique and undisturbed record of past and recent changes in the composition and state of the atmosphere. Studies on deep ice cores drilled in Antarctica provide some of the best temporal accounts of the close correspondence between air temperature and greenhouse gas concentrations, and allow the reconstruction of climate changes over the last 500,000 years. In contrast to well-mixed greenhouse gases, the incorporation into snow layers of persistent atmospheric pollutants deposited through snowflakes or aerosols is governed by largely unknown processes. However, as will be discussed in Chapter 4, significant changes in pollutant concentrations in snow and ice on extended temporal scales can be used to infer similar changes in the composition of the atmosphere. These changes can be used to detect the impact of anthropogenic activities in Antarctica and elsewhere in the world.

For more than 25 Ma, Antarctica has had no terrestrial connection with any of the continental landmasses in the Southern Hemisphere and, during this time, the continent has been almost completely covered with permanent snow and ice. Even in the present warm interglacial period, less than 3% (about 331,700 km^2) of the continental area (including the islands) is permanently or seasonally free of ice and snow. Most of these ice-free areas are characterised by low temperatures and precipitation (cold deserts) and, like deserts in warmer regions, they show dry kettles, ventifacts and surface salt encrustations, with scarce biota. Antarctic terrestrial and freshwater ecosystems play a minor role in global water and carbon cycles; nevertheless, they are very important for research on the environmental and ecological effects of global climate changes. The remoteness of these ecosystems from human civilisation and the extreme environmental conditions make Antarctica a unique laboratory for studying cold adaptation and colonisation processes by organisms which are often located very far from sources of propagules. By virtue of the reduced number of species and interactions among organisms, Antarctic ecosystems allow the identification of critical factors operating in the environment, assessment of the flux of nutrients and pollutants between abiotic and biotic components, easy identification of accidentally introduced alien organisms, and of colonisation processes in old or newly exposed substrata. This knowledge is very useful for better understanding of more complex ecosystems elsewhere and predicting their possible response to climate and environmental changes. Besides contributing to ecological science as a whole, pristine Antarctic ecosystems offer a unique opportunity to detect changes in the amount and composition of global atmospheric pollutants. Knowledge of pollutant concentrations in the Antarctic environment allows the establishment of global baselines and proper management of polar ecosystems in view of the progressive expansion of field research, tourism and human activity in scientific stations.

In the previous chapter, we saw that areas of the Antarctic Peninsula are experiencing one of the largest warming trends in the Southern Hemisphere and probably, of the world. Simplified biotic communities in cold desert ecosystems are strongly affected by climate forcing, and the responses of terrestrial and freshwater ecosystems are already tangible in the Antarctic Peninsula. Antarctic ecosystems are expected to provide a better indication of the effects of climate change than ecosystems at lower latitudes, where the response of biotic communities to external forcing is buffered by more complex biological interactions and feedback processes.

2.2 Glacial Systems

According to a general classification scheme (e.g. Armstrong et al. 1973; Sugden and John 1976), ice sheets (or ice caps, if the surface area is less than 50,000 km^2) are glacial systems which submerge the landscape, at least in their central portion. These systems are unconstrained by topography and the ice flow is largely independent of sub-ice undulations. Ice domes are approximately symmetrical, upstanding areas of ice sheets or ice caps usually characterised by slow-moving ice (sheet flow). Faster-moving ice (stream flow) occurs at the margin of ice sheets in outlet glaciers (bound by ice-free terrain) and/or ice streams (flanked by slow-moving ice) discharging ice towards the periphery (Bentley 1987). However, some fast-moving ice systems, such as the world's longest (Lambert Glacier, East Antarctica, about 700 km long) and fastest (Jacobshavn Glacier, Greenland; 8,360 m year^{-1}) glaciers, are ice streams along some stretches and outlet glaciers along others (Benn and Evans 1998).

Many terms have been introduced to describe the forms of glaciers constrained or controlled by topography (e.g. ice fields, valley glaciers, cirque glaciers, piedmont glaciers; Embleton and King 1975; Sugden and John 1976). Although some of these glaciers occur on mountain slopes and in coastal ice-free areas of Antarctica, glacier ice shelves, resulting from the flotation of glacier tongues, are much more important for their extension (about 7 % of the surface area of ice sheets and 44 % of the coastline).

Glacial systems receive snow and ice from direct snowfall, blown snow and snow avalanching from slopes above the ice surface. In Antarctica, the climatic boundary of permanent snow lies below sea level and, because of very limited or no summer melting, the entire continent is an area of net snow accumulation. The main exceptions are probably local surfaces scoured by strong katabatic winds, which redeposit snow at sheltered lee-side locations and/or on the sea. Snow precipitation accumulates year after year in successive accumulation layers, and the increasing load of solid precipitation determines a reduction of air-filled spaces. The snow progressively transforms into a mass of loosely packed ice crystals with interconnecting air passages (firn). The density increases from 100–200 kg m^{-3} in freshly fallen snow to values above 400 kg m^{-3} in firn. The transition to ice occurs when interconnected air passages become sealed, the air is closed in individual bubbles, and the density is 830 kg m^{-3} (Paterson 1994). Glacier ice may reach a density of more than 900 kg m^{-3} through further compression of air bubbles.

The transformation of snow into ice occurs at different times and depths, depending on the climate. The development of ice is maximum and takes place after a few years, at shallow depths, in areas such as the Antarctic Peninsula or in sub-Antarctic islands where high snowfalls and the pressure of overlying snow cause crystals to move relative to one another, producing rather rapid compaction and an increase in density. Another factor accelerating the transformation of snow into ice is occasional melting at the surface or within the snow pack. The meltwater percolates downwards through the snow and then refreezes at the contact with snow or ice (superimposed ice). Because heat is released during freezing, the formation of superimposed ice warms the surrounding snow and produces further melting, until all the water is refrozen when the weather turns colder (Benn and Evans 1998). Melting does not occur on the Antarctic plateau, and the snow accumulation rate is usually very low. Winds increase the density of snow by blowing it across ice surfaces, breaking up snowflakes into smaller crystals and depositing them in drifts. Although these drifts have a much higher density than snow deposited in still air, on the Antarctic plateau the transformation of snow into ice occurs at depths of 70–100 m and after 2,000–2,500 years (Barnola et al. 1987). This depth and time are typical of climatic conditions in the present interglacial phase, but the process may require up to 6,000 years during glacial periods (Baroni et al. 2001). In the McMurdo Dry Valleys (Victoria Land) and other very dry, ice-free areas of East Antarctica, snowfall is not abundant enough to promote the development of an ice sheet.

The ice accumulated in glacial systems moves from accumulation areas in the continental interior to coastal ablation zones. In order to maintain a steady state, the accumulation of ice and snow in glaciers must compensate the losses through wind ablation, runoff, evaporation, sublimation, the calving of icebergs and avalanching of ice blocks from terminal ice cliffs. About 75 % of glacier ablation in Antarctica is achieved by iceberg calving into the Southern Ocean, and basal melting occurs where the snouts are floating.

The mass balance is the difference between gains and losses, expressed in terms of water equivalent and measured over a specified time period. Direct measurements of annual snow accumulation can be performed in pits excavated in the snow pack, while ablation is measured through a network of stakes in the ice. However, as discussed in the previous chapter, these methods are unsuitable on Antarctic ice sheets and more reliable estimates of mass balance can be obtained through remote sensing methods (Jacobs et al. 1992; William and Hall 1993) and from meteorological data (Braithwaite and Olesen 1989). Annual ablation and accumulation vary with altitude (mass-balance gradients) in most glaciers, and the equilibrium line altitude (i.e. where total annual accumulation exactly balances total annual ablation) is an important indicator of glacier response to climate change (Paterson 1994).

Glaciers move when the forces exerted by the weight and surface slope of the ice overcome the strength of the glacier or its bed, allowing the ice to slide past obstructions on the bed or to deform (Benn and Evans 1998). Detailed analysis and description of why and how glaciers move and their interactions with the landscape are reported in several publications (e.g. Hutter 1983; Twiss and Moores 1992; Paterson 1994; Benn and Evans 1998). Essentially, movement is the cumulative effect of three processes acting singly or in combination: permanent internal deformation of ice itself, deformation of the bed underlying the glacier, and sliding at the ice–bed interface.

Glacial systems are not uniformly cold, and temperature variations have profound implications because ice moves efficiently when it is at or close to the melting point. The temperature at which ice melts is not always 0 °C, but decreases according to pressure at a rate of 0.072 °C per million Pa (Pascal) of pressure. Moreover, the geothermal heat flux and frictional heat generated by the flow may also increase basal ice temperatures. Thus, although the temperature of surface ice on the Antarctic plateau is always well below 0 °C, the base of very thick ice sheets may be at the melting point. Ice streams and tide-water glaciers with warm basal ice and efficient basal sliding may reach velocities of some thousands of metres per year. Due to topographic funnelling and very high mass turnover, the San Rafael Glacier in North Patagonia, for instance, has a velocity of 7,000 m year[-1] (Warren 1993). In general, the flow of Antarctic ice systems ranges from tens to hundreds of metres per year; it is faster in Antarctic Peninsula glaciers, given the high accumulation of snow and relatively high surface temperature which penetrates to the bed.

There are piedmont and local glaciers in the Antarctic coastal belt and mountains emerging from the ice. Although these glacial bodies are small and their volume is negligible compared to that of major Antarctic glacial systems, they are important for glaciological and palaeoenvironmental studies because they usually have lower inertia and react more promptly to climate variations. These glaciers are often frozen to the bed (i.e. their basal temperature is inferior to the melting point related to pressure). Sliding and abrasion are negligible or nonexistent beneath these cold-based glaciers, and motion (a few metres per year) is exclusively due to internal deformation (Siegert 2001). Accumulation and ablation are mainly controlled by the wind and, although localised superficial melting can occur in summer, it is believed (Baroni 1991) that ablation is mainly due to deflation and sublimation.

2.2.1 Antarctic Ice Sheets

Antarctica is covered by a dome-shaped ice sheet which reaches altitudes above 4,000 m. With a surface of about 12×10^6 km^2 and a volume of 26×10^6 km^3, the Antarctic ice is an important component of the terrestrial hydrosphere; through complex energy and mass exchanges, it interacts with the atmosphere and global climate system. The reflectivity and altitude of the Antarctic ice sheet ensure low temperatures, thus playing a fundamental role in its perpetuation. The ice sheet strongly limits the poleward propagation of cyclones, thereby helping to maintain its present configuration and the balance between accumulation of snow and ice and their ablation in the Southern Ocean.

Although Antarctica is apparently covered by a single ice body drained at its margins by numerous outlet glaciers and ice streams, Antarctica comprises two distinct ice bodies (the East Antarctic and West Antarctic ice sheets; Fig. 11) separated by the Transantarctic Mountains. Several glacial bodies, domes and an ice cap occur across the Antarctic Peninsula. In the main ice sheets there is no melting, and the ice flows from the interior domes by slid-

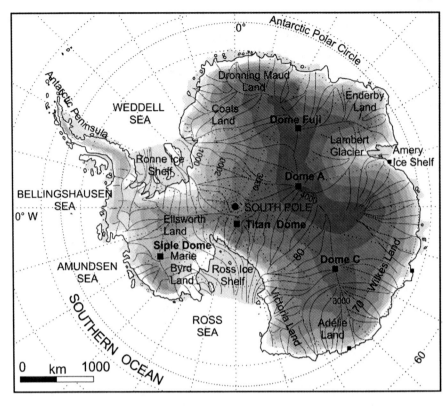

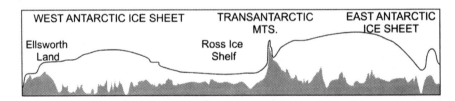

Fig. 11. Surface topography and major ice divides of Antarctic ice sheets

ing and/or basal sediment deformation. Flow velocities increase towards the coast, and ablation mostly occurs by calving into the sea. In contrast, Greenland ice sheets are land-based and, due to progressive melting and reduction of ice volume, their velocity decreases towards the coast.

The East Antarctica ice sheet has an area of about 9.9×10^6 km^2 (excluding floating ice shelves) and an average ice thickness of about 2,500 m. Its external edge is crowned by several coastal ice-free areas and nunataks (rocks emerging from the ice). On the Antarctic plateau, the ice culminates in large domes (Fig. 11). The enormous mass of ice rests on continental rocky substrata and, if the ice were to be removed from East Antarctica, the bedrock surface would be above sea level. Thus, in addition to a surface topography characterised by ice and a few emerging mountains on the coastal slopes, Antarctica has a very complex sub-glacial topography. The latter shows an extensive continental shield marked by relatively gentle convexities and depressions, sub-glacial mountain chains, basins and valleys. Some depressions lie below sea level; the most extensive is Wilkes Basin, along the Transantarctic Mountains (Fig. 12). The deepest depression (–2,341 m) is in the Astrolabe Basin, between Adélie Land and Wilkes Land. As a result of subglacial morphological irregularities, the thickness of the ice is extremely variable; it is greatest (4,776 m; Baroni 2001) over the Astrolabe Basin.

The flow of ice from the Antarctic plateau may be impeded and channelled by the Transantarctic Mountains and coastal ranges. In areas where the ice flow is blocked by mountains and the snow is blown away by katabatic winds, the exposed ice sublimates, giving rise to "blue ice". These ablation zones at the edge of the Antarctic plateau are among the most important meteorite

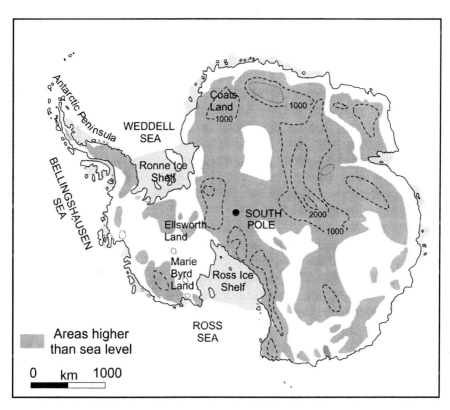

Fig. 12. Schematic representation of sub-glacial topography and sea level in Antarctica (*non-shaded areas* are below sea level)

traps on Earth (Cassidy et al. 1992; Zolensky 1998). The ice flow carries mete-orites many kilometres away from impact sites and, in ablation zones, they cannot escape because the mountains or nunataks act as a barrier to the fur-ther flow of ice. Concentrations of meteorites in Antarctic ice were first dis-covered by the Japanese in 1969 at Yamato Mountains (Queen Maud Land), and since then at least another 30 sites have been found, especially along the Transantarctic Mountains. For instance, since its discovery in 1984, the Fron-tier Mountains blue ice field in northern Victoria Land has yielded 472 mete-orites weighing more than 200 g each (Folco et al. 2002).

Compared to meteorites collected in temperate regions, those from cold and dry Antarctic ice sheets are much better preserved and more representa-tive of all meteorite types, with no organic contamination. Most of them are thought to come from asteroids, but some may have originated on the Moon and one class of meteorites, such as ALH84001, appears to derive from impact events on Mars (McSween 1997, 1999). Besides allowing a better understand-ing of material formed in the early solar system, Antarctic meteorites will probably settle the long-standing debate on the presence of life on Mars (McKay et al. 1996). Moreover, meteorites in space absorb and record cosmic radiation, and the time elapsed since they impacted the Earth can be deter-mined through laboratory studies. The terrestrial residence age of meteorites can provide additional information on Antarctic ice sheets.

Lambert Glacier, about 700 km long and 50 km wide, is the world's largest glacier. It drains about 14 % of the East Antarctica ice sheet into the Amery Ice Shelf, terminating as a calving wall in Prydz Bay. The centreline velocity near the ice front in Prydz Bay is 1,200 m year^{-1}. Rapidly moving (up to 2 m day^{-1}), often heavily crevassed ice streams (30–80 km wide and 300–500 km long) also flow from the West Antarctica ice sheet into the Ross Ice Shelf. One of the main factors involved in this rapid flow is probably the presence of basal water (Clark 1995). Liquid water is rather widespread under Antarctic ice sheets. Since the 1970s, airborne radar profiles have shown that water can collect in small sub-glacial lakes (Oswald and Robin 1973). To date, more than 70 sub-ice lakes have been found under Antarctic ice sheets. These lakes, which may have existed for millions of years, are attracting the attention of the scientific community and the general public because of the nature of possible resident biota, tectonic forces responsible for their formation, and the record of Antarctic climate history which the underlying sediment probably contains. In 1993, altimetric data from the ERS-1 satellite confirmed the presence of a very large sub-ice lake (a surface of more than 14,000 km^2, depths sometimes greater than 500 m, and a vol-ume of 1,800 km^3) near Vostok Station. Although the bed of Lake Vostok lies below sea level, radio echo-sounding indicates that it is largely composed of freshwater; it has a thick layer of sediments and is expected to support a res-ident microbial population (Ellis-Evans and Wynn-Williams 1996; Kapitsa et al. 1996). From 3,590 m below Vostok Station, the ice was accreted from liquid water associated with the lake. Priscu et al. (1999) found that this ice contains microbes (ranging from 2.8×10^3 to 3.6×10^4 cells ml^{-1}) with phylotypes closely related to alpha- and beta-Proteobacteria and actinomycetes. Karl et al. (1999) reported a lower concentration of bacteria ($2-3 \times 10^2$ cells ml^{-1}) in the accreted ice above Lake Vostok. They also found lipopolysaccharides which suggest the prevalence of Gram-negative bacteria and respiration of ^{14}C-labelled acetate, and glucose substrata which suggest the presence of viable microorganisms. As the 4-km-thick ice sheet goes afloat as it crosses the lake, significant quantities of water are probably exchanged between the base of the ice sheet and Lake Vostok. Such exchanges would enrich this extreme environment (pressure~350 atm, temperature~3 °C, permanent darkness) with air hydrates and sediments released from the melting basal glacier, and would also promote the circulation of lake water (Siegert et al. 2000, 2001).

There is ever greater scientific interest in sub-glacial lakes under the East Antarctic ice sheet, and a number of ongoing research projects aim to study their origin, palaeoclimate and possible microbial life, which has undoubtedly important evolutionary and astrobiological implications. The National Aeronautics and Space Administration (NASA) is interested in exploring these extreme Earth environments, in part to prepare planetary missions to Europa, the largest Jovian moon, which is the smoothest object in the solar system. Thus, exploration of Antarctic sub-glacial lakes and underlying sediments would satisfy several scientific interests and allow important technological testing of sterile recovery systems used to access and retrieve samples. Ice coring at Vostok Station stopped at 3,623 m (about 120 m above the lake surface) to prevent contamination of one of the most pristine environments in the world. The challenge for the international science and engineering community is the development of a scientifically sound research programme which, when implemented, will not contaminate the lake and will minimise damage to the environment. However, there is no doubt that exploration of the lake will affect the environment. Although the lake is perhaps millions of years old, radar data have identified a significant along-lake flow component of the ice sheet; the rate at which ice is frozen (accreted) to the base of the ice sheet is greatest at the shoreline, and the accreted ice layer is subsequently transported out of the lake (Bell et al. 2002). According to these authors, every 13,300 years the overlying ice sheet removes all the water, which is replaced by other sources. As the lake water captured by the moving ice sheet is deposited as layers of ice along the eastern shoreline of Lake Vostok, it might be possible to search for evidence of life in the eastern shore ice; this further fuels the debate on what should be done with Lake Vostok.

The build-up of glaciers in the Antarctic mountains commenced about 40 Ma ago (Robin 1988), but the development of the East Antarctica ice sheet probably began after the opening and deepening of the Drake Passage and the establishment of westerly atmospheric and oceanic circulation (between 30 and 22 Ma ago). The development of an oceanic polar front blocked the flow of warm waters to high latitudes and, together with the increasing albedo, helped cool the continent. The build-up of the East Antarctica ice sheet was completed around 14 Ma ago (Sugden 1996), and geomorphological evidence and offshore oxygen isotopic records show that it has since been rather stable (Robin 1988; Warnke et al. 1996; Marchant and Denton 1996).

To foresee how ice sheets might behave in the future, it is important to understand how they behaved in the past. However, research must clarify whether till from the Transantarctic Mountains containing fragments of Pliocenic marine diatoms (Webb et al. 1984; Barrett et al. 1992) and glacial lake sediments near the top of Mount Murphy (Marie Byrd Land, West Antarctica; LeMasurier et al. 1994) testify to the repeated growth and decay of Antarctic ice sheets.

The West Antarctic ice sheet is much smaller (surface of about 2.2×10^6 km^2 and average thickness of 1.7 km) than the eastern one. It is a marine-based ice sheet anchored to a series of islands and archipelagos. Most sediments and rocks under the ice lie below sea level (Fig. 12); if the ice were removed, even considering isostatic rebound (Siegert 2001), the bedrock would remain below the modern sea level. Although there is still controversial debate on the stability of Antarctic ice sheets during the last 30 Ma (e.g. Barrett et al. 1992; Marchant et al. 1994; Sugden 1996) and on evidence of glacial systems in several Antarctic and sub-Antarctic regions between 30 and 45 Ma ago (e.g. Barron et al. 1988; Birkenmajer 1988), the West Antarctica ice sheet probably first developed 10 Ma ago and reached its maximum size about 5 Ma ago (Robin 1988). Since then, the ice sheet has probably undergone minor fluctuations.

The mass balance of the Antarctic ice sheets as a whole has been determined through estimates of accumulation and ablation (e.g. Oerlemans 1989; Bentley and Giovinetto 1991). However, there is uncertainty as to whether calculations yield an approximate, or a slightly underestimated mass balance. Besides the fragmented nature of glacial systems and topography, reliable mass-balance estimates are hindered by geographic variations in climatic conditions and by the presence of very sensitive areas such as the Antarctic Peninsula. Moreover, the Antarctic ice sheets are now responding to changes determined in the past, such as during the end of the last glacial period. This lag effect makes it very difficult to predict the possible retreat or advance of glacial systems in response to present-day climate changes. For most of the Antarctic ice sheet, the response time will probably be in the order of several thousand years.

2.2.2 Ice Core Records of Climate and Environmental Change

In culminating ice caps near the centre of the Antarctic ice sheet, the horizontal flow of ice is negligible and annual snow layers accumulate without melting, growing thinner and thinner with no discontinuities. As ice deformation occurs by vertical compression rather than shearing, the age of the ice is a function of depth and the snow accumulation rate. The physical transformation of snow into ice does not change the chemical composition of the snow or aerosols, or of volcanic and cosmic dust deposited on the surface. Owing to very low accumulation rates and vertical compression, the individual layers are not apparent in most Antarctic ice cores. Dating should therefore be done through indirect methods such as the analysis of isotopes and other components of ice and air bubbles, ash and acidic aerosols from known volcanic eruptions, and cosmogenic radionuclides (Paterson 1994; Legrand and Mayewski 1997; Petit et al. 1999; Archer et al. 2000; Wagner et al. 2000).

The balance of a number of isotopic constituents within the atmosphere, cryosphere and oceans is peculiar to particular climate conditions and varies according to climate change. During the evaporation of seawater, the water molecules composed of light isotopes (e.g. ^{16}O, ^{17}O or $^{1}H_2$) vaporise more easily than those containing ^{18}O or deuterium (D). The evaporation rate is also related to water temperature (cold water allows relatively more ^{16}O to evaporate than ^{18}O). In equilibrium conditions, atmospheric water vapour contains 1 % less ^{18}O and 10 % less D than average ocean water. During vapour condensation, molecules containing heavy isotopes precipitate more readily, and the remaining vapour is therefore depleted in heavy isotopes. The progressive cooling of water vapour, such as during its passage from the Southern Ocean to the colder Antarctic landmass, will result in precipitation with increasing concentrations of lighter isotopes. Thus, although many factors can affect the isotopic composition of precipitation, the effect of temperature is remarkably predictable. A mass spectrometer can be used to determine variations in $^{18}O/^{16}O$ and/or $^{1}H/D$ ratios in ice melted from different layers of a core; the difference between the measured ratios and those of Standard Mean Ocean Water (SMOW) are expressed as $\delta^{18}O$ and δD. Low values of $\delta^{18}O$ and δD indicate low palaeotemperatures, because snow is enriched in ^{16}O and the ocean in ^{18}O during periods of glaciation.

The first ice core from the Russian Vostok station (2,083 m) was obtained during a series of drillings in the early 1970s and 1980s. Lorius et al. (1985) performed the first isotopic analysis of the ice core, and two years later Jouzel et al. (1987) analysed an ice record spanning a full glacial–interglacial cycle. Drilling continued at Vostok until January 1998, reaching a depth of 3,623 m. By measuring the continuous deuterium profile along the ice core, the Vostok temperature record was extended to the past 420,000 years and four glacial–interglacial cycles (Petit et al. 1999; Fig. 13). Although the third and fourth climate cycles in the ice core show a shorter duration than the first two

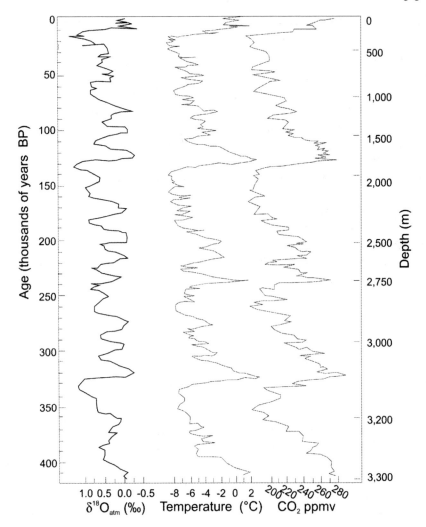

Fig. 13. Plot showing the Vostok ice core record of [18]O levels, reconstructed air temperature and CO_2 levels. (Data from Petit et al. 1999)

cycles, all cycles are characterised by a similar sequence of a warm interglacial followed by a cold glacial period, which ends with a rapid return to an interglacial period. The overall glacial–interglacial temperature change in surface temperatures is about 12 °C. Climate cycles deduced from the Vostok ice core appear to be more uniform than those in deep-sea core records (Petit et al. 1999). However, more recent measurements (Petit et al. 2000) show that the Vostok climate record may be disturbed below 3,311-m depth.

At Dome C (75° 06′S and 123° 24′E, about 3,233 m above sea level), the EPICA research team (a consortium of European countries) drilled the ice to a depth of more than 3,100 m during the 2002–2003 summer season. The ice is believed to be older than 500,000 years, and a longer historical climate record will probably be obtained through further drilling down to a depth of about 3,300 m.

Cosmic rays and solar irradiation impinging on the upper atmosphere produce [10]Be and [36]Cl. After their formation, these isotopes become quickly attached to aerosols and are removed from the atmosphere by precipitation. Their concentrations in ice cores can thus be used both as stratigraphic markers to compare different ice cores and as markers of long-term changes in the amount of snow deposition (Paterson 1994).

Isolated ice bubbles trapped in ice contain "fossil air"; by placing a thin slice of ice core in a vacuum chamber and cracking the ice, the concentrations of escaping gases, such as CO_2 and other greenhouse gases, can be determined. Since the finding by Barnola et al. (1987), many ice cores from Greenland and Antarctica have shown remarkable similarities between the greenhouse gas curve and that of $\delta^{18}O$ and δD. Peak CO_2 concentrations occur during warm periods and low concentrations mark glaciations. This trend probably reflects feedback mechanisms between glacial, oceanic, atmospheric and biological systems.

According to Siegert (2001), whether CO_2 records are synchronous or occur before or after $\delta^{18}O$ variations remains unresolved. This knowledge is necessary to understand the effective role of greenhouse gases in forcing glacial activity and/or their development as a result of glaciation. Once the role of CO_2 greenhouse gases in the behaviour of ice sheets is established, it will be possible to evaluate the relative importance of feedback mechanisms involving greenhouse gases in recent climate change. A study of CO_2 glacial/interglacial cycles recorded in ice cores (Archer et al. 2000) indicates that this gas potentially forced climate change in the last two glaciations (i.e. its concentrations increased prior to the decay of ice, as shown in the $\delta^{18}O$ signal). The most likely driver of CO_2 change over a glacial cycle is the ocean, but it is still not clear what processes (e.g. increased solubility of CO_2 in cold water, biological productivity, pH variations) are responsible for CO_2 uptake and delivery across glacial cycles. Moreover, numerical modelling of the role of CO_2 variations in the last glacial–interglacial cycle (from the Vostok ice core) in forcing the Earth's climate show that CO_2 variations alone cannot reproduce the ice-age cycle (Loutre and Berger 2000). This result seems to indicate that CO_2 variations may not themselves be forcers of climate change, but that they may be influential as part of a feedback mechanism.

Ice impurities can be analysed to obtain information about deep ice sheets and their former environment. Glacial periods are characterised by an increase in windborne concentrations of fine sand, silt and clay particles. In contrast to atmospheric circulation in Greenland, which was probably affected by rapid change during the last glacial cycle (Svensson et al. 2000) – which in Antarctica was rather stable – dust concentrations in the

Vostok ice core show a strong periodicity of 100,000 and 41,000 years (Petit et al. 1999).

Large volcanic events can be recorded in ice cores as ash layers or as increases in acidity (tephra horizons) produced by the transformation in the atmosphere of sulphur dioxide into sulphuric acid aerosols. Ash and acidity peaks in ice core layers constitute useful stratigraphic markers because they match with historical eruptions (e.g. Hammer et al. 1980; Francis 1993). Three ash layers 3.3 km below the ice-sheet surface have been detected in the Vostok core (Petit et al. 1999) and, together with acidic layers detected in deep ice sheets by airborne radar sounding (Millar 1981), these records reveal no obvious relationship between major volcanic events and ice-age cycles.

The chemical composition of the atmosphere has been dramatically altered by human activity. Lead isotope measurements in Greenland ice cores indicate that early large-scale atmospheric pollution of the Northern Hemisphere by mining of this metal in Spain began between 150 B.C. and 50 A.D. (i.e. during the Carthaginian and Roman civilisations; Rosman et al. 1997). In the last 200 years, the world population has increased by more than 500 % and ice cores worldwide contain higher concentrations of CO_2, CH_4, N_2O and persistent pollutants from atmospheric nuclear bombs, and industrial and agricultural activity. As will be discussed in Chapter 4, some of these pollutants have a global distribution, and the chemical composition of Antarctic snow and ice cores reflects the impact of heavy metals, radionuclides and persistent organic pollutants (POPs) from remote anthropogenic sources and/or human activity in Antarctica. Radionuclides and persistent pollutants also provide useful stratigraphic marker horizons which can be used to date snow and ice cores and to reconstruct the mass balance of glaciers (e.g. Lefauconnier et al. 1994). Moreover, ice cores can be used to study the possible relationship between changes in atmospheric composition and past global changes. A strong correlation between concentrations of Na^+, Ca^{2+}, SO_4^{2-} or the value of the ratio Cl^-/Na^+ and the $\delta^{18}O$ signal has been found in Antarctic ice cores (Legrand and Mayewski 1997). As discussed in the next chapter, ocean–atmosphere interactions and the sulphur cycle are among the processes linking atmospheric chemistry to global climate change. Measurements of soluble and insoluble constituents in snow and ice over Antarctica are valuable not only as indicators of changes in their source strength but also as tracers of atmospheric circulation over the continent. In the simplest case, for instance, marine versus continental air masses can be differentiated on the basis of sea salt (e.g. NaCl) versus continental dust (e.g. Al, $CaSO_4$).

2.2.3 Ice Shelves

Land-based glaciers or ice streams flowing into the Southern Ocean may originate floating glacier tongues or ice shelves which accumulate snow on their surfaces. The two largest embayments of Antarctica are occupied by the Ross Ice Shelf in the Ross Sea and by the Filchner-Ronne Ice Shelf in the Weddell Sea. These two shelves and the Amery Ice Shelf, which is fed by the Lambert Glacier, drain a combined area of more than 60 % of the Antarctic continent, with a flow of 0.8–2.4 km/year. Together with many other smaller shelves occurring along the coast and comprising about 47 % of the coastline, Antarctic shelves have a surface area of about 1.72×10^6 km^2. Their seaward edges are marked by cliffs rising 30–50 m above sea level, with an overall thickness of about 200 m. As shelves are grounded and constrained by promontories and islands, their thickness increases landwards and in places where ice streams enter the floating ice mass. The grounding line of the Ross Ice Shelf, for instance, is about 1,000 m thick. Based on basal melting and freezing patterns, this shelf can be divided into three zones (Souchez and Lorrain 1991): an area of enhanced bottom freezing near the grounding line, one with slow bottom freezing, and an outer zone affected by stronger circulation, greater heat exchange and net basal melting.

The stability of ice shelves depends on ice discharge from feeding glaciers, the morphology of the coast, location of bedrock "pinning points", net snow accumulation at the surface, and freezing at the base. Ablation is due to basal melting of the lower surface (about 3 m year^{-1} in the ice front of the Ross Ice Shelf; Jacobs et al. 1986), net surface ablation and, above all, to the calving of icebergs. Jacobs et al. (1992) estimated that calving accounts for 77 % of ice lost from the Antarctic ice sheet. Calving from Antarctic shelves is generally episodic, releasing large tubular icebergs which can be tracked over vast distances and for several years.

The Antarctic ice shelves are regularly monitored through satellite imagery, and the calving of giant icebergs is increasingly attracting the attention of mass media, contributing to a growing general concern over the possible effects of global warming. Too much importance is probably attributed to calving events in Antarctica. The breaking-off of icebergs as large as small countries from the Ronne-Filchner and Ross ice shelves may be part of their normal lifecycle. The edge of a shelf may retreat on one side and advance on the other. The B-15 iceberg (about 290 km long and 37 km wide), calved from the Ross Ice Shelf in March 2000, was the largest iceberg recorded by the US National Ice Centre, and the images made the rounds of the world. However, the ice shelf was in a northerly advanced state, and even greater bergs were probably calved from its edge in the past. For instance, in 1956 during an expedition to Antarctica, the US icebreaker *Glacier* reported a berg about 330 km long and 95 km wide in the Ross Sea.

As discussed in the previous chapter, the Antarctic Peninsula is experiencing an enhanced warming trend, and the progressive retreat of its shelves (overall about 13,000 km^2 since 1974) is probably linked to changing climate conditions. In contrast to the calving of large icebergs from the Ronne-Filchner and Ross ice shelves, in 1995 the northern Larsen Ice Shelf (A; about

2,000 km²) disintegrated into thousands of small icebergs. Further south, the Larsen (B) Ice Shelf and other shelves began to break up, receding past their historical minimum extent. Satellite imagery shows that the retreat is continuing, and that the northern section of the Larsen (B) Ice Shelf has completely shattered and separated from the continent; in February 2002 about 3,250 km² disintegrated in a plume of thousands of icebergs adrift in the Weddell Sea.

2.3 Life in Snow and Ice

Like in other polar and alpine regions, Antarctic snow and ice contain suitable habitats for microbial growth. In general, Antarctic snow is deep frozen and dry, and conditions for life are much more favourable in the sea ice than on the continent. However, patches of brightly coloured snow algae can develop in snowfields or in melt holes on glacier surfaces in islands and warmer coastal sites such as those in the western Antarctic Peninsula. Microbial colonisation may also occur in water-filled bubbles within freshwater lake ice and in large liquid water systems which form each summer over parts of the major ice shelves. Colonised sites are usually within or adjacent to ice-free areas, because liquid water within the snow can persist for some weeks, and because the wind and birds can carry soil particles and nutrients to these zones. Soil or sediment particles are important for enhancing the melting process (through the absorption of solar radiation) and the supply of soluble ions, and as microbial inoculum. Marine aerosols are further sources of nitrate, ammonium and other nutrients for snow algae. Although relatively rare in comparison to snowfields in temperate latitudes, snow algae from the South Orkney Islands were studied by Fritsch in 1912, and other papers were published in the 1960s (e.g. Llano 1962; Fogg 1967; Kol and Flint 1968) and more recently (e.g. Ling 1996; Mataloni and Tesolin 1997; Ling and Seppelt 1998). The most widespread colour in Antarctic snow is red, usually due to the presence of spores of green algae such as *Chlamydomonas nivalis* and *Chlorosphaera antarctica*, which accumulate photo-protective red astaxanthin esters (Bidigare et al. 1993). The green colour is due to the filamentous green alga *Hormidium subtile*, while yellow algae communities are dominated by chlorotic green algae or chrysophytes. Glacial ecology is in a pioneer stage (Yoshimura et al. 1997; Jones et al. 2001), and very few data are available on primary productivity and interactions between organisms. According to Fogg (1998), productivity is low (few mg C m² day⁻¹) even under favourable conditions. Moreover, there is evidence that algae promote the development of heterotrophic snow bacteria through the extra-cellular release of about 10% of their photosynthetic production (Thomas and Duval 1995).

Algae need light, nutrients and liquid water, and are therefore more common in the sub-Antarctic islands and Antarctic Peninsula where the air temperature reaches or approaches 0 °C for several weeks during the summer. Kol and Flint (1968) gave a detailed description of green algae from the Balleny Islands (67° S, 163° E), but snow algae have rarely been reported at higher latitudes in coastal Victoria Land (Vincent 1988). During several expeditions in this region, Bargagli found only one patch of red snow on Apostrophe Island (73° 32′S, 167° 25′E) in January 1996. The summer air temperature on the small island is often below 0 °C, and organisms experience frequent freeze–thaw cycles. The presence of red snow was probably due to the concomitance of several factors, such as the ablation of snow which determines a concentration of algal cells, and the presence of an ice-free area (about 50 m away) with nesting skuas. Soil dust and algae pigments probably absorbed enough solar energy to allow the production of liquid water, which in turn has a lower albedo and higher heat capacity than snow.

The positive feedback effects which allow the growth of algae in snow also promote the development of cryoconite (rock dust windblown onto ice) communities. Local melting produces holes in glacier ablation zones, which may contain microorganisms growing as mats and films, with algae (mainly diatoms and chlorophytes) and cyanobacteria (especially *Phormidium frigidum* and *Nostoc*) dominating the rich microflora. Cryoconite holes tend to reach a relatively constant depth which depends on local climatic and environmental conditions. They can be connected by surface or subsurface streams and may form pools.

Much more extensive systems of lakes, pools and streams (up to several tens of kilometres) may form in the larger ice shelves. In the ablation areas of the McMurdo Ice Shelf, melting continues for 1–2 months during summer, and meltwater systems at the ice shelf grounding zone often contain large amounts of sediments derived from the seafloor (Kellogg and Kellogg 1984). Physico-chemical properties of meltwaters vary enormously (from freshwater to waters 2–3 times saltier than seawater), and under favourable conditions the biotic community is largely composed of benthic mats of cyanobacteria and diatoms (Vincent 1988).

In contrast to sea ice, that overlying Antarctic lakes and pools lacks brine channels and other structures which allow the formation of liquid water habitats. Liquid water is essential to life on Earth, and arguably to any form of life in the solar system. In recent years mass media have spread the news of liquid water existing elsewhere in the solar system, fuelling people's imagination. Antarctica, the continent with many environmental extremes, has therefore become one of the most important places in which to look for microorganisms thriving in extreme habitats (extremophiles; Macelroy 1974). The discovery of extremophiles throughout Antarctica, Siberia and other polar regions (e.g. Abyzov 1993; Vorobyova et al. 1996; Nealson 1997; Priscu et al. 1998; Staley and Gosink 1999; Gilichinsky et al. 1999) is prompting the search for life in the polar caps of Mars, Europa, Callisto and other icy moons of the solar system, and for possible panspermia (the transport of life from one planet to another). Interest in exobiology and astrobiology has grown astoundingly during the past decade. Antarctic bacteria, which thrive below ice in the coldest and driest desert on Earth, have become of primary interest to research programmes (US National Science Foundation and NASA) on life in extreme environments and to those of the European Union in biotechnology (Aguilar et al. 1998).

In Antarctic lakes covered by ice, gas bubbles may become trapped in the ice and gradually move to the surface by the freezing of water below and the ablation of ice in the upper surface. These air bubbles are rather common in ice overlying lakes with benthic mat communities of cyanobacteria. Pieces of mat are buoyed and detached by gas from the sediments; they float upwards and accumulate beneath the ice cap, which incorporates them during the autumn. The following spring, mats absorb the radiation penetrating through the ice and promote the formation of gas bubbles, water and biotic communities.

The McMurdo Dry Valleys site is one of the coldest and driest ice-free areas on Earth. Here most lakes have a permanent ice cover (3–6 m thick) and contain a layer of windborne soil particles, which allow the formation of liquid water inclusions during summer. Priscu et al. (1998) studied viable microbial assemblages in ice samples from six lakes and found that ice sediment particles serve as nutrient for microorganisms. The ice-embedded biotic community was capable of photosynthesis, nitrogen fixation and decomposition; it was therefore suggested that this habitat may serve as a model for life on Mars and Europa. Compared to assemblages of autotrophic and heterotrophic organisms described in the ice of several alpine lakes, those in the permanent ice of Dry Valley lakes grow and reproduce under much more extreme physico-chemical conditions and originate only from the atmosphere, rather than from lake waters or sediments (Psenner and Sattler 1998). However, microbes inhabiting Antarctic ice sheets and permafrost meet even more extreme conditions. These are probably communities of "survivors" (Rothschild and Mancinelli 2001; i.e. they have been trapped in the ice and are more resistant than others which suffered a similar fate). Active organisms live in water films between grains of ice and/or soil, while viable but inactive forms are frozen (anabiosis) and many are dead. Owing to their preservation in ice, however, cell membranes, organelles, proteins, DNA and RNA from the latter forms are important for molecular (palaeontological) research. Through microbiological studies on an ice core, collected under sterile conditions at Vostok Station, Abyzov (1993) found yeasts, fungi, bacteria and a new species of actinomycetes (*Nocardiopsis antarcticus*) capable of synthesising a melanoid pigment, a UV-absorbing substance. Microorganisms were rather rare along the ice core and showed a random distribution independent of the sampling depth. Fungal spores, especially bacterial spores, are the more resistant forms, and some are capable of reactivation after 8,000–12,500 years of anabiosis.

2.4 Ice-Free Areas and Terrestrial Habitats

During the austral summer only about 330,000 km^2 of the Antarctic surface is free of ice and snow. These ice-free areas are mainly located in the western Antarctic Peninsula; on the continent they occur only in scattered coastal areas, on the steep slopes of the Transantarctic Mountains or in nunataks. Although the term oasis originated in hot deserts to describe areas with groundwater and vegetation, it is also used to indicate ice-free areas of continental Antarctica. Pickard (1986) distinguished Antarctic oases (minimal surface area 10 km^2) from smaller ice-free areas (i.e. nunataks, beaches and moraines), and identified the Victoria Land Dry Valleys (mountain oasis) and six low coastal oases: Bunger Hills, Vestfold Hills, Windmill Islands, Schirmacher Oasis, Soya Coast and Thala Hills. Most of these areas are cold deserts, with very sparse biota and a small number of cryptogamic and (terrestrial and aquatic) invertebrate species. They therefore play a minor role in the global carbon cycle and in driving global climate trends. Nevertheless, Antarctic ice-free areas are a unique laboratory for understanding processes of cold adaptation, the spread and development of pre-adapted organisms, and colonisation processes in extreme environments very far from propagule sources. These areas play an important role as breeding grounds for seabirds and as sites in which to study the geological and glacial history of the continent. Besides contributing to ecological and geological science as a whole, Antarctic ice-free areas offer a unique opportunity for detecting and predicting the effects of global climate changes on terrestrial ecosystems (Bargagli 2000, 2001).

By virtue of their reduced complexity and largely unpolluted and pristine conditions, Antarctic terrestrial ecosystems allow the assessment of fluxes of nutrients and other elements between biotic and abiotic components of ecosystems, and the detection of persistent pollutants derived from human activity in Antarctica and elsewhere in the Southern Hemisphere.

Weather fluctuations and the physical environment exert a determining influence on the survival and development of Antarctic organisms. Life history strategies are often based on long lifecycles, slow growth rates, low reproductive output and high energy investment in the ability to respond rapidly to adverse changes. They are adapted to extreme environmental conditions but may be highly sensitive or intolerant to changes exceeding pre-existing thresholds. The biological colonisation of old or newly exposed substrata is easily recognisable in desert environments, and knowledge of the response of terrestrial organisms to climate and environmental change is essential for the management and protection of these sensitive ecosystems. Such knowledge is also useful for a better understanding of climate-induced changes in more complex ecosystems elsewhere (Block 1994).

Many ice-free areas have emerged during the past few thousand years from the retreating ice in Antarctica, and glacial erosion is the dominant land-forming factor. Several coastal areas show evidence of isostatic uplift, which has produced raised beaches and inland cliffs and, sometimes, freshwater or brackish small lakes. The surface of these periglacial environments is characterised by scattered erratic boulders and suites of glacial till and unsorted rock rubble. In general, moraines are rare and limited in size compared to those in the Alps or other mid-latitude regions. On the contrary, sediment cores from continental shelves around Antarctica show a widespread occurrence of glacio-marine deposits (Anderson JB 1991; Barrett et al. 1991).

Most ice-free areas in East Antarctica are typical cold desert environments. The term cold desert was first applied to largely unvegetated lands in the Russian High Arctic and to the western and northern portions of the Canadian Arctic Archipelago, with annual precipitation ranging from 100 to 250 mm and the warmest monthly mean air temperatures<5 °C (Aleksandrova 1988; Vincent 1997). Antarctic cold deserts are much smaller than Arctic ones, but their annual precipitation is often well below 100 mm and the warmest mean air temperatures are <0 °C. The Antarctic cold desert undoubtedly represents a climatic and environmental extreme relative to the rest of the biosphere.

Due to low temperatures and arid conditions, chemical weathering and many other water-based rock decay processes are ineffective. Most snowfalls ablate, and in summer even snow meltwaters (especially near or over large, dark, north-facing boulders sheltered from the wind) may suddenly evaporate. Thus, processes such as frost wedging, frost cracking and the transport of products of disintegration play a marginal role in continental Antarctica; features such as talus or scree slopes are therefore rather uncommon. The main processes involved in rock disintegration and the formation of regolith are glacial and wind action, salt weathering, insolation, ice formation, and biotic exfoliation by endolithic communities. These weathering processes determine very slow transformations. Glacial deposits from the Miocene or earlier may show well-preserved landscapes and, for the distinctive features of soil and landscape, Antarctica can be regarded as a distinct morphogenetic region (Campbell and Claridge 1987).

As shown in Fig. 14, the wind armed with fine sand, snow and ice causes faceting and polishing of rocks (ventifacts), and the redistribution of fine materials with the formation of coarse armoured lag (desert pavements). Strong evaporation determines the accumulation of soluble salts on the surface of regolith, or encrustations and efflorescences just beneath surface boulders, cobbles and pebbles (Ugolini and Anderson 1973; Keys and Williams 1981; Gore et al. 1996). These salts consist largely of chlorides, nitrates and sulphates of sodium, potassium, calcium and magnesium. Although salts may originate from different sources, most ions come from the marine environment through marine aerosols, snowfall and seabirds (Bockheim and Wilson 1992; Rankin and Wolff 2000; Bargagli et al. 2001). The crystallisation of salt within rock pores and cracks contributes to rock decay and the formation of

Fig. 14. The effect of wind on ice-free areas. *Above* The faceting and polishing of rocks. *Below* Salt encrustations (northern Victoria Land)

heavily pockmarked surfaces. The cold desert soils of continental Antarctica contain large amounts of soluble salts, and soils are underlain by permanently frozen ground (permafrost) even in coastal fringes. Permafrost occurs in ice-free areas with an average annual temperature lower than −1 °C (Bockheim 1995). Although an active surface layer melts above the permafrost during summer, it seldom results in the formation of liquid water. The depth of the active layer (usually a few decimetres) mainly depends on temperature and the characteristics of surface materials; freezing and melting cycles determine the formation of ground patterns such as block fields and cracks.

In maritime Antarctica, climatic and environmental conditions are much less extreme, and ice crystallisation and chemical weathering are more important in rock disintegration. The increased leaching allows the formation of clay-sized materials and reduced amounts of salt encrustations. The organic content of soils may be low, but pads of humus can develop beneath isolated plant cushions. At poorly drained sites, peat-like material can accumulate under moss turf. This material is often cemented by ice crystals, and humic compounds only penetrate slightly into the underlying mineral soil (Fogg 1998).

2.4.1 Antarctic Soils

In the past, some soil scientists were unwilling to identify as soil the weathered surficial deposits in ice-free areas of continental Antarctica. However, as discussed by Bockheim (1982), these materials fit most classical definitions of soils and may support microflora and cryptogams. Since the 1960s, many studies have been published on soils of the Victoria Land cold desert (e.g. Ugolini 1963; Calkin 1964; Claridge 1965; Tedrow and Ugolini 1966; McCraw 1967; Campbell and Claridge 1968; Ugolini 1970). Most of these studies conclude that, owing to low temperatures, low moisture availability and negligible biological activity, the main soil-forming processes on parent materials are oxidation and salinization. In this region, soil development and weathering range from virtually nil on young land surfaces (<50,000 years) to soils with distinctive pedogenetic features on older land surfaces (Campbell and Claridge 1987). The latter soils show a desert pavement usually containing more than 90 % rocks and coarse fragments, with coarse sand filling the voids. The desert pavement is underlain by a reddish oxidised zone due to the presence of free Fe oxides. Soils lack cohesion and structural development, contain abundant water-soluble salts and negligible amounts of organic matter. Bockheim (1997) reported analytical data and properties of eight pedons in the Dry Valleys region representing three climatic zones and parent materials, and ranging in age from Holocene to Pliocene. The soluble salts increased linearly with soil age, ranging from salt encrustations in Holocene soils to indurated salt pans in early Quaternary soils. The salts (mainly NaCl in coastal areas, $NaSO_4$ in inland regions and $NaNO_3$ along the polar plateau) reflected the composition of snow and originated primarily from atmospheric deposition. Chemical weathering was generally restricted to oxidation of Fe-bearing minerals and clay authigenesis. Some soils contained ice-cemented permafrost, but older soils had dry permafrost, with <5 % moisture content. A classification was attempted, although the question of whether Antarctic cold desert soils with dry permafrost and abundant soluble salts should be classified as a special suborder or subgroup of Aridosols, rather than as soils containing permafrost (Gelisols), remains an open issue. On the basis of temperature, moisture, geomorphic and pedological characteristics, Campbell et al. (1998) differentiated five groups of soils within the McMurdo Dry Valleys.

Throughout Antarctic coastal ice-free areas, ornithogenic soils develop around rookeries of pygoscelid penguins (Tedrow and Ugolini 1966). Large quantities of organic matter are derived from droppings, feathers and bird remains, and high concentrations of P and N are deposited and incorporated into these soils. The wind helps spread nutrients to neighbouring areas.

General theories for soil formation in Antarctica, according to geo-pedological features and latitude, were developed by Bockheim and Ugolini (1990). However, the results of pedological research performed during the last decade in coastal ice-free areas of East Antarctica, excluding the cold desert of Victoria Land (e.g. Blume and Bölter 1993; Blume et al. 1997), suggest that these theories should be extended. The Casey area (Wilkes Land) lacks the ahumic red soils described in Victoria Land, and soil formation and chemical weathering in this area may occur to a greater extent than predicted in former models. Beyer et al. (2000) summarised pedogenetic findings in this area based on a large-scale database, and suggested the importance of podzolization and of the accumulation of organic matter in soil-forming processes along the desert margins. Podzols represented 20 % in total of the soil landscape in the Casey area (one-third is ornithogenic and two-thirds non-ornithogenic in origin; Beyer and Bölter 2000). The low soil salinity suggests that during the thawing period, free water determines downward leaching which exceeds upward transport by evaporation. On the other hand, the migration of organic acids, non-humified carbohydrates and N-bearing compounds is considered the main podzolisation mechanism. Some relationships between soil development and vegetation patterns were suggested. Soil formation processes in Wilkes Land are thus rather similar to those in other cold areas of the world, although they require longer timescales to form visible structures. Obviously, soils develop faster (albeit with a high degree of cryoturbation) in the milder, wetter conditions of the Antarctic Peninsula, which determine more active chemical weathering processes, abundant clay-sized materials and low concentrations of soluble salts.

2.4.2 Terrestrial Ecosystems

One of the most impressive aspects of Antarctic ecosystems is the startling contrast between the richness and high biomass of marine organisms and the extreme biological impoverishment of terrestrial ecosystems. In the marine environment, steady environmental conditions and evolutionary processes in isolation contributed to the development of ecophysiological adaptations and a high degree of endemism among fish and invertebrate species. In contrast, many species of cryptogams and terrestrial invertebrates in Antarctic ice-free areas show scarce evidence of evolved adaptation to extreme variations in environmental conditions. Most biotic communities occur as small, isolated patches containing broad-niche genotypes highly tolerant of severe variations in water availability, temperature, light, and UV radiation.

Although the taxonomic revision of Antarctic flora has not be completed, recent studies have considerably reduced the number of species previously described as new. Dodge (1973), for instance, described more than 400 species of Antarctic lichens and claimed that most of them were narrow-ranging endemisms new to science. According to a preliminary estimate by Castello and Nimis (1997), the total number of lichen species in Antarctica is probably about 260; the percentage of endemic species falls from more than 90 % to less than 40 %, and that of bipolar and cosmopolitan species increases from about 2 % to more than 40 %. This trend suggests that most of the flora in continental Antarctica is young, and probably originated from postglacial immigrants. After the Pleistocene glaciation, the re-colonisation of ice-free areas was likely accomplished through long-distance dispersal by birds, winds and ocean currents (Lindsay 1977), rather than expansion from Antarctic "refugia" such as ice-free nunataks (Dodge 1973). Although Antarctica has its own microflora (e.g. yeasts and the cryptoendolithic alga *Hemicloris antarctica* in Dry Valleys soils and rocks; Vincent 1988) and microfauna (e.g. oribatid mites of the family *Maudheimidae* in Dronning Maud Land; Marshall and Pugh 1996), there is evidence that viable propagules are constantly introduced to the continent (Schuster 1979; Smith 1991; Linskens et al. 1993). Despite the remoteness of Victoria Land, and the altitude and very limited extent of recently exposed fumarolic grounds close to the summit of Mt. Erebus (3,794 m), Mt. Melbourne (2,733 m) and Mt. Rittmann (2,370 m), exotic species of algae and bryophytes have colonised microenvironments with suitable temperature (usually ranging from 15 to 60 °C), shelter under ice and snow hummocks, and with a regular supply of moisture from both steam condensation and snowmelt (Broady et al. 1987; Melick et al. 1991; Bargagli et al. 1996a). Genetic analyses of mosses growing on the fumarolic grounds of Mt. Melbourne (*Campylopus pyriformis*) and Mt. Rittmann (*Pohlia nutans*) show that these extremely isolated populations have low levels of genetic diversity and probably derived from a single immigration event, followed by vegetative growth (Skotnicki et al. 2002).

Through increasing aircraft and shipping movements between temperate latitudes and many Antarctic sites, mankind has become an important source of alien microorganisms (Cameron et al. 1977), and propagules of cryptogams and higher plants (Linskens et al. 1993). In just a few hours or days, viable propagules may reach suitable environments such as manned stations or geothermally heated grounds (Broady 1993). The present magnitude and success of alien introductions to Antarctica is completely unknown (Vincent 1988), but thermophilic bacteria and common fungi from temperate regions have often been detected by microbiologists near research stations or field camps (e.g. Cameron 1972; Mercantini et al. 1989).

Although it is not easy to discriminate between the influence of cold and desiccation (both temperature and moisture gradients show parallel north–south and coast–inland variations), the bioavailability of liquid water, rather than biogeographical isolation or low temperatures, is the main factor limiting the distribution and abundance of terrestrial organisms in continental Antarctica (Kennedy 1993; Block 1994; Bargagli et al. 1999). The water supply depends on the moisture content of supporting substrata, but it is also affected by the high salt burden of xeric soils which lowers the water potential. Moreover, at temperatures<0 °C, hydrated organisms may continue to lose water by evaporative transfer to neighbouring ice crystals (Worland 1996). Thus, the survival of many organisms in Antarctic ice-free areas largely depends on their resistance to dehydration. Cyanobacteria, lichens, xerophytic mosses and microinvertebrates such as nematodes and tardigardes are desiccation-tolerant.

In winter, microhabitat temperatures are close to air temperatures, but in summer the temperature can rise up to 40 °C (i.e. seasonal range of 60–70 °C and diurnal range of 25–50 °C), depending on insolation, exposure, colour, moisture and wind conditions on the surface of soils, rocks and cryptogams. In order to tolerate frequent freeze–thaw cycles and such thermal regimes, life history strategies of Antarctic terrestrial organisms are characterised by long lifecycles, slow growth rates and low reproductive output, with most energy resources dedicated to survival adaptations. For instance, most Antarctic invertebrates cannot tolerate tissue ice formation; they therefore produce antifreezes such as polyols and sugars, and mask and/or eliminate potential ice nucleators (Block 1990).

Features of Antarctic terrestrial ecosystems, the distribution and ecology of organisms, their adaptations and evolutionary processes have been reviewed in several books (e.g. Llano 1977; Kushner 1978; Laws 1984; Longton 1988; Vincent 1988; Stonehouse 1989; Friedmann 1993; Ross et al. 1996; Lyons et al. 1997; Fogg 1998; Priscu 1998; Øvstedal and Smith 2001; Beyer and Bölter 2002). In general, two different biogeographical zones are distinguished: continental Antarctica and maritime Antarctica (Fig. 15).

Continental Antarctica includes the cold and dry coastal fringe of East and West Antarctica, the east coast of the Antarctic Peninsula and islands close to

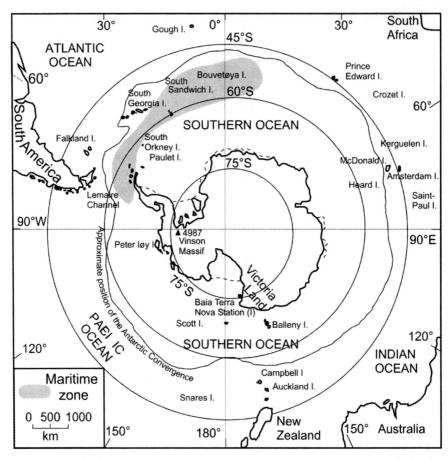

Fig. 15. Outline of the South Polar regions showing the maritime and continental zones

the continent. These areas usually have the mean temperature of the warmest month below 0 °C and an annual precipitation<20 cm. A range of habitats can be found, from very hostile environments, such as those of isolated nunataks in the polar plateau, to milder climatic conditions in coastal ice-free areas. During the summer, rocks and soils may provide moistened surfaces with dilute mineral solutions (mainly from precipitation, salt encrustations, runoff and splash; Bargagli et al. 1999, 2001) on which living propagules may settle and grow. Bacteria and blue-green algae can fix atmospheric nitrogen; they are among the main settlers and the most widespread life forms in Antarctic ice-free areas. Some species form crusts over soils and rocks, and modify substrata in ways which favour the growth of settling spores of macroscopic cryptogams. Lichens are the most widespread organisms on rock and boulder surfaces (epilithic), while mosses prefer sand and gravel substrata. Thus, in sparse areas facing north, with rather stable substrata, low salt contents and an almost neutral pH, and with maximum duration and frequency of insolation, moisture availability, and protection from wind, restricted communities of algae, mosses and lichens can proliferate to form surface mats. These communities mainly occur in coastal ice-free areas, where the sea mitigates temperatures and increases snow precipitation and, together with contributions from nesting seabirds, enhances the availability of nutrients (Bargagli et al. 2001). Soil fauna (protozoans, rotifers, tardigrades, nematodes, mites and springtails) may be locally abundant, although often with a limited number of species. Although in some frigid inland areas occasional algae, lichen and moss vegetation may occur, these environments are mainly characterised by bare rocks and soils. However, endolithic and/or sublithic communities can develop in apparently lifeless translucent rocks such marble, granite and sandstone. These communities usually include microalgae, cyanobacteria, fungi and heterotrophic bacteria, with phototrophs and fungi often associated symbiotically as a microscopic form of lichen (Vincent 1988). Endolithic communities are well insulated from the external environment and have a very simple trophic structure, with eukaryotic and prokaryotic algae as phototrophs supporting heterotrophic bacteria and fungi. There is evidence that some endemic species evolved under the unique conditions of Antarctic endolithic habitats (Vishniac 1985; Nienow and Friedmann 1993), and these ecosystems are attracting considerable interest as models for possible life forms on other planets.

Milder and wetter maritime Antarctica includes the Antarctic Peninsula and neighbouring islands such as the South Orkney, South Shetland and South Sandwich islands (Fig. 15). In these regions the mean monthly temperature exceeds 0 °C for 1–4 months in summer and rarely falls below –15 °C in winter. Annual precipitation (very occasionally as rain in summer) is usually greater than 30 cm water equivalent. Terrestrial ecosystems are therefore characterised by a greater abundance and variety of organisms than those in continental Antarctica; however, south of 68° S there is a rapid increase in aridity and a sharp decline in the diversity of plant and invertebrate species. This southern part of the west coast of the peninsula represents the transition from communities of maritime Antarctica to those in coastal continental Antarctica (Smith 1984). The vegetation is dominated by cryptogams and, in wetter coastal habitats, mosses and cyanobacteria may form closed stands, which locally accumulate peat in the northern province. Lichens predominate in drier habitats and inland. The small-scale topography and disruption by periglacial activity are important in the development and distribution of organisms. In areas protected from winds and exposed to solar radiation, bryophytes, algae and lichens receive moisture and warmth and grow in close stands; however, small-scale gradients in morphological and environmental features break the continuity of plant communities, leading to the development of barren areas. The moister soils may also support two species of native flowering plants: Antarctic pearlwort, *Colobanthus quitensis*, and Antarctic hair grass, *Deschampsia antarctica*. These plants are pioneer colonists of deglaciated grounds, and there is evidence of a rapid increase in population in some areas of the Antarctic Peninsula, probably in response to the enhanced warming of this region (Fowbert and Smith 1994). Soils associated with vascular plants support a larger and more diverse population of bacteria, microfungi and invertebrate fauna. Besides protozoans, rotifers, tardigrades and nematodes, the microarthropod fauna comprises at least eight species of springtails, 30 species of free-living mites, and the widespread wingless midge endemic to Antarctica, *Belgica antarctica* (Wirth and Gressit 1967; Smith 1996).

2.4.3 Freshwater Ecosystems

Although small rivers occur at various sites around Antarctica, the continent does not have stream–river drainage systems. Ephemeral streams and small seeps, fed by ice fields, glacier ice and/or melting snow banks, only exist in coastal regions during summer. They usually have a small discharge (few litresper second) and are subject to large variations in the duration of stream flow (from a few weeks to some months), depending on latitude. The Onyx River in the McMurdo Dry Valleys (about 77° 30′S), Antarctica's longest known water course (over 30 km), normally flows during mid–late December through January (Howard-Williams et al. 1997), while the Tierney River complex in the Vestfold Hills (about 68° 30′S) may flow for some months (from November to mid-May; Vincent 1988). All Antarctic streams show large interannual fluctuations in the timing and magnitude of discharge. As on the continent, most streams originate from glaciers, and solar radiation is the main factor controlling short-term and seasonal variations in discharge. When melting snow banks are the main source of water, the spring peak and duration of the discharge are also affected by the amount of atmospheric precipitation during winter. Waters flowing in Antarctic ice-free areas often have a large suspended sediment load, and ecosystems contain sparse microbial communities. Due to freeze–thaw cycles, nutrient concentrations in a stream may show considerable spatial and temporal variations. The highest concentrations usually occur during the first flows of the season and at maximum distance downstream. Figure 16 shows a statistically significant increase in conductivity and K^+, Mg^{2+} and SO_4^{2-} concentrations in water samples collected from a small stream at Edmonson Point (northern Victoria Land) at increasing distances from the source (melting ice and snow banks). The stream ends at a large beach with a penguin rookery. In addition to seabirds, many other factors probably contribute to the progressive increase in ion concentrations with distance downstream, such as water evaporation, marine aerosols, leaching from streambed sediments, rocks and biota. The slow-moving, clearer and nutrient-rich waters in the terminus of the stream favour the

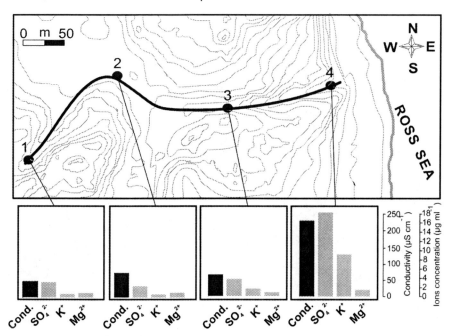

Fig. 16. Increasing concentrations of major ions (μg ml^{-1}) and conductivity (μS cm^{-1}) in water samples collected along a small stream at Edmonson Point (northern Victoria Land)

development of extensive mats and films of cyanobacteria, phytoflagellates, chlorophytes and moss communities along margins (Fig. 17). Phototrophs support a wide range of heterotrophic bacteria, fungi, protozoans and microinvertebrates (rotifers, nematodes and tardigrades). Although the species composition varies widely, these communities are widespread in flowing water ecosystems of continental and maritime Antarctica. However, there is evidence that the overwintering characteristics of filamentous chlorophytes and cyanobacterial mats in southern Victoria Land are different from those at Signy Island (Vincent et al. 1993). In the southernmost ecosystems of continental Antarctica, epilithic mats and films remain deep-frozen and dry during the winter, but retain a high level of viability and provide an important inoculum for the next season of stream flow (they begin photosynthesis and respiration within 30 min of rewetting; Vincent and Howard-Williams 1986). On the contrary, at Signy Island the chlorophytes are almost completely destroyed by the winter freeze, and the development of the algal community in the following season when streams begin to flow is due to a few, highly resistant cells which initiate the new cycle.

In Antarctica there are many sub-glacial and sub-aerial lakes, and their morphological, thermal and salinity characteristics are so unusual and varied

Fig. 17. The growth of algae and mosses along the margins of a small stream at Edmonson Point (northern Victoria Land)

that they have interested scientists since the International Geophysical Year in the late 1950s (e.g. Hawes 1963; Goldman 1970; Burton 1981; Hobbie 1984; Heywood 1984; Priddle 1985; Vincent and Ellis-Evans 1989; Simmons et al. 1993; Green and Freeman 1993; Spigel and Priscu 1998). Many lakes and pools formed by glacial retreat are located in poorly developed rocky catchments and usually lack outlets. They are therefore the main sink for water and solutes, and reflect the biogeochemical characteristics of their surrounding catchment. Many lakes in high-latitude regions are frozen solid to the bottom or permanently ice-covered. Waters gained by lakes during summer (through temporary moats or groundwater) supply latent heat and replace the water lost by sublimation of surface ice. The ice cover prevents stirring of waters by surface winds, and reduces light penetration and gas exchange between the water and atmosphere. The waters of ice-covered lakes are usually stratified, supersaturated with gases, and contain biota growing in a rather stable, shaded environment. Lake Vanda in the Wright Valley (77° 32'S, 161° 32'E; McMurdo Dry Valleys), probably the most studied freshwater ecosystem in continental Antarctica, is fed by the glacial meltwaters of the Onyx River during summer. The lake is about 69 m deep and has a smooth, permanent, 3-m-thick ice cap. The water above 45 m, cool, oligotrophic and rich in dissolved oxygen, is one of the clearest natural waters on Earth (Vincent and Vincent 1982). Below 45 m the lake is strongly stratified, chlorinity and temperature increase abruptly, and near the bottom the water becomes anoxic, remarkably tepid (about 25 °C) and 3 times saltier than seawater (Spigel and Priscu 1998). However, physico-chemical characteristics of Antarctic lakes vary widely and some, such as the Don Juan Pond (McMurdo Dry Valleys), have salt-rich brine which does not freeze in winter. Although many lakes with no ice cover in summer freeze almost solid during winter, bottom waters may remain liquid due to high salt concentrations. The solutes may arise from seawater trapped during the isostatic uplift of marine bays such as in Ace Lake, Deep Lake and other lacustrine ecosystems of the Vestfold Hills. In lakes which did not originate from the sea, salt accumulation is mainly due to strong evaporation which concentrates ions from melting snow and ice, leaching from catchment rocks, soils and widespread salt encrustations. Figure 18 shows the increase in conductivity and average concentrations of some anions and cations in four lakes at Andersson Ridge (73° 44'S, 162° 42'E, northern Victoria Land) during January 2002.

With respect to terrestrial organisms, those in freshwater ecosystems are generally more protected against desiccation, solar radiation and temperature variations; most of the biomass of Antarctic non-marine ecosystems therefore accumulates in lakes and ponds. Owing to the wide range of physico-chemical and environmental characteristics, the trophic status of Antarctic lakes may vary widely from hypooligotrophy to mesotrophy or even eutrophy, especially in coastal areas receiving nutrients from the marine environment through seabirds. The main phototrophs in Antarctic inland waters are cyanobacteria (*Phormidium*, *Oscillatoria* and *Lyngbya*), flagellates such as *Cryptomonas* and *Chroomonas* (Cryptophyceae), *Ochromonas* (Chrysophyceae), *Clamydomonas* (Chlorophyceae), *Pyramimonas* (Prasinophyceae), and chlorococcaleans such as *Ankistrodesmus* and *Schroederia* (Vincent 1988). Terrestrial bryophytes such as *Bryum pseudotriquetrum* may grow as benthic phototrophs in lakes near Syowa Station and Schirmacher Oasis (Imura et al. 1999). Photosynthetic

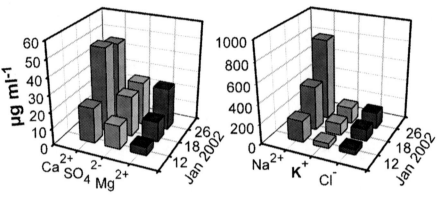

Fig. 18. Increase in major ion concentrations (µg ml⁻¹)in lake waters during January 2002 at Andersson Ridge (northern Victoria Land)

sulphur bacteria occur in lakes derived from seawater, where they generate high concentrations of H_2S through sulphate reduction. Cyanobacteria are the most widespread organisms in all Antarctic freshwater ecosystems. Although they may sometimes dominate plankton, they typically occur in lacustrine benthos where they form widespread, thick films and mats. Cyanobacterial mats overlying sediments are one the main feature of Antarctic lakes. The mats may have different macromorphologies, colour, and communities of cyanophytes, diatoms, bacteria, yeasts, protozoans, rotifers, nematodes and tardigrades (Parker and Wharton 1985; Fumanti et al. 1997).

Although data on phototrophs and heterotrophs in freshwater ecosystems of continental and maritime Antarctica are only available for a few locations, the trophic structure of these ecosystems increases northwards. There are no crustaceans in Victoria Land, and the main grazers are protozoans, rotifers, nematodes and tardigrades, while the aquatic fauna in Vestfold Hills also comprehends a cladoceran and some species of copepods (Vincent 1988). The greatest species diversity is found in the plankton and benthos of lakes and pools of the Antarctic Peninsula. In addition to the common *Philodina gregaria*, the rotifer fauna probably contains at least seven to eight taxa, and the arthropod mesofauna comprehends some species of crustaceans (Anostraca, Branchiopoda, Copepoda Calanoida, Copepoda Harpacticoidea, Cladocera and Ostracoda; Smith 1996).

2.5 Antarctic Ecosystems as Indicators of Change

Global climate models predict that polar regions will be affected by air temperature increases and changes in the amount and pattern of precipitation. Changes in temperature (melting and freezing) and precipitation in Antarctic ice-free areas are likely to strongly affect water availability and the distribution and appearance of new, uncolonised areas. Climate amelioration and increased water bioavailability may act alone or synergistically to enhance colonisation and the establishment of local invasive or exotic pre-adapted species (Walton 1990; Watson 1999).

In highly desiccated saline soils, liquid water controls biogeochemical processes, and changes in chemical weathering, salt solubilisation, substrate stability and terrestrial productivity will strongly affect lacustrine ecosystems, which are multiyear integrators of changes in landscape processes. Furthermore, climate changes are likely to affect the limnology of polar desert lakes through changes in the duration and thickness of snow and ice cover (i.e. underwater light availability), and in thermal and chemical regimes, with unpredictable effects on organism abundance and diversity.

Biotic communities in Antarctic cold desert ecosystems are rather simple and fragile, with many species either living at the limit of their range or adapted to cope with extreme environmental conditions. The low species diversity and lack of several functional groups, such as angiosperms in terrestrial ecosystems and macroinvertebrates in aquatic ecosystems, means that the loss or gain of even a single species may strongly affect the integrity and productivity of these ecosystems. Species of sub-Antarctic invertebrates and plants possess sufficient physiological characteristics and ecological plasticity to allow survival in maritime or continental Antarctica, and climate amelioration will increase the pool of potential colonists (Convey 2000). Undoubtedly, Antarctic ecosystems are more sensitive indicators of regional climate changes or colonisation processes than ecosystems at lower latitudes, where responses of biotic communities to external forcing and potential colonists are buffered by more complex biological interactions and feedback processes.

Atmospheric circulation patterns are responsible for the deposition of persistent airborne pollutants from lower latitudes into Antarctic ecosystems (Bargagli 2000). Global warming will probably enhance the contraction of sea ice and the penetration of air masses and clouds into the continent. It cannot be excluded that these changes will increase the deposition of atmospheric pollutants on the continent. Xenobiotic compounds may more severely impair organisms growing in extreme environments than related species growing under less stressful conditions in temperate regions.

All these aspects make Antarctic ice-free landscapes sensitive indicators of regional climate and environmental change, and models of change on a global scale (Walton et al. 1997). In looking to the future of Antarctic ecosystems, the Scientific Committee on Antarctic Research (SCAR 1989, 1993; SCAR/COMNAP 1996) has identified the possible implications of enhanced global warming and the impact of human activity on Antarctic ecosystems as research priorities. As a successor of the SCAR-BIOTAS (Biological Investigations of Terrestrial Antarctic System) programme, a new international research programme SCAR-RiSCC (Regional Sensitivity to Climate Change in Antarctic Terrestrial Ecosystems) has been developed to study responses of terrestrial and freshwater ecosystems to climate change. As an analogy for future climatic and environmental change, research is focused on a variety of sites along a latitudinal gradient, covering about 40° (from the peri-Antarctic islands to southern Victoria Land). To develop a predictive understanding of ecological change, research must address the ways in which ecosystems interact with their environment, the influence of abiotic variables on population performance, and relationships between species abundance and richness.

2.5.1 Climate-Change Indications

Among scientists there is little doubt that there are already tangible changes in Antarctic ice-free areas, especially on the peninsula and sub-Antarctic islands (e.g. Smith 1990; Hall and Walton 1992; Kennedy 1995; Convey 1997, 2000). The Antarctic Peninsula is a transition zone from a relatively temperate north to a more polar-influenced south. Its role as a physical barrier to atmospheric circulation makes it highly sensitive to climate changes. As discussed in the previous chapter, in the last decades the west coast of the peninsula has been affected by a remarkable atmospheric warming trend and an increase in atmospheric precipitation. The spectacular collapse of ice shelves has attracted public and scientific interest in the region (e.g. Ross et al. 1996; Vaughan and Doake 1996; Lucchitta and Rosanova 1998; Rott et al. 1998);

moreover, there is evidence that the warming is inducing glacier retreat (Kejna et al. 1998; Park et al. 1998). The enhanced melting of snow, ice and permafrost increases water availability, leaching and drainage processes, and the rate of soil formation.

Experiments using small Perspex cloches (Fig. 19; increasing moisture, temperature, and changes in other microenvironmental factors) over barren fell-field sites at Signy Island (Smith 1993; Wynn-Williams 1993) showed that spore lying dormant within the soil may germinate. After three years some species of mosses and cyanobacteria had covered the mineral substratum. On the same island, from repeated observation of terrestrial ecosystems over 25 years, Smith (1990) detected the expansion of lichen and moss communities on previously uncolonised glacial moraines. He suggested that rates of successional change may be faster than expected. Moreover, melting snow and ice, along with wind, birds and humans, may help release of viable propagules onto newly exposed substrata. Small colonies of two species of

Fig. 19. Long-term experiments on colonisation processes with Perspex cloches (Edmonson Point, northern Victoria Land)

moss (*Polytrichum piliferum* and *P. longisetum*) were found at a site on Signy Island which had become ice-free in the previous decade (Convey and Smith 1993). Neither species was previously known in the South Orkneys, which probably have the most intensely studied bryoflora in the Antarctic. The nearest known occurrence of *P. piliferum* was 700 km to the west–southwest (King George Island), and that of *P. longisetum* was over 1,700 km (southern Patagonia).

There is only one record of alien plants persisting more than a few years at a site on the Antarctic Peninsula: some *Nothofagus* saplings with associated soil and ground flora from Ushuaia were transplanted at Ciervia Point in 1955. The trees did not survive, but some plants of *Poa pratensis* became established and were still surviving in 1995 (Smith 1996). The reason why *Deschampsia antarctica* (Poaceae) and *Colobanthus quitensis* (Caryophyllacea) are the only two native species of vascular plants growing in the Antarctic Peninsula is still obscure. However, climatic factors seem to play an important role in their survival and colonisation. Due to regional warming, the two species have become rapid colonisers in many recently exposed grounds adjacent to receding ice fields (Smith 1996; Grobe et al. 1997). From 1964 to 1990, on the Argentine Islands there was a 25- and 5-fold expansion of *D. antarctica* and *C. quitensis* respectively (Smith 1990).

As discussed in the previous chapter, warming trends in temperature records from continental Antarctica are less marked than in those from the peninsula, and it is impossible to discern long-term trends from inter-annual variability in ecosystem processes. Freshwater ecosystems should be key monitoring sites for early detection of global-change effects; however, little can be substantiated on the basis of available data from continental Antarctica. Wharton et al. (1993) reported ice thickness records for Lake Vanda since 1960. The data showed a thinning from approximately 4 to 3 m between 1961 and 1982, while there was no clear trend after 1982. However, the level of the lake has increased by about 10 m since 1973 (Chinn 1993). This rise is likely due to the increase in the annual discharge of the Onyx River (from 2.5×10^6 m^3 in 1970 to 5.5×10^6 m^3 in 1992), which is the only significant inflow to the lake. Since the 1970s, the levels of several lakes in the McMurdo Dry Valleys (Chinn 1993), Lake Wilson (at about 80° S in southern Victoria Land; Webster et al. 1996) and the Vestfold Hills (Fulford-Smith and Sikes 1996) have increased, and their ice cover has thinned. These variations have generally been ascribed to regional climatic changes which increase meltwater flows with respect to ablative losses.

2.6 Future Research in Antarctic Periglacial Areas

Climate change is expected to produce faster and greater changes in high-latitude regions, because it is likely to be amplified by alterations in albedo, atmospheric composition and permafrost. Terrestrial ecosystems in polar regions are strongly regulated by temperature and precipitation; they are less affected by human activities and will likely show the strongest signal-to-noise ratio. In the context of the International Geosphere-Biosphere Program (IGBP), the Global Change and Terrestrial Ecosystems (GCTE) research programme for the Arctic provides a useful scheme for studies pertaining to the prediction of the effects of climate change on terrestrial ecosystems. The primary aim of GCTE is to understand and model the effects of primary ecosystem processes such as the exchange of energy, water and trace gases with the atmosphere, element cycling and storage, and biomass accumulation or loss. Through intensive studies in key biomes along biogeochemical transects, the programme aims to model changes in species distribution and composition, and consequent changes in ecosystem function, in order to predict patterns of change in ecosystem composition and structure. Arctic research over the past 20 years has shown that climatic change has produced a change in the function of terrestrial ecosystems, which are now net sources of CO_2 to the atmosphere (Oechel and Vourlitis 1996). This change is thought to be transient, but there is no evidence as to how long it will last. Moreover, the results of simulations based on climate-change scenarios are mixed with regard to whether C sequestration will increase or decrease within the next two decades. Plant productivity seems rather unresponsive to the observed warming, and the expected northward migration of vegetation will probably occur in the long term, in contrast with the relatively instantaneous response of soil microorganisms. According to Oechel and Vourlitis (1996), despite the development of the GCTE programme, we are still far from being able to predict the response of Arctic ecosystems to climate change. They maintain that much of this ignorance stems from the multivariate nature of natural systems, and the multitude of interactions and feedback which are difficult to define through field observations, experimentation, or even modelling approaches.

Antarctic terrestrial ecosystems have poorly organised communities whose dynamics are controlled by rather simple biological interactions and cybernetic feedback processes. The desert environment of Antarctic ice-free areas is strongly affected by changes in water availability, which in turn affects soil processes, hydrology and biogeochemistry, plant productivity, survival, and colonisation processes. Any reduction in ice cover will probably result in the exposure of new substrata, enhanced mechanical and chemical weathering of rocks, and increased rates of soil formation. There is therefore a need for continuous, improved instrumental monitoring of the physico-chemical and biological characteristics of periglacial areas in order to understand and model the effects of global change on water, permafrost, soil, and primary ecosystem processes. Monitoring should include remote sensing and in-situ measurements, mapping the extent of vegetation, and biological characterisation at community and population levels (both floristically and faunistically) to detect changes in community structure and extent, and possible biological invasions. Although Antarctic terrestrial ecosystems are rather "simple", the results of studies within the framework of the GCTE indicate that it is difficult to predict their response to climate and environmental changes. Research will be necessarily speculative, as predictions will be based on extrapolation of environmental and biogeographical data, palaeobiogeographical reconstructions, ecosystem modelling and ecophysiology (Adamson and Adamson 1992; Hansom and Gordon 1998). Although a short-term effect of warming and ice melting in maritime Antarctica is an increase in new uncolonised grounds, it cannot be excluded that enhanced snowfalls will reduce habitats for terrestrial and freshwater organisms in the future and/or in other Antarctic regions. The potential impact of exotic immigrant species may be overstated.

As discussed by Walton et al. (1997), some aspects of research on Antarctic ice-free landscapes cannot be ignored, for example, logistic costs and potential environmental impact, the paucity of long-term environmental data for many locations, the high sensitivity to climatic forcing, which may make it difficult to distinguish long-term trends from interannual variability, and the presence of many generalist species which may provide a buffer against change. The impact of logistic support to research activities and tourism could confuse the signals of global climate change and those of long-range transport of persistent pollutants. On the other hand, there is no doubt that global-change research in Antarctica will help identify environmental management options, and suitable approaches to minimise the impact of human activity and to ensure long-term protection and conservation of freshwater and terrestrial ecosystems.

2.7 Summary

Antarctica is covered by two ice sheets which have coalesced to form the largest ice body on Earth. There is no melting in the continental interior, and the slowly accumulated ice flows towards the coast in outlet glaciers and ice streams along the margins of the plateau. In areas where the flow is blocked by mountains, ablation of the ice favours the concentration of well-preserved meteorites. The scientific community is also greatly interested in the origin and palaeoclimate of subglacial lakes, in the possible microbial life they contain, and in microbes inhabiting the Antarctic ice sheets and permafrost. The most striking findings of research on Antarctic is the assessment of relationships between changes in atmospheric composition and past climate change. Ice records extend to the past four glacial–interglacial cycles, and progressively longer records are recovered through coring of different domes.

A very small percentage (about 2 %) of the Antarctic surface is free of ice and snow in summer. Most ice-free areas have emerged from the retreating ice in the past few thousand years. Apart from nunataks or the slopes of the Transantarctic Mountains, most of these areas are distributed along the coast; they often lack moraines and glacial deposition typical of deglaciated areas of the Northern Hemisphere, because the deposition of materials in Antarctica occurs in the sea beneath ice shelves or at glacier snouts. The marked latitudinal and altitudinal (inland from the coast) gradient of temperature and water availability determines strong spatial variations in the chemical weathering of rocks and soil formation processes. The southernmost ice-free areas are cold deserts with few developed soils, often encrusted with water-soluble salts. Podzols with low salinity and accumulation of organic matter occur at higher latitudes in Wilkes Land, and active chemical weathering processes with abundant clay-sized materials occur in soils of the Antarctic Peninsula. Ephemeral streams fed by melting ice and/or snow banks flow for some weeks or months in summer, and they may feed small ponds or subglacial and sub-aerial lakes. Most lakes lack outlets and are therefore the main sink for water and solutes from the surrounding catchment area. The climatic gradient and particularly the bioavailability of water influence the abundance and diversity of terrestrial organisms. Although there are few species of cyanobacteria, diatoms, chlorophytes, fungi and lichens in the southernmost rock outcrops of continental Antarctica (at about 86–87° S), well-developed cryptogamic communities and some species of terrestrial microinvertebrates occur further north in sheltered and wetter coastal habitats of the McMurdo Sound region. Cyanobacteria are the most widespread organisms in freshwater ecosystems, and the mats overlying sediments host communities of cyanophytes, diatoms, bacteria, yeasts, protozoans, rotifers, nematodes and tardigrades. Greater species diversity and abundance of organisms are found in ecosystems of continental Antarctica at lower latitudes. Milder and wetter maritime Antarctica (Antarctic Peninsula) not only contains cryptogamic vegetation but also two species of native flowering plants, an endemic wingless midge, and several other species of terrestrial and freshwater invertebrates.

In the first chapter we saw that Antarctica plays a very important role in global climate, and that available data and models suggest that polar regions will be particularly affected by warming and changes in the patterns and amounts of precipitation. This chapter emphasises the role of deep ice core records from the Antarctic plateau in providing one of the best temporal accounts of climate change over the last 500,000 years, and the unique opportunities provided by Antarctic terrestrial and freshwater ecosystems in detecting and predicting the possible effects of change, both climate-based and human-induced. Antarctic ecosystems are deemed among the most sensitive indicators of regional climate change and reliable models of change in more complex ecosystems elsewhere. Some changes, such as the collapse of ice shelves, enhanced melting of ice, and colonisation of newly exposed substrata by cryptogams and two species of vascular plants, are occurring in zones of the Antarctic Peninsula more affected by the regional warming trend.

In conclusion, Antarctic ice sheets and glaciers, sub-glacial lakes, and terrestrial and freshwater ecosystems are very important research sites for a variety of scientific investigations and for addressing global-change issues. Several international research programmes have been developed with the conviction that the value of the expected research findings will compensate for the potential environmental impact and high costs. Increasing difficulties in obtaining funds for complex research beyond the interests of single nations, and concern about the impact of human activity in Antarctica require the development of collaborative multidisciplinary research in a few key areas of Antarctica.

3 The Southern Ocean Environment: Anthropogenic Impact and Climate Change

3.1 Introduction

The Southern Ocean, one of the largest, oldest and coldest deep-water marine systems, encircles Antarctica in a 2,500-km-wide, semi-closed belt. Like the Antarctic continent, it has distinctive physico-chemical and biological features and is important to global processes. Waters with temperatures near the freezing point (about $-1.9\,°C$), produced along the Antarctic continental margin, contribute to the formation of bottom waters and spread through the global ocean. In winter a belt of sea ice with an area larger than that of the continent itself rings Antarctica, with profound effects on the sea/air exchange. The surface albedo, lack of heat and moisture exchanges between the ocean and atmosphere affect the formation of clouds and the stability of the atmosphere over large areas of the Southern Hemisphere. Sea-ice extent and thickness are very sensitive to global changes and may provide an early indication of warming due to increasing greenhouse gas concentrations.

Zonally uniform water masses with unique physical characteristics began to develop in the Southern Ocean in the Early Miocene, when circum-Antarctic currents largely unaffected by landmasses began to flow around the continent. Marine organisms able to adapt to the new environment evolved for more than 20 million years in a semi-closed system, and several taxa assumed a circumpolar distribution. The upwelling of warmer waters from lower latitudes affects the stability of the water column; in spring and summer it favours the growth of phytoplankton, with the establishment in the seasonal pack-ice zone of the characteristic short food chain diatoms–krill–whales (or seals or penguins). Although ocean phytoplankton mostly consists of unicellular organisms with a one-day lifespan, it seems to play a very important role in controlling global warming through the uptake of CO_2 and the production of dimethylsulphide (which affects the number of cloud condensation nuclei available in remote regions). There is evidence that a positive phytoplankton–climate feedback system developed in the Southern Ocean during the glacial–interglacial transition.

Contrary to organisms in Antarctic terrestrial ecosystems, those in the Southern Ocean do not experience extreme variations in temperature, water availability and solute concentrations. The most important changes are due to the formation of the sea-ice cover, which determines marked seasonal variations in light penetration and primary productivity. Sea ice provides a platform on which birds and mammals can live and breed, and from which they can make foraging forays into the water. Compared to land ice, sea ice is usually warmer, with a much more heterogeneous and variable structure. Biotic communities may develop within the sea ice and on its underside. The boundary between sea ice and the open sea is one of the ecologically most interesting regions in the Southern Ocean because, like ecotones in terrestrial ecosystems, the sea-ice edge has a variety of niches, with great species diversity and abundance.

Reduced competition and long isolation in the Southern Ocean presumably provided notothenioid fish and several taxa of marine invertebrates with the opportunity to speciate and fill ecological niches occupied by other species at lower latitudes. Owing to their unique evolutionary history, Antarctic marine biota show a much higher percentage of endemic species (e.g. up to 97 % for notothenioid fish) with respect to organisms in continental terrestrial and freshwater ecosystems. However, like cryptogams and terrestrial microinvertebrates, most marine organisms have slow growth rates and can survive long periods of starvation at negligible metabolic cost. In the neritic province (near the coast), owing to the low zooplankton biomass, most phytoplankton and ice algae sink in the summer, becoming food for the very rich benthic fauna. Benthic invertebrates are mostly suspension feeders such as sponges and echinoderms, and are important sources of food for fish. The latter are eaten by other fish, birds and mammals, which are in turn the prey of leopard seals or killer whales. Thus, in contrast to the short food webs in the oceanic province, those in coastal ecosystems are rather complex and long.

This chapter examines the main characteristics of the Southern Ocean, interactions with global atmospheric and oceanic circulation, and the structure and functioning of the main biological systems. Contrary to popular belief, the Southern Ocean is far from being a pristine environment. The exploitation of living marine resources during the last two centuries has had dramatic impact on seal and whale populations and provides important lessons for regulating the exploitation of current and future marine resources.

Antarctic marine organisms have unique ecophysiological adaptations due to their long evolutionary history in isolation; compared to related species from lower latitudes, they are probably more sensitive to the potential effects of anthropogenic pollutants and to climate and environmental change.

Although several scientific reports on the Southern Ocean were made by naturalists who accompanied several 19th century expeditions, the modern era of ecological research in the Southern Ocean began in 1972 with the establishment of a subcommittee of the SCAR Working Group on Biology, the Committee on Marine Living Resources of the Southern Ocean. The BIO-MASS (Biological Investigation of Antarctic Systems and Stocks) Programme (1977–1991) was a major collaborative effort of scientists from many nations. It aimed to gain a better understanding of the Antarctic marine environment and to provide a sound basis for the management of living resources. The development of this programme facilitated the establishment in 1980 of the international Convention on the Conservation of Antarctic Marine Living Resources (CCAMLR) as part of the Antarctic Treaty System. The results of oceanographic and ecological research were archived in the BIOMASS Data Centre and have been reported in many publications. El-Sayed (1994) provided a comprehensive record of the accomplishments of the BIOMASS Programme, and other books on oceanography and ecosystems of the Southern Ocean have been published recently (e.g. Knox 1994; Johannessen et al. 1994; Ross et al. 1996; Lizotte and Arrigo 1998; Jeffries 1998; Jacobs and Weiss 1998; Hobbs et al. 1998; Karoly and Vincent 1998; Spezie and Manzella 1999; Faranda et al. 2000).

3.2 The Southern Ocean

As reported in the first chapter, the opening of the Drake Passage and separation of the South Tasman Rise from Victoria Land allowed the Antarctic Circumpolar Current (West Wind Drift) to establish in the Late Oligocene. Although the ocean surrounding Antarctica is a continuation of the Atlantic, Pacific and Indian oceans, 22 Ma ago a major physico-chemical and biological barrier developed between cold Antarctic surface waters and warmer sub-Antarctic waters (the Antarctic Polar Front, sometimes called the Antarctic Front or Antarctic Convergence, at latitudes between 47 and 61° S; Fig. 20). The establishment of this barrier, which is still operating, favoured the progressive cooling of the continent and the formation of Antarctic ice sheets. Isostatic equilibrium adjustment, in response to massive accumulation of ice on the continent and scouring by ice shelves, lowered the continental shelves surrounding Antarctica. The transition from the continental shelf and continental slope ("shelf break") usually occurs at depths of about 400–500 m (i.e. 3–4 times deeper than in any other continent). The mean width of Antarctic continental shelves is about 200 km, and at the foot of the continental slope there are three deep basins (4,000–6,500 m) partially bounded by a series of ridges and a plateau (Fig. 21). The mid-ocean ridge systems between Antarctica and the continents to the north represent seafloor spreading sites which determine the evolution of Southern Ocean morphology and circulation.

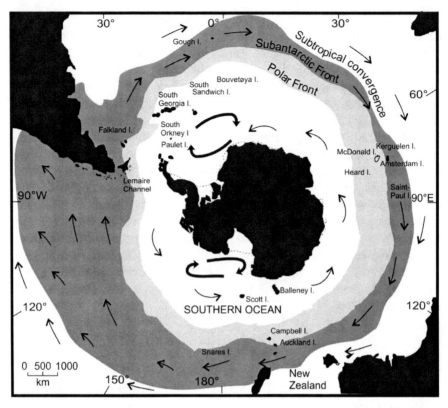

Fig. 20. Map of surface currents and the mean position of principal fronts in the Southern Ocean

3.2.1 Water Masses and Circulation Patterns

The area covered by the Southern Ocean (about 77×10^6 km^2) extends from the Antarctic coasts to a northern limit established by water mass characteristics. The isopleth lines in this area are approximately parallel to the lines of latitude, although gradients are more concentrated in frontal zones. The Antarctic Polar Front (APF) is the zone where cold Antarctic Surface Water sinks below warmer Sub-Antarctic Water and continues to flow northwards at intermediate depths as Antarctic Intermediate Water. The latter water mass can be traced through its low salinity and temperature of 4–5 °C; it can be detected in coastal waters emerging both near the continents of the Southern Hemisphere and north of the equator. Antarctic Intermediate Water thus plays a crucial role in the interchange of heat and minerals between the Southern Ocean and other oceans. Antarctic Circumpolar Deep Water, a warmer (about 1–2 °C), more saline (about 34.75 parts per thousand), nutrient-rich water

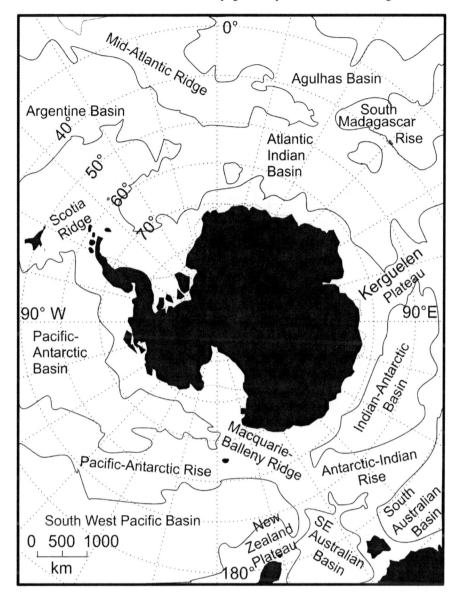

Fig. 21. Map of principal deep basins and submarine canyons around Antarctica

mass which lies below Antarctic Surface Water (Fig. 22), is the largest water mass in the Southern Ocean; it flows southwards from the world's warmer oceans at a depth of some thousand metres, rises at the APF and upwells at the Antarctic Divergence. Antarctic Bottom Water, another important water mass with slightly lower salinity and temperature (about −1 °C) and higher density, forms near the continent; it flows down along the continental slopes and

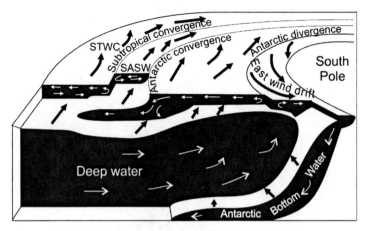

Fig. 22. Major water masses in the Southern Ocean

spreads northwards into the lowermost oceanic basins, cooling and ventilating a large proportion of deep sea (Whitworth et al. 1998). Antarctic Bottom Water mainly forms in the Weddell and Ross Seas and other coastal areas of East Antarctica as a result of winter freezing and particular regional conditions, such as the presence of deep ice shelves and polynyas. According to Godfrey and Rintoul (1998), the volume of dense water spreading northwards from near Antarctica is equal to or a little higher than the volume of dense water produced in the North Atlantic and, as a whole, dense water masses formed in the Southern Ocean account for more than 50 % of the volume of the world ocean.

The APF is actually a broad zone of transition with meanders and eddies, rather than the narrow band schematised in Fig. 20. Crossing the Polar Front in a southward direction, surface water temperature sharply decreases (about 3–8 °C in summer and 1–5 °C in winter). In the Weddell Sea–Drake Passage region, the APF has a seasonal variability of 1–2° in longitude and long-term variability of up to 4° in latitude (King and Turner 1997). Despite the variability in its position, width and temperature gradient, the APF divides the Southern Ocean into a sub-Antarctic region to the north and an Antarctic one to the south, with great differences in weather conditions and in physico-chemical and biological features. To the north of the APF, an increase in water salinity (about 0.5‰) marks the Sub-Tropical Convergence or Sub-Antarctic Front. This front encircles Sub-Antarctic Water and is regarded as the boundary between the Southern Ocean and three other oceans to the north. Its position varies widely, especially off the coast of Chile (Fig. 20). Antarctic Surface Water to the south of the APF and up to a depth of about 100–250 m is cold (from 1 to –1.9 °C in winter), with small seasonal variations and salinity usually less than 34.5‰. However, salinity (and therefore density) increases when sea ice forms, and decreases when it melts.

Ocean circulation is driven not only by temperature and salinity (thermohaline circulation) but also by winds. The two forcing mechanisms coexist and, although we consider them separately, they cannot be separated geographically or dynamically. The winds in the Southern Ocean are among the most intense and constant and are characterised by an easterly flow from the continental margin to 65° S and by a wide zone of westerlies, which extends northwards to more than 40–35° S. In the latter zone the strong westerlies drive the largest current system of the world, the Antarctic Circumpolar Current (ACC) or West Wind Drift. The ACC flows clockwise around the continent at a relatively slow velocity (about 20 cm s^{-1}, compared to more than 200 cm s^{-1} for the Gulf Stream), but it transports a water mass larger than that of any other oceanic current system – more than 100×10^6 m^3 s^{-1} (i.e. 100 Sv, Sverdrup; Nowlin and Klinck 1986). The flow is strongly affected by bottom topography, because the current may reach depths of 3,000 m. The Antarctic Peninsula and southern part of South America constrain the flow of the ACC, determining a convergent flow in the Drake Passage, with an increase in speed (up to 1 m s^{-1}) and water volume (up to 150 Sv).

In marine areas south of 65° S and close to the coast of Antarctica, there is a westward-flowing coastal current (the Antarctic Coastal Current, water volume of about 10 Sv) driven by easterly winds (i.e. katabatic drainage winds from Antarctic ice sheets which are deflected leftwards by the Earth's rotation). The westward coastal current in the Weddell and Ross Seas is diverted northwards by the Antarctic Peninsula and Victoria Land, and this results in a cyclonic circulation of the sea (Fig. 20). In addition to these two large, permanent-flowing gyres, another one occurs east of the Kerguelen Plateau (Gordon and Molinelli 1982; Deacon 1984), and several eddies, current rings and meanders have been reported in the Drake Passage and other regions of the Southern Ocean (e.g. Joyce and Patterson 1977; Gordon 1988).

In the Southern Hemisphere, the Coriolis force deflects ocean currents to the left, thereby driving the ACC to the north and the Antarctic Coastal Current to the south. This divergent flow (the Antarctic Divergence; Fig. 22) promotes the upwelling of warm and nutrient-rich subsurface Circumpolar Deep Water in the region between the ACC and the Antarctic Coastal Current, except east of the Drake Passage (Deacon 1984).

3.2.2 Air–Sea Exchanges

As discussed in the first chapter, the global climate system is driven by a gross radiative imbalance at the top of the atmosphere between the tropics and the poles. The poles absorb less short-wave solar radiative energy than is returned to space by long-wave terrestrial radiation, leading to an energy deficit. The reverse occurs at the tropics and, in both hemispheres, 38° latitude approximately marks the boundary between the zone of net radiative cooling and that of net radiative heating. This radiative imbalance would steadily warm the tropics and cool the poles if it were not for a poleward flow of heat in both the atmosphere and ocean. This transfer of heat exists because there are temperature differences between poleward and equatorward flows. Not only heat, but also carbon, persistent pollutants and biological constituents of the atmosphere are transported southwards. Only a brief account of air–sea interactions will be given here; readers interested in pursuing this topic further are referred to specific books, such as those by Gill (1982), Trenberth (1992), Jakeman et al. (1993), Hobbs et al. (1998) and Karoly and Vincent (1998).

The atmosphere and ocean are coupled through exchange of heat, moisture and momentum at the sea surface (i.e. fluxes at the sea surface both depend on and influence the state of the atmosphere and ocean). More than 200 years ago researchers discovered that deep ocean water in the tropics has a low temperature; it was deduced that this water must have originated at the poles and that, in order to conserve mass, there must be an accompanying poleward flow of warm water at the surface. The heat gained by oceans in the tropics through radiative heating is advected to higher latitudes by the strong poleward flow of subtropical currents. This pattern is largely determined by wind-driven circulation in the upper ocean, although thermohaline circulation also plays an important role. Waters at high latitudes receive relatively small amounts of radiative heat and lose large amounts of heat to the atmosphere. The evaporation of moisture extracts latent heat from the sea, making surface water colder and saltier (i.e. more dense), and combines with other heat flux processes and mixing to drive the ocean's thermohaline circulation. Given that water from the Southern Ocean is denser than that at lower latitudes, it flows equatorwards beneath the lighter waters, determining the vertical stratification of mid- and low-latitude waters. Wind-driven divergence of surface waters along the equator and in some land boundaries in the Southern Hemisphere causes upwelling of cold water and the consequent cooling of the atmosphere above it.

The majority of the world's deep water is formed in the North Atlantic (Norwegian/Greenland Sea and Labrador Basin; Hall and Bryden 1982) and Southern Ocean (mainly in the Weddell and Ross Seas), while cold-to-warm water conversion occurs above all in the subtropical Indian and Pacific Oceans. Thermohaline and wind-driven circulations combine to produce a global conveyor belt system which redistributes heat throughout the oceans. However, in the Northern Hemisphere the oceans are separated by landmasses, and they are linked only at their southern extremity. The massive eastward flow of the ACC, the only free-flowing oceanic current which completely encircles the Earth, is the primary means by which mass, heat and solutes are carried from one ocean to another (Godfrey and Rintoul 1998). The impact of climate change in the Southern Ocean is thus likely to trigger large-scale changes in the world's oceans and in the global climate system.

These will be long-term changes because, in contrast to the atmosphere which usually responds in a few weeks to changes in forcing, seawater has large thermal inertia, and changes in Southern Ocean processes affect the global climate on timescales of years to centuries.

There is evidence that climate change has probably already altered the physico-chemical characteristics of some Antarctic water masses (Bindoff and McDougall 2000). Wong et al. (1999) found temperature and salinity changes in the Pacific which are consistent with surface warming and freshening in the region of the Southern Ocean where water masses form. However, reliable predictions require coupled ocean–atmosphere models. Although over the past decade there has been marked improvement in the simulation of Southern Ocean responses to global warming (e.g. Budd and Wu 1998; Hirst 1999), the present generation of coupled models cannot give a satisfactory explanation of sea-ice and polynya distributions. This is a serious drawback, because most simulations show that Southern Ocean thermohaline circulation may be very sensitive to climate changes, and that a large proportion of Antarctic dense water forms in leads and polynyas between the sea ice.

Most ocean-climate models predict reduced sea-ice formation in the Southern Ocean. This will decrease the rate of formation of downwelling dense water masses, with a consequent stabilisation of the water column. The effects of these changes will probably be amplified by the fact that the ACC is partly driven by thermohaline circulation associated with the formation of Antarctic Bottom Water. Modelling by Cai and Gordon (1998), for instance, shows that an increase in CO_2 concentrations and global warming would promote a decrease in the magnitude of the ACC. The El Niño–Southern Oscillation phenomenon could also contribute to changes in seawater circulation and in the climate of Antarctica. However, as discussed in the first chapter, possible connections between ENSO and the Southern Ocean remain to be explored. Whether the magnitude or frequency of ENSO will change in a changing climate is still an open issue. Although this is a topic of intense research, it cannot be addressed here, and the reader can refer to the textbook by Philander (1990) or more recent publications by Knutson et al. (1997), Gillespie and Burns (2000), and Grimm et al. (2000).

3.2.3 Sea Ice

The seasonal formation of vast areas of sea ice (a 400–2,000 km wide belt with an area of up to 19×10^6 km^2, which completely encircles the continent in September; Gloersen et al. 1992) is a predominant feature of the Southern Ocean. Unlike in the Arctic, more than 85 % of Antarctic sea ice melts in summer; most of it is therefore usually much thinner than Arctic ice. Ice formation in the Southern Ocean increases in autumn (from April to June) and progressively extends northwards until September–October. Melting is relatively rapid (with a rate more than double that of freezing) in spring, and the most extensive areas of ice remaining at the end of summer are those along the coasts of the Bellingshausen and Amundsen Seas and in the western Weddell Sea (Fig. 23). The sea–air heat exchange and upwelling of relatively warm water (about 2 °C), which resides below the Southern Ocean pycnocline (Gordon 1981), probably contribute to the rapid break-up and melting of Antarctic sea ice. The sea-ice zone in the Arctic varies by less than 25 %, and freezing is much faster than the retreat of ice. According to Gordon (1981), this reverse pattern is due to the large input of freshwater, which induces a strong pycnocline in the seawater column and consequent rapid freezing mainly determined by the ocean–atmosphere heat flux.

Surface seawater in the Southern Ocean has a salinity of about 34‰ and reaches the freezing point at –1.91 °C. When the surface layer of the sea reaches this point, additional heat loss determines a slight supercooling of

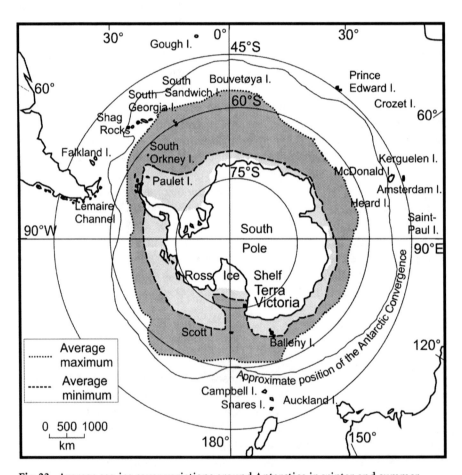

Fig. 23. Average sea-ice cover variations around Antarctica in winter and summer

water; ice crystals of pure water begin to form and salt is expelled into the surrounding water. At first, a floating suspension of small ice crystals or platelets (called frazil ice) occurs at or near the water surface to produce a soupy, unconsolidated mixture (grease ice). If freezing continues and ice formation exceeds 30–40 %, the transition to a solid cover begins. Wave fields break this thin layer of ice into small, circular pieces (termed pancake ice) with raised rims due to rubbing against other pieces. Pancake ice eventually coalesces to form a composite and continuous sheet (Maykut 1985; Wadhams 1991). As seawater freezes, brine-rich streamers form and sink, carrying away heat from the ice formation zone. Long, columnar ice crystals begin to form below (congelation ice) in response to conductive heat loss along the temperature gradient. Less saline water rises up, and frazil ice or platelets may form within the zone of convection; these float up, accumulate and fuse together beneath congelation ice to form a porous layer (bash ice) containing pockets of brine. Platelets may also form at great depth in supercooled water streaming out from under ice shelves or at the seafloor in inshore zones of convective circulation (anchor ice). Although anchored to the bottom, the anchor ice can move in response to thermal or mechanical stress and, where it adjoins land or ice sheets, it shows typical parallel tide cracks. The annual ice sheet forms early in the autumn near the coastline (fast ice); although it usually begins to detach from the shoreline in late summer, it may also remain for some years, giving rise to multiyear ice with a thickness of several metres. The ice sheet formed later and seawards can break up under the influence of storms, and pack ice can move considerable distances before re-freezing into solid pack (Fig. 24).

Depending on physical and hydrodynamic conditions, sea-ice formation and structure can vary considerably according to location and from one year to the next. The ice formed in leads and polynyas often consists of frazil crystals quickly piled by wind to form thick layers. In the Weddell Sea, frazil ice layers are often sandwiched between congelation ice (Knox 1994). Sea ice is usually overlain by a variable amount of snow, which ensures that a large proportion of incoming solar radiation is reflected. The snow–ice interface may become flooded with seawater along leads in the pack ice, especially in the marginal ice zone. The ice edge is the transitional zone between the ice cover and open waters, and forms a complex frontal system affected by dynamic interactions among ice, seawater and the atmosphere. The ice edge is characterised by large spatio-temporal variations – while the boundary between consolidated pack ice and open water is sometimes sharply delineated (hundreds of metres or a few kilometres), at other times the transition zone with loose pack may be tens or some hundred kilometres wide.

In springtime, sea ice is fragmented by the wind and waves, and melting is favoured by the north-easterly component of the ACC, which constantly pushes the ice to lower latitudes. In areas of the Southern Ocean such as the Weddell and Ross seas, where semi-enclosed gyres occur, sea ice is obviously

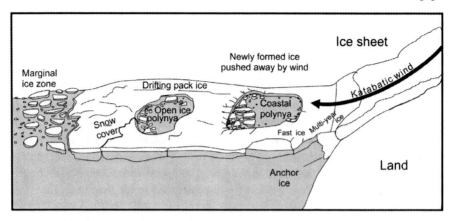

Fig. 24. Schematic representation of coastal and open ice polynyas and sea-ice formations in the Southern Ocean

more likely to become multiyear ice. Ice melting releases water with a low salt content, which floats to the sea surface and promotes the stability of the water column. Like sea ice, great tabular icebergs calved from the fronts of ice shelves or the snouts of tidewater glaciers melt below the waterline as they float away from the continent. However, they seem to have an opposite effect on the water column because they generate vertical convection and promote the upwelling of nutrient-rich deep water (Allison et al. 1985).

3.2.4 Antarctic Sea Ice and the Global Climate System

At its greatest extent, Antarctic sea ice covers a marine area larger than that of the continent. Its seasonal formation and melting greatly affects the climate of marine and coastal areas, essentially through changes in radiative energy and mass exchange processes. As the albedo of sea ice covered by fresh snow can be as high as 90 %, in the Southern Ocean during winter most incoming solar radiation is reflected back (King and Turner 1997). The capping of the ocean prevents the exchange of heat, gases and moisture between the sea and atmosphere, and has important effects on the stability of the atmosphere, cloud formation and precipitation. Much lower albedos and significant fluxes of heat and moisture occur in areas with open water surrounded by sea ice and/or land ice, such as leads and polynyas.

Even when the ice cover in the Southern Ocean is at its maximum, about 20 % of the marine area is ice-free (Gloerson et al. 1992). Persistent coastal polynyas are usually formed by the drainage flow of katabatic winds and by the continuous freezing of water and advection of formed ice by the wind, which removes a large amount of latent heat from surface waters, thereby producing cold, saline, dense waters. These water masses have important implications in deep-water formation and the primary productivity of shelf areas (Massom 1988). Bromwich (1989) estimated that a small (about 50×50 km) and recurring coastal polynya in Terra Nova Bay produces about 10 % of sea ice in the entire Ross Sea region. Polynyas can also form in the open ocean (Fig. 24). In the eastern Weddell Sea, for instance, a polynya of about 2.5×10^5 km^2 developed roughly at the same location for three consecutive winters (1974–1976; Zwally and Gloersen 1977). Carsey (1980) found upwelling of water in the centre of this large ice-free area, while the edge was characterised by thermohaline convection, formation of ice and its transport away from the polynya. It was thought that these characteristics of the polynya were related to the upwelling of warm subsurface water over a topographic high in the seabed (the Maud Rise; Comiso and Gordon 1987). It was later suggested (Enomoto and Ohmura 1990) that polynya formation was probably due to their location beneath the circumpolar trough, in an area with variable position of surface wind fields. The formation of coastal and open ocean leads within sea ice seems essentially due to interactions between ocean surface and atmospheric circulation patterns, especially the strength and persistence of wind (Bromwich et al. 1998). A regional warming and/or change in the extent of sea ice will affect air mass circulation, wind regimes and the formation of polynyas, with consequent changes in ice production, formation of dense water masses, spring disintegration of sea ice, and marine productivity.

Antarctic sea ice is not confined by land margins, and it exhibits large inter-annual and regional variability throughout the year (Gloersen et al. 1992). The annual cycle of expansion and contraction in a roughly concentric zone around Antarctica is controlled by the equilibrium between air temperature, upper ocean structure, pycnocline depth, wind direction and strength, leads, and variation in the ACC and other major currents. Another factor affecting the nature and extent of sea ice is the depth and thermal properties of overlying snow, which controls most of the radiative exchange and sub-ice primary productivity. The exact role of overlying snow is largely unknown, and this parameter is difficult to model (Iacozza and Barber 1999). Besides the lack of reliable data on the thickness of snow, a further complication for modelling studies is the timing of snowfalls in relation to the formation of sea ice (Barber and Nghiem 1999).

Most scientists believe that Antarctic sea ice is very sensitive to climate change, and that it will provide an early indication of global warming. A reduction of the area covered by sea ice will increase the absorption of solar radiation, determining a further increase in temperature and change in atmosphere–sea coupling. However, at present there is no firm evidence indicating any significant long-term trend in the extent of Antarctic sea ice (King and Turner 1997), although it cannot be excluded that an early effect of warming could be a reduction of the total mass of Antarctic sea ice, rather than its extension.

Through the analysis of whaling records, de la Mare (1997) suggested that the Antarctic summer sea-ice edge moved 2.8 degree of latitude southwards between the mid-1950s and early 1970s. Based on an atmosphere–ocean sea ice model, Wu et al. (1999) concluded that the extent of sea ice in the Southern Ocean decreased by 0.4–1.8° latitude over the 20th century. Broad-scale records from passive microwave imagers flown on polar orbiting satellites have shown a decrease in the late winter maximum (between 1973 and 1977) and summer minimum (especially in 1980 and 1981; Gloersen et al. 1992). However, the situation reversed during the 1980s, with an increase in the winter maximum and summer minimum. At present, in spite of the growing satellite database (about 30 years), no statistically significant overall trends have been detected (Gloersen et al. 1992; Johannessen et al. 1994; Jacka and Budd 1998); only some local trends, such an increase in sea ice in the sector from 0 to 40° E and a larger decreasing sector from about 65 to 160° W in the Bellingshausen and Amundsen seas, have been detected.

As for the future, according to a CSIRO coupled model (Gordon and O'Farrell 1997), a reduction of about 25–45 % of the sea-ice volume is predicted for a doubling of CO_2 concentrations and a global warming of 2.1 °C. Through a quite similar coupled atmosphere–sea ice model, Wu et al. (1999) calculated that a global warming of 2.8 °C and higher albedo feedback by surface snow would reduce the extent of Antarctic sea ice by about 2° latitude. However, interactions among sea ice, atmosphere and ocean are complex and largely unknown. The scarce knowledge of a number of parameters such as the thickness and density of sea ice and that of the overlying snow makes it impossible to depict reliable scenarios. In current models it is very difficult to represent even detailed aspects of sea-ice distribution and timing.

In addition to brine released by the freezing of seawater, especially in coastal polynyas, the formation of dense water masses on the Antarctic continental shelf is also due to the cooling of seawater by submerged ice flowing off the continent, especially in the large Ross and Filchner-Ronne ice shelves. When flowing under deep ice shelves, seawater may reach temperatures below –1.95 °C (i.e. below the typical surface freezing point of –1.91 °C; Foldvik and Gammelsrød 1988). This "supercooled" water spills off the shelf, descends the continental slope and mixes with Warm Deep Water to form Antarctic Bottom Water. Significant changes in the extension of Antarctic ice shelves will thus affect water circulation in the Southern Ocean.

As discussed in previous chapters, it is not yet clear whether the warming of the Antarctic Peninsula is in response to global changes or whether it is a natural, exaggerated local response to regional climate variations. Whatever the case, the Antarctic Peninsula shows a remarkable atmospheric warming trend, and its most notable effect is the progressive and continuous retreat of the Wordie, Müller, George VI and Wilkins ice shelves on the west coast, and of the Larsen (A and B) Ice Shelf on the east coast. These events are attracting considerable media coverage and scientific interest, but available data are still inadequate to identify significant changes in ocean circulation and impacts on the marine environment. The Antarctic Peninsula is not well resolved in the present generation of coupled models, and the models cannot help forecast possible changes in the near future. As ice shelves float, their melting has probably no effect on sea levels; moreover, as shelves are usually replaced by the formation of a sea-ice cover, there is probably no dramatic change in the overall albedo. There is, however, evidence of relationships between the warming trend and regional changes in sea-ice extension (Stammerjohn and Smith 1996) and sedimentary processes. Domack and McClennen (1996) found a particularly high sedimentation rate of terrigenous material in two cores collected in the northern region of the western Antarctic Peninsula. This process was consistent with historical records of increasing surface air temperature and was due to the enhanced input of glacial meltwater, which is usually laden with large amounts of terrigenous materials.

Analyses of bones and other organic remains in abandoned penguin rookeries in the near vicinity of Palmer Station show that they were inhabited by Adélie penguins for at least the past 600 years. No remains of chinstrap or gentoo penguins were found (Emslie et al. 1998). The latter two species currently inhabit the Palmer region, where they are believed to have established in the last 20–50 years (Parmelee 1992). Given that Adélie penguins are obligate associates of winter pack ice, whereas their congeners, the chinstrap penguins, occur almost exclusively in ice-free waters (Ainley et al. 1994), it has been hypothesised that changes in seabird distribution are related to regional warming in the Antarctic Peninsula (Fraser et al. 1992; Smith et al. 1999).

In contrast to Antarctic Peninsula ice shelves, those on the continent will probably not be significantly affected by global warming in the next century (Vaughan and Doake 1996); in any case, long-term responses of ice shelves to warming are uncertain, and possible consequences for Southern Ocean circulation are unpredictable. A model study by Warner and Budd (1998) shows that with a global warming of 3 °C, it will take at least several centuries for Antarctic ice shelves to disappear. The warming will initially increase the melting of ice, thereby introducing a layer of freshwater (Jenkins et al. 1997; O'Farrell et al. 1997); however, this layer will probably mitigate melting if it is not flushed away by currents or tides.

3.2.5 Biogeochemical Cycles of C, Fe, S and Other Elements in the Southern Ocean

About 3.5 % (by mass) of seawater consists of solutes, most derived from weathering of terrestrial ecosystems, which are carried by rivers to the sea. As in Antarctica, there are only very small ice-free areas and no rivers; the input of organic matter and soluble ions of continental origin in the Southern Ocean is therefore negligible. The main inputs are from the atmosphere (especially volcanic gases and particulate materials, which become condensation nuclei and fall out in precipitation), and hydrothermal vents which release He, Ca, Mn, K, H and other elements from the oceanic crust. The loss of elements mainly occurs through biological processes (such as organisms forming siliceous frustules or calcareous shells), the formation of sea spray, or through the adsorption of ions to clay particles and new minerals in sediments. Thus, in the Southern Ocean many elements such as Al, Bi, Ce, Co, Pb, Mn and Th, which have prevailing terrestrial sources and are highly reactive in seawater, are probably scavenged by particulate matter and may occur in lower concentrations than in other seas.

The Southern Ocean plays a key role in the oceanic cycle of Si. The downward flux of highly silicified frustules of diatoms (the so-called silicate pump, Dudgale et al. 1995) is a very important transfer mechanism of Si, C and other elements of the upper water column which are incorporated into diatoms (e.g. Ra, Ba and Ge; Holm-Hansen 1985). Diatomaceous ooze accumulated on the Southern Ocean floor represents a large proportion of total recent Si deposition throughout the world ocean. An increasing body of evidence (e.g. Nelson et al. 2001; Sigmon et al. 2002) indicates that silicic acid concentrations are very low in very productive areas of the Southern Ocean, and that Si may constitute a limiting factor for primary production.

Because of its high solubility and chemical reactivity, CO_2 is taken up by the oceans much more effectively than most other gases released by anthropogenic sources. The total amount of carbon in the ocean is about 50 times higher than in the atmosphere; however, because CO_2 solubility is temperature dependent, net fluxes show regional and seasonal patterns. As the cooling of surface waters tends to drive CO_2 uptake, while warming drives outgassing, the Southern Ocean plays a very important role in the uptake of C, producing vertical gradients and its transport from polar to tropical regions in dense bottom waters. The seasonal and regional distribution of CO_2 is also driven by primary production (about 100 petagram C year^{-1}; petagram, Pg; Falkowski et al. 1998). Part of this C is transformed into dissolved inorganic carbon through autotrophic respiration, while the remainder is the net primary production (on the basis of global remote sensing data, estimated to be about 45 Pg C year^{-1}; Falkowski et al. 1998). According to these authors, the global export production (i.e. the sinking of dead organisms, detritus and dissolved organic C) ranges from 10 to 20 Pg C year^{-1}, but only a small fraction of C sinks into the sediments. Heterotrophic respiration at depth converts the remaining organic matter back into dissolved inorganic C, which is transported by deep-water masses to other locations where it upwells and re-equilibrates with atmospheric CO_2. The presence of dissolved inorganic C at depth contributes to lower concentrations of atmospheric CO_2 (about 200 ppm; Maier-Reimer et al. 1996).

The formation of calcium carbonate shells by marine organisms depletes surface carbonate ions, and reduces alkalinity and uptake of CO_2 from the atmosphere. Thus, although overall ocean productivity is largely determined by nitrate and phosphate (and silicon for specific types of phytoplankton) supplied from deep water, it is also affected by $CaCO_3$ formation in surface waters.

In the Southern Ocean the production of $CaCO_3$ by marine organisms is much lower than in temperate and tropical seas, and the water column is rather unstable due to dominant upwelling, ice formation and melting. The distribution pattern of chemical elements is therefore less "structured" than at lower latitudes. The seasonal cycle of primary productivity is mainly limited by insufficient light penetration during the austral spring (Smith and Gordon 1997), and by Si or micronutrient availability in summer (Boyd et al. 1999). The depletion of nitrate and phosphate concentrations in surface waters, a regular feature of many marine areas, rarely occurs in the Southern Ocean. This is partly the result of upwelling and of the high rate of nutrient recycling by microbe populations in the euphotic zone (N, for instance, is recycled six to seven times before settling to deeper waters as particulate N; Holm-Hansen 1985). Like in the equatorial Pacific, the surplus of nutrients in the Southern Ocean (Coale et al. 1996) is often associated with a relatively low phytoplankton biomass (the Antarctic paradox; Tréguer and Jacques 1986). There is evidence (Martin et al. 1990; Boyd et al. 2000) that High-Nutrient, Low-Chlorophyll (HNLC) regions are characterised by low concentrations of certain biolimiting trace elements such as Fe and Mn. The artificial addition of Fe to Southern Ocean seawater stimulates phytoplankton growth and increases the uptake of atmospheric CO_2 (e.g. Timmermans et al. 1998; Boyd et al. 2000). A recent study by Hiscock et al. (2003) in the Pacific sector of the Southern Ocean (from 54 to 72° S) shows that in zones where photosynthetic performance was low, Fe-enrichment response was high; on the contrary, where performance was high (low-silicic acid waters), the Fe-enrichment response was low. Based on these results and on silicic acid limitation (Nelson et al. 2001; Sigmon et al. 2002; Hiscock et al. 2003), there is reason to believe that the region south of the southern boundary of the ACC and north of the continental margin would respond positively to the addition of Fe; the zone between the APF and the southern boundary of the ACC (the Seasonal Ice Zone) would give the same response in spring but not in summer (during the latter season the zone has little silicic acid, while Fe is sufficient) and, finally, the region north of the APF would not respond to the addition of Fe.

As will be discussed in Chapter 5, on Antarctic continental shelves the concentration of most trace elements in seawater are similar to or higher than those reported elsewhere and seem adequate to sustain biological growth. In contrast, offshore surface waters receive scarce quantities of lithophilic elements from shelves, melting icebergs, rivers or aeolian transport. Owing to very low concentrations of Fe and other trace elements, phytoplankton can use <10% of available major nutrients (Martin et al. 1990; Westerlund and Öhman 1991a; Fung et al. 2000). Long-term series data and global surveys indicate that oceanic nitrogen fixation varies spatio-temporally and is sensitive to climatic conditions (Hansell and Feely 2000). It has been hypothesised that, over glacial–interglacial timescales, Fe can indirectly influence the nitrate content of oceans (Falkowski et al. 1998). Martin et al. (1990), for instance, postulated that 50-fold higher aeolian Fe supplies to the Southern Ocean during the last glacial maximum could have stimulated enough new productivity to reduce atmospheric CO_2 from 280 to less than 200 ppm. However, more recent circulation and mixing models predict that the effect of increased nutrient uptake on atmospheric CO_2 concentrations would be much lower, and that a more efficient Southern Ocean biological CO_2 pump alone cannot explain lower CO_2 levels during ice ages (Watson et al. 2000). Stephens and Keeling (2000) suggested that low atmospheric CO_2 levels during glacial periods may result from reduced deep-water ventilation associated with year-round Antarctic sea-ice coverage or wintertime coverage, combined with ice-induced stratification during summer.

Although the ongoing debate on nutrient uptake in the Southern Ocean underlines the importance of studying how it is regulated and whether it has changed in the recent geologic past, it is now believed (IPCC 2001) that the role of marine organisms in driving the C cycle in oceans is probably close to the steady state, and that the oceanic uptake of CO_2 is above all a physically and chemically controlled process. Indeed, in deep ocean water, the concentrations of nutrients and dissolved organic carbon are closely correlated and their ratios match the nutritional requirements of phytoplankton. Primary production may thus have little potential to drive a net air–sea transfer of C.

Hypothetically, oceans can incorporate most of the C released by anthropogenic activity, but the uptake of atmospheric CO_2 mostly occurs in oceanic regions where waters which have spent many years in the ocean upwell (i.e. are re-exposed to the contemporary atmosphere which contains a greater amount of C due to human activity; Doney 1999). Owing to the finite rate of exposure of "older" and deeper waters to the atmosphere, the uptake process will take several hundred years to complete; therefore, even assuming that there is no further increase in anthropogenic emissions of C, atmospheric concentrations of CO_2 will increase. Deep ocean sediments may also contribute to reducing atmospheric C contents through CO_2 reaction with $CaCO_3$, but a response time of about 5,000 years has been estimated (Archer et al. 1997). Studies on the Southern Ocean by Caldeira and Duffy (2000), for instance, have shown high fluxes of anthropogenic CO_2 but very little burial of organic carbon in sediments. Notwithstanding this, the Southern Ocean is very important in determining the uptake of CO_2 from the atmosphere (Fig. 25). Most "in-situ" measurements performed during the last decade have shown undersaturation (e.g. Robertson and Watson 1995; Stoll et al. 1999), although local source areas are also present (Bakker et al. 1997). Through the application of a one-dimensional model to quantify the distribution of CO_2

resents the largest biogenic flux of S to the atmosphere (Andreae 1990; Kettle and Andreae 2000). Once in the atmosphere, DMS forms sulphur dioxide, sulphates and methane sulphonic acid. Sunlight-scattering sulphur aerosols and cloud condensation nuclei can potentially affect the radiative balance and global climate system. Charlson et al. (1987) hypothesised a phytoplankton-climate link, and that the global warming trend could be mitigated to some extent by increased DMS emission stimulated by the warming. However, the existence of this homeostatic feedback and the strength of any cooling effect by increased DMS emission are still uncertain (Andreae and Crutzen 1997). Moreover, any such feedback acting through the S cycle must include other climatic effects. Global warming, for instance, decreases the uptake of CO_2 by oceans, leading to its accumulation in the atmosphere and producing an opposite effect with respect to DMS (Kiene 1999).

In spring and summer, Antarctic coastal waters and marine areas at the retreating pack-ice edge are characterised by phytoplankton blooms, with concentrations higher than 100 nM, i.e. much greater than the average 3 nM in the world's ocean (Gibson et al. 1990; di Tullio and Smith 1995). There is evidence that higher values of the DMSP:chl a ratio are often associated with phytoplankton assemblages dominated by the colonial prymnesiophyte *Phaeocystis antarctica*. Research by di Tullio et al. (1998) in the southern Ross Sea found that ice diatoms are sometimes as important as *P. antarctica* in DMSP production. The Southern Ocean plays a prominent role in the total global flux of DMS into the atmosphere, because diatoms and *P. antarctica* represent a large proportion of Antarctic phytoplankton, and areas with seasonal and intense algal blooms extend for several million square kilometres. Projections based on the assumption that DMS fields and ice cover will not change between the year 2000 and 2100 indicate that the global flux of DMS will increase in the next century from 26.0 to 27.7 tg S $year^{-1}$, and localised increases are foreseen in areas of the Southern Ocean immediately adjacent to the continent (IPCC 2001). However, the depletion of the ozone layer increases the amount of ultraviolet radiation reaching the ocean surface, and there is evidence that this radiation inhibits DMPS production by *P. antarctica* and increases the oxidation of DMS, thus reducing the flux of DMS to the atmosphere (Hefu and Kirst 1997).

3.3 Pelagic Ecosystems

The sea surrounding Antarctica represents one of the largest and most dynamic environments on Earth. The unrestricted flow of currents circulating around the continent and the seasonal pulse of sea-ice freezing and break-up are the main factors controlling the Antarctic marine ecosystem. Views on the productivity of pelagic communities have changed greatly over the years.

High primary production near pack ice and the short food chain (diatoms–krill–whales) were taken as an indication of rich, potentially exploitable resources. It is now recognised that phytoplankton biomass and overall primary production in the oceanic part of the Southern Ocean are much lower than supposed. Besides the previously discussed lack of some micronutrients, other factors limit primary production, these being (1) the vertical instability of the water column, which reduces light availability and prevents algae from making full use of the rich supply of nutrients; (2) low temperatures, which depress the growth rate of algae; and (3) the biomass of herbivores such as krill, copepods and salps, which can further limit the growth of the phytoplankton population.

In general, biogeographical subdivisions in the pelagic zone of the Southern Ocean reflect the different water masses, fronts and seasonal ice cover. In spite of within-zone and longitudinal variations, especially in areas of major oceanographic disturbance such as in the vicinity of the Antarctic Peninsula, three main circumpolar pelagic zones are usually identified (Hempel 1985): the zone of ice-free open waters, the seasonal pack-ice zone, and the permanent sea-ice zone. In general, the ice-free zone of the ACC is rather rich in nutrients but, apart from a few areas, relatively poor in primary production. The zooplankton is rather similar to that of the northern North Atlantic and is dominated by copepods, salps and small euphausiids. Squid, myctophids, small juveniles of benthic fish species and some petrels and whales also occur, but they play a minor role compared to the biomass in the boreal zones of the North Atlantic and North Pacific.

The seasonal pack-ice zone occupies most of the East Wind Drift; it is mainly ice-free in spring and summer, and seeding by ice algae contributes to a high primary production. Although large amounts of phytoplankton may sink to the bottom, the pelagic food web comprises large communities of salps, copepods, fish larvae, chaetognaths and, above all, krill (*Euphausia superba*). Shoals of krill provide the food base for baleen whales, crabeater seals (*Lobodon carcinophagus*), penguins and other seabirds.

In polynyas and the permanent sea-ice zone, primary production is very intense but limited to a short summer period. Krill is often replaced by the smaller *E. crystallorophias*, and the zooplankton biomass is low; therefore, most algae are not consumed and fall to the bottom to sustain a rich fauna of benthic feeders. Many species of benthic invertebrates exploit organic matter accumulating at the water–seabed interface during summer and survive long periods of starvation at negligible metabolic cost. These invertebrates are food sources for crustaceans and many fish species on which Emperor penguins (*Aptenodytes forsteri*) and Weddell seals (*Leptonychotes weddellii*) feed.

In general, mere survival in the Southern Ocean is a feat, and organisms in pelagic communities are often characterised by late and low reproduction, long lifespans, large body size and high total biomass with low net production. Models of food web dynamics and the management of living resource exploitation have to take into account the low efficiency of the food chain and the possible effects of climate change on the extent of sea ice and on ocean circulation.

3.3.1 Primary Productivity

The study of Antarctic marine phytoplankton dates back to the James Clark Ross expedition (1839–1843), when the botanist and surgeon J.D. Hooker reported that diatoms "occurred in such countless myriads, as to stain the Berg and the Pack-ice" (Hooker 1847). He sent some samples to the German botanist C.G. Ehrenberg, who published the first book on Antarctic diatoms in 1844. Since then many papers have been written on the distribution and biogeography of phytoplankton in the Southern Ocean (e.g. Hart 1934; El-Sayed 1968; El-Sayed et al. 1979; Sakshaug and Holm-Hansen 1984; Sullivan et al. 1988; Jacques and Fukuchi 1994; Priddle et al. 1994; Arrigo et al. 1998a; Saarhage 1998; Smith et al. 2000a). The seasonal cycle of phytoplankton biomass and productivity is characterised by large variations, which remain poorly resolved despite the importance of the Southern Ocean in the marine C cycle (Sarmiento et al. 1998). During the last decade, large-scale distribution and spatio-temporal variations in phytoplankton observed by satellites have been used to understand relationships between primary productivity and environmental forcing such as wind and sea ice (e.g. Banse 1996; Moore et al. 1999). However, satellites are often of relatively little use in the Southern Ocean, because they cannot estimate phytoplankton abundance for extended periods in areas of extensive cloud and/or ice cover. On the other hand, the large amount of data collected over the past decades through research vessels shows a rather scarce geographical and temporal coverage, because most studies are confined to short, often ice-free summer periods.

Studies on net phytoplankton or microplankton (i.e. plankton 20–200 μm in size) show that algae mainly consist of colonial or chain-forming diatoms. More than 100 diatom species, often belonging to the genera *Chaetoceros*, *Odontella*, *Thalassiosira*, *Rhizosolenia* and *Nitzschia*, have been reported from Southern Ocean waters. Dinoflagellates (about 60 species; Knox 1994), chiefly the genera *Ptotoperidium* and *Dinophysis*, silicoflagellates, and especially unicellular motile and colonial prymnesiophytes of the genus *Phaeocystis* are other important algal groups (e.g. Holm-Hansen and Huntley 1984; Estep et al. 1990; Caron et al. 2000). If the phytoplankton sampled with nets appears dominated by quite large diatoms, since the 1970s the use of other sampling approaches revealed that ultraplankton (microorganisms<20 μm long) may account for a large proportion (up to 80–90 %) of total estimated phytoplankton carbon in many regions of the Southern Ocean. These microorganism communities include nanoplankton (size 2–20 μm, mostly small diatoms belonging to the genera *Chaetoceros, Fragilariopsis* and *Nitzschia*, and unicellular green flagellates with calcareous or siliceous skeletal elements) and picoplankton (size 0.2–2 µm, including unicellular eukaryotes such as coccoid green flagellates and, above all, cyanobacteria and prochlorophytes). Viruses, bacteria, phagotrophic flagellates and ciliates provide heterotrophic activity which balances the phototrophic activity of ultraphytoplankton. The average density of bacterioplankton in the Southern Ocean (about 10^6 cells ml^{-1}) usually corresponds to that in temperate seas and correlates with phytoplankton biomass. Archaebacteria, once thought to be restricted to hypersaline, extremely hot or anoxic habitats, may be particularly abundant in Antarctic waters (more than 30 % of prokaryotic biomass; de Long et al. 1994).

Although most species of Antarctic marine phytoplankton are circumpolar, there are large spatio-temporal variations in species composition. The algal bloom in coastal Antarctic waters is often dominated by a single opportunistic species (e.g. *Thalassiosira tumida* in the south-western Weddell Sea or the prymnesiophyte *Phaeocystis* in the southern Ross Sea; El-Sayed and Fryxell 1994; Sweeney et al. 2000). Diatoms are undoubtedly the most important phytoplankton class in the Southern Ocean, but there is evidence (di Tullio et al. 2000) that huge blooms of *Phaeocystis antarctica* determine an early and rapid C export to deep water and sediments in the Ross Sea. Sequence data from 18S small subunit ribosomal DNA have shown that *Phaeocystis antarctica* is genetically distinct from *P. pouchetii*, the northern cold water form, and *P. globosa*, the warm water species, from which the former two seem to have evolved (Medlin et al. 1994). In general, the most important taxa of Antarctic phytoplankton are cosmopolitan but, in addition to the monospecific genera (*Charcotia* and *Micropodiscus*), about 80 % of Antarctic dinoflagellate species and 37 % of diatom species seem to be endemic (Fogg 1998).

A low standing crop of phytoplankton (average chlorophyll concentrations of about 0.5 mg m^{-3}, with maximum values at 50–70 m depths) is usually reported in oceanic waters, the Drake Passage, and the Bellingshausen Sea. Average productivity in these areas is low (about 0.1–0.2 g C m^{-2} day^{-1}) and corresponds to that of oligotrophic seas such as the eastern Mediterranean or the North Pacific gyre (El-Sayed and Fryxell 1994). In contrast, high standing crop and primary production are usually reported in coastal waters and in the vicinity of Antarctic and sub-Antarctic islands. In the southern Ross Sea, for instance, concentrations of chlorophyll and particulate organic carbon in the surface layer may reach values exceeding 15 µg l^{-1} and 0.85 µmol l^{-1} respectively. These values are twice the maximum concentrations reported in the Peruvian upwelling system or in the shelf of the Bering Sea (Smith et al. 2000b).

In general, phytoplankton growth follows the seasonal cycle of radiation fairly closely – biomass increases rapidly during the austral spring, and primary productivity reaches a maximum in December/January (average values about 2.6 g C m^{-2} day^{-1}) and declines in February (Smith et al. 2000a). Quite similar trends, with productivity peaks above 3 g C m^{-2} day^{-1}, have been reported in many other coastal regions of Antarctica (El-Sayed and Fryxell 1994).

Ultraplankton in Antarctic waters has a seasonal periodicity similar to that of microplankton, although the peak occurs later in summer and the curve of the standing crop is flatter than the bell-shaped curve of microplankton. In winter, the standing crop of ultraplankton is about five times that of microplankton and may play a very important role for planktonic and benthic consumers (Clarke and Leakey 1996).

3.3.2 Effects of UV-B on Phytoplankton and Primary Production

Measurements in polar firn air (Butler et al. 1999) have shown that gases with the largest potential to deplete ozone ($CFCl_3$, CF_2Cl_2, $CF_2ClCFCl_2$) are definitely released by anthropogenic sources. Current surface measurements of these compounds show that concentration growth rates have decreased since the Montreal Protocol (WMO 1999; Prinn et al. 2000). However, initial recovery of stratospheric ozone will not be detected much before 2010, and depletion due to halogens will probably recover during the next 50–100 years (Hofmann and Pyle 1999). These forecasts are uncertain due to the increasing consumption of O_3-depleting substances by developing countries. Moreover, with respect to chlorine compounds, increasing concentrations of CO_2, CH_4 and N_2O and the impact of climate change on stratospheric temperature and circulation may cause even larger changes in stratospheric O_3 (IPCC 2001).

In this scenario, considering that C fixation plays a central role in ecological and climatic processes, the biological effects of springtime stratospheric O_3 depletion (as much as 50 % during the last decade) and the increase in UV-B radiation (290–320 nm) reaching the surface of the Southern Ocean are of particular concern. During the austral spring and summer, high nutrient and sunlight availability at the receding edge of the pack ice promotes phytoplankton blooms which account for a large proportion of total primary production in the Southern Ocean. Sea-ice meltwaters form an upper layer of relatively fresh water over a saltier deeper one, and this stratification concentrates algal blooms in near-surface waters. This highly productive upper layer is therefore most at risk from enhanced UV-B radiation, especially during the period of maximum O_3 depletion.

The Antarctic O_3 hole was first reported in 1985, but observations at Halley Station have shown that depletion began in the 1970s (Farman et al. 1985). Research on the biological and ecological consequences of O_3 depletion for Antarctic ecosystems began some years later (e.g. Bidigare 1989; Voytek 1990; Karentz 1991) and were therefore hindered by the lack of baseline data; the response of Antarctic organisms was tempered by about 20 years of adaptation and species selection under enhanced seasonal UV-B radiation (Karentz 1994).

As the biological effects of UV radiation strongly depend on wavelength, and even differences of a few nanometres are important (Cullen et al. 1992; Helbling et al. 1994), small changes in O_3 concentrations may disproportionate harmfulness of incident UV-B radiation. As a rule, DNA is the primary lethal target of UV-B but RNA, proteins and other molecules are also adversely affected by exposure. UV radiation may catalyse photochemical reactions in seawater and within algal cells, causing oxidative stress and impairing nutrient uptake, membrane transport and photosynthesis, thereby inhibiting growth and reproduction, and ultimately leading to death (Vincent and Roy 1993). The sensitivity of algae show large inter- and intraspecific variations, depending on avoidance strategies, number and efficiency of repair systems, physiological state and genotypic differences (Karentz 1994). During the last decade, several studies on the Southern Ocean have concentrated on O_3-dependent shifts in in-water spectral irradiance and on alterations to spectrally dependent phytoplankton processes (photo-inhibition, -reactivation, -protection, and -synthesis; Smith et al. 1992; Helbling et al. 1994; Arrigo 1994; Boucher and Prézelin 1996; Neale et al. 1998; Bracher and Wiencke 2000). The results of these studies have often been used to estimate the loss of primary production. However, spatio-temporal variations in ozone depletion, cloudiness, sea-ice cover and vertical mixing make it difficult to reliably estimate the temporal pattern of phytoplankton exposure. The reported decreases in annual primary production therefore span quite a large range (from 0.1 to 12 %).

Recent research suggests several factors and processes which can contribute to a relatively small loss of primary production as a result of O_3 depletion. Helbling et al. (1994), for instance, found that flagellates were much more sensitive to UV than diatoms, and that the latter tended to dominate the phytoplankton crop in areas with a shallow upper mixed layer, while flagellates dominated crops at stations with deep mixed layers (more than 40 m). Mycosporine-like amino acids (MAAs), which occur in several algae taxonomic groups, play a photo-protective role against UV-B exposure. Bracher and Wiencke (2000) studied the effects of spectral exposure at normal and depleted stratospheric O_3 concentrations on photosynthesis and MAA contents in natural phytoplankton communities; they found that only samples outside the phytoplankton bloom showed a significant decrease in the photosynthetic production rate due to enhanced UV-B radiation. Models describing UV-influenced photosynthesis in the presence of vertical mixing (Neale et al. 1998) show that O_3 depletion can inhibit primary productivity in Southern Ocean open waters, but the natural variability in exposure of phytoplankton to UV radiation, associated with vertical mixing and cloud cover, can enhance or diminish the effect on water-column photosynthesis. Interactions between vertical mixing and UV radiation have direct effects on photosynthesis and can also influence the acclimatisation and selection of phytoplankton. Thus, despite the small loss in primary production, there is no doubt that O_3 depletion and UV-B radiation in the Southern Ocean constitute significant environmental stress. Possible cumulative impacts on interspecies variations and the pelagic food web cannot be excluded, although these impacts are difficult to foresee in communities which were tempered by about 20 years of adaptation and species selection under increased seasonal UV-B radiation.

3.3.3 The Ecological Role of Sea Ice

The presence of algae in Antarctic sea ice was discovered by Hooker (1847), and several reviews (e.g. Horner 1985; Vincent 1988; Garrison 1991; Palmisano and Garrison 1993; Ackley and Sullivan 1994; Lizotte and Arrigo 1998; Fogg 1998) address the biology and ecophysiology of rich microbial assemblages in Antarctic sea ice. Owing to the extension of the ice in the Southern Ocean in October (up to 19×10^6 km^2), organism assemblages growing on the surface, interior and bottom of the ice constitute one of the largest ecosystems on Earth. Sea-ice biota interacts with and contributes to the high productivity and dynamics of the food web in the ice-edge zone (Smith and Garrison 1990).

Sea-ice surfaces and microhabitats such as brine inclusions or interstices, which develop when sea ice forms and ages, may be colonised by airborne propagules as well as planktonic organisms. The bulk of biota in the ice (microalgae, bacteria, protozoans and small metazoans) stems from the sea, and the composition of interior biotic assemblages of recently formed ice usually reflects that of the plankton beneath. Marine organisms are collected and concentrated by ice crystals floating to the sea surface or by water drawn through aggregations of frazil ice (Fogg 1998). As a rule, this composition changes with time, and differences exist between pack ice and land-fast ice. In the latter, the lower margin of the columnar ice (skeletal layer) is frequently inhabited by abundant ice microalgae; these communities are absent or much less developed in pack ice (Legendre et al. 1992). In land-fast ice the growth of sea-ice microalgae in the skeletal layer is often limited by salinity and light attenuation due to the thickness of the overlaying snow cover. On the contrary, pack-ice algae usually grow at or near the sea-ice surface and have about an order of magnitude higher photosynthetic capacity than land-fast ice algae (Arrigo et al. 1998b). The availability of nutrients is the main limiting factor for the development of surface or near-surface pack-ice assemblages. Although snow cover reduces the amount of light available for algal growth, it also directly and indirectly provides nutrients through surface flooding caused by snow loading and submersion of the pack ice. Seawater flooding on depressed ice floes can form a layer of infiltration ice along the snow-ice

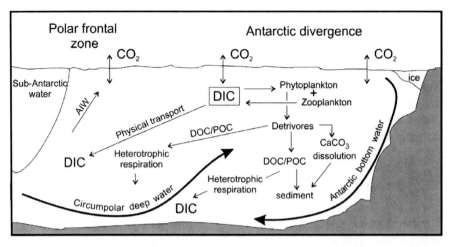

Fig. 25. Schematic illustration of carbon cycling in the Southern Ocean (*DIC* dissolved inorganic carbon, *POC* particulate organic carbon, *AIW* Antarctic Intermediate Water)

sources and sinks in the Southern Ocean and to simulate its variability over the period 1986–1994, Louanchi and Hoppema (2000) calculated a mean uptake of 0.53 Pg C year^{-1}, with an increase between 1986 and 1994. The inter-annual variability (0.15 Pg C year^{-1}) was related to the Antarctic circumpolar wave, which affects the sea surface temperature, wind speed and sea-ice extent. These estimates agree with the results of other studies based on either field data or models (e.g. Poisson et al. 1994; Bakker et al. 1997), and indicate that in the early 1990s the Southern Ocean may have helped decrease the atmospheric CO_2 growth rate by about 3–5 Pg C year^{-1}. Global warming experiments using coupled ocean–atmosphere models show that under climate-change forcing, the CO_2 uptake rate is likely to decrease more in the Southern Ocean than in any other ocean (Sarmiento and Le Quere 1996). The increased stratification of the upper water column will cause a decrease in vertical mixing along isopycnals, vertical transport of C, convective overturning, and upward mixing of warm waters from below (Sarmiento et al. 1998). The temperature of Southern Ocean surface waters will thus increase less than elsewhere, and may even decrease in some regions (Manabe and Stouffer 1994). The increased stratification will also cause a gradual collapse of thermohaline circulation in the entire deep ocean, which has a major role in the C cycle on century timescales.

The surface layer of the ocean influences the climate system not only through the exchange of greenhouse gases but also through the release of dimethylsulphide (DMS), which is thought to exert a cooling influence. DMS is a breakdown product of dimethylsulphoniopropionate (DMSP), a constitutive osmoprotectant of algal cells which is released by exudation and through autolysis, grazing or bacterial and viral attacks. Oceanic emission of DMS rep-

interface. Physical processes, such as ice rafting, pressure-ridge formation, the break-up of ice floes by waves and sea swell, and the formation of brine cells, contribute to the heterogeneous structure of sea ice. The presence of algae during seasonal warming may increase the absorption of solar radiation, thus contributing to localised melting within floes. Consequently, the volume of brine increases and adjacent cells tend to fuse and migrate downwards, forming brine channels in late-season ice or multiyear ice. Figure 26 illustrates the main habitats in Antarctic sea ice and the usually identified, three broad categories of surface, interior and bottom assemblages. In general, the most common algae in sea ice are small diatoms (over 100 species), nano- and microphytoflagellates, and the prymnesiophyte *Phaeocystis*. The majority of studies have concentrated on these organisms, which account for most of the biomass. However, bacteria (free-living, or attached to algae or detritus), fungi, autotrophic and heterotrophic protozoans (mostly ciliates, flagellates, amoebae and foraminifers), and a few metazoans co-inhabit the three kinds of assemblages. A detailed list of organisms from Antarctic sea ice is reported by

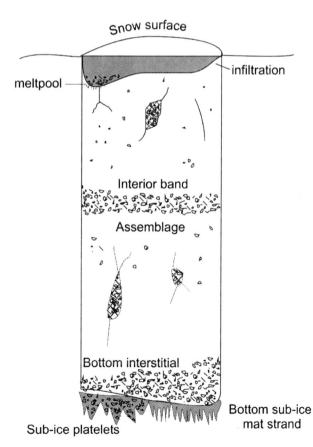

Fig. 26. Schematic representation of habitats and algal assemblages in Antarctic sea ice

Palmisano and Garrison (1993). Algae photosynthesis drives the ecosystems and provides organic matter for heterotrophic organisms. Bacteria usually reach higher densities in sea ice than in seawater, and are consumed by heterotrophs which regenerate mineral nutrients for autotrophs. These close and self-sustaining systems can reach primary production rates greater than 1.0 g C m^{-2} day^{-1} in pack-ice surface-layer assemblages (Garrison and Buck 1991), and as high as 2.1 g C m^{-2} day^{-1} in land-fast ice bottom-layer assemblages (Grossi et al. 1987). A cryopelagic community of metazoans such as adult and larval zooplankton (amphipods, copepods, larval euphausiids) and fish grazes off the algae on the undersurface of the sea ice. In shallow regions some benthic species may seasonally colonise and graze on ice-associated biota; in any case, benthic organisms on the continental shelf will receive considerable amounts of living and/or detrital material released from the melting sea ice.

The seasonal melting of pack ice is characterised by the release of ice biota into the waters of the progressively receding ice-edge zone. There is evidence (e.g. Sullivan et al. 1990; Garrison and Buck 1991) that algae, bacteria and protozoans in the sea-ice assemblages could be a source of "seed" populations to developing ice-edge phytoplankton blooms.

3.3.4 Ice-Edge Processes and Communities

At the edge of the receding pack ice, the sea ice and associated biota disappear and are replaced by open water and pelagic organisms. This zone is a dynamic oceanographic front marked by distinct physico-chemical properties, biological processes and communities, and by brief periods of intense productivity. A comprehensive treatment of the ice-edge environment is beyond the scope of this chapter. The Antarctic Marine Ecosystem Research at the Ice-Edge Zone (AMERIEZ) programme was a multidisciplinary investigation of the structure of pelagic ecosystems. It gave substantial results on microbial rates and processes, and papers such as those by Hanson and Lowery (1983), Sullivan and Ainley (1987), Nelson et al. (1987), Cota et al. (1990) and Palmisano and Garrison (1993) should be consulted to supplement the following brief review of the topic.

Research carried out in Arctic and Antarctic seas essentially shows that the melting of sea ice during spring and summer produces a layer of low-salinity water, with increased phytoplankton biomass and productivity, and elevated abundance of marine birds and mammals. Because of the short fetch of wind over the water floes, lenses of nearly fresh water (salinity as low as 0.03‰) form near the ice edge, determining a strong vertical gradient of salinity and temperature, vertical circulation and horizontal currents parallel to the front. The stabilisation of the water column enhances the availability of light for algal photosynthesis by decreasing the extent of cell mixing below the euphotic zone. The area affected by these short-lived conditions is not a sharply delimited belt between sea ice and the open sea. The pack ice extends at least 100 km outwards from the edge of the continuous sea ice, with wide areas of open water and concentrated ice floes. The stabilised area may thus extend up to 250 km into the open sea (Smith and Nelson 1985), as a mosaic of patches of stratified and mixed waters. As the concentrations of nutrients are nearly always high in Southern Ocean waters, opportunistic algae such as *Nitzschia curta* and *Phaeocystis* dominate the plankton community until mixing brings in competitors. The algal bloom is coupled with an increase in bacterial density and, together with protozoans, they can contribute 20–30 % of the total microbial biomass (Fogg 1998).

Average values of productivity measured in the receding ice edge range between 556 and 962 mg C m^{-2} day^{-1} (Jennings et al. 1984; Smith and Nelson 1986), and the estimated total annual production (approximately 380×10^{12} g C) constitutes a large proportion of total photosynthetic production in the Southern Ocean. Since the enormous availability of biomass at the ice edge is restricted in space and time, it mostly settles on the ocean floor. Many pelagic grazers and benthic invertebrates have developed life strategies which allow them to exploit this momentary but immense availability of food. Some species of pelagic tunicates, salps, euphausiids, copepods, gammarids and hyperiids are among the most represented organisms in ice-edge zooplankton. Comparatively large forms such as krill (*Euphausia superba*) can benefit from concentrations of protozoans, such as tintinnids and choanoflagellates, and from refuges provided by decaying ice floes. The abundance of plankton at the pack-ice edge attracts some species of myctophid fish, seabirds, seals and baleen whales, whose migrations and reproductive cycles depend on the availability of prey. Seabirds include communities associated with the pack ice, such as Emperor (*Aptenodytes forsteri*) and Adélie (*Pygoscelis adeliae*) penguins, and snow (*Pagodroma nivea*) and Antarctic (*Thalassoica antarctica*) petrels (Fraser and Ainley 1986). North of the ice edge, in waters generally free of ice but still under its influence, the most widespread seabird species include macaroni penguins (*Eudyptes chrysolophus*), the southern fulmar (*Fulmarus glacialoides*), cape pigeon (*Daption capensis*), mottled petrel (*Pterodroma inexpectata*), and the light-mantled sooty albatross (*Phoebetria palpebrata*) and black-browed albatross (*Diomedea melanophris*). The crabeater seal (*Lobodon carcinophagus*), leopard seal (*Hydrurga leptonyx*), Ross seal (*Ommatophoca rossii*) and minke whale (*Balaenoptera acutorostrata*) are among the mammals which concentrate in the marginal ice-edge zone.

3.3.5 Krill and Pelagic Food Webs

Although the term krill derives from the Norwegian "kril", used by whalers to denote small fish, it is now applied to crustaceans eaten by baleen whales in the Southern Ocean, and it is often reserved for the largely dominant species *E. superba*. This macroplanktonic species plays a central role in the seasonal pack-ice region of the Southern Ocean due to its abundance (probably one of the most abundant and successful animal species on Earth), large size (up to 6 cm long) and fresh weight of about 1 g. It largely affects the dynamics of pelagic food webs and biogeochemical fluxes of macro- and micronutrients, and therefore attracted much attention from biologists in the British *Discovery* expeditions (e.g. Fraser 1936; Marr 1962), the BIOMASS Programme (e.g. Miller and Hampton 1989; El-Sayed 1994), and in recent years (e.g. Knox 1994; Trathan et al. 1995; Hagen et al. 1996; Daly 1998; Perissinotto et al. 2000; Reid 2001). In spite of the large number of papers on krill biology, ecophysiology, behaviour and adaptations, its distribution, life cycle and biomass are still not fully understood. They generally grow to 5 cm in two years and have a lifespan of 5–8 years. While adults can tolerate starvation for over 200 days, the larvae can survive without food for only about 2 months. Thus, larvae usually overwinter beneath the ice, normally feeding off phytoplankton and small zooplankton filtered through a basket formed by fringed thoracic limbs adapted to capture algae scraped off the undersurface of the sea ice. Although sea ice allows the overwintering of adults and provides a nursery for larvae, it is still unclear whether the under-ice krill population represents a significant proportion of the total Southern Ocean population or whether it is only part of the more general ocean current-driven life cycle of krill (Hansom and Gordon 1998). It is still unknown to what extent krill populations take advantage of overwintering under the pack ice. For instance, there is evidence that, in summer and winter, krill in coastal waters near the South Shetland Islands (Ligowski 2000) mainly feed on benthic diatoms.

   In general, *E. superba* shows a very patchy distribution. It mainly occurs in zones between the summer and winter limits of the pack ice, at a depth ranging from 20 to 150 m, where it takes advantage of phytoplankton blooms. It rises to the surface at night and sinks during the day, concentrating in swarms which may be very large (super-swarms may reach a density>1,000 g m^{-3}, thickness of 100–200 m and length of up to several kilometres; Knox 1994). With respect to other zooplanktonic organisms, krill can swim at speeds greater than 10 cm s^{-1}; the movement of individuals in swarms is coordinated and synchronised, and there is evidence that swarms can move against currents. However, *E. superba* adults are carried towards the APF in Antarctic Surface Water, where eggs released at depths of about 100 m begin to sink, hatch at around 1,000 m, and are carried back southwards together with young larvae in Circumpolar Deep Water. As larvae develop, they slowly ascend to the surface and begin a new cycle, allowing the distribution of *E.*

superba at different latitudes between the APF and the continent. Like assemblages of phytoplankton and of other Antarctic marine organisms, krill has a circumpolar distribution and a latitudinal zonation (Fig. 27). There is evidence that two main circumpolar currents, and the transport and collection of larvae and young adults in gyres and eddies provide a framework for high krill biomass in the Scotia Sea, around South Georgia Island, in the Bellingshausen Sea, and in several areas off the coast of East Antarctica, such as north of the Ross Sea, off Wilkes Land, Enderby Land and Queen Maud Land.

The short food chain based on krill represents the most outstanding product of a long evolutionary process in isolation, in a cold and nutrient-rich environment with a seasonal ice cover and light availability (Hempel 1985). While krill has a rather wide food spectrum, most of its consumers are specialised to feed on it alone. Like other highly productive pelagic areas which host short food chains with a large herbivore biomass (e.g. anchovies and sardines), in the marginal ice zone *E. superba* is the main species channelling organic matter from phytoplankton to cephalopods, fish, birds, seals and baleen whales. The giant squid (*Mesonychoteuthis hamiltoni*, over 5 m in length and 150 kg in weight) and other species of cephalopods are among the most important krill consumers, and they are in turn important in the diet of

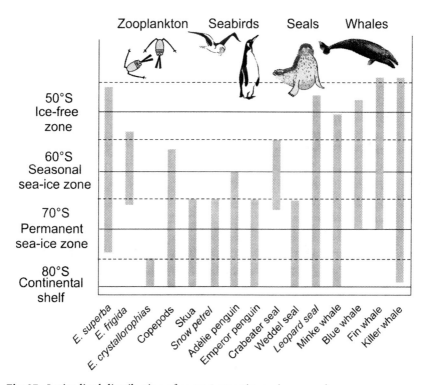

Fig. 27. Latitudinal distribution of some Antarctic marine organisms

sperm whales. Thousands of squid beaks, many belonging to species rarely or never caught in nets, are sometimes found in the stomachs of these whales (Nemoto et al. 1988). Wandering albatrosses, Emperor penguins, seals and killer whales are among the other consumers of Antarctic squid.

There is still uncertainty about the total standing stocks of krill, and the most reliable estimates are probably those based on krill consumption by major predators. As a general indication, krill consumption by fish, seabirds, squid and seals has been estimated to be 15, 30, 40 and 130×10^6 tonnes year^{-1} respectively (Knox 1994; tonnes, t). Current consumption by reduced whale stocks has been estimated to be about 40×10^6 t year^{-1} (Armstrong and Siegfried 1991), i.e. about five times lower than the amount consumed before exploitation. However, it is wrong to suppose that the reduction in whale stocks has determined a "surplus" of krill, because there are no data on krill biomass during the period of whale exploitation and because other krill-consuming populations have probably increased as a result of the reduced competition with whales. As suggested by Berkman (1992), pelagic ecosystems in the Southern Ocean appear to have already adjusted to a constant level of krill consumption ($300-400 \times 10^6$ t year^{-1}, which roughly corresponds to the total annual production; Knox 1994).

Owing to its wide distribution and abundance, since the 1960s Antarctic krill has attracted the interest of fishing fleets searching for new species as traditional ones are fished out. According to CCAMLR estimates, krill harvesting peaked at more than 500,000 t in 1981–1982, and during the last decade was about 100,000 t year^{-1} (Fig. 28). Overall, more than 5.74×10^6 t has been harvested to date. Krill fishing is causing much concern, and a better understanding of *E. superba* biology, distribution and swarm composition is essential for the proper management of fisheries. Although the present exploitation rate in relation to the total stock is probably not excessive, it cannot be excluded that localised stocks at the northern edge of the pack ice are already overexploited (Everson and Goss 1991). In addition to krill, squid are also fished commercially in some northern areas of the Southern Ocean, such as offshore Patagonia, the Falkland Islands and New Zealand. As in the case of krill exploitation, squid exploitation would pose problems for a number of predator species which, especially in the APF, depend almost exclusively on cephalopods for their breeding season diet.

Besides *E. superba*, the Southern Ocean contains other important herbivorous (e.g. protozoans, copepods, euphausiids, salps) and carnivorous (e.g. copepods, chaetognaths, pelagic polychaetes) zooplankters. These organisms are eaten by *Parathemisto gaudichaudii* and, in contrast to *E. superba*, they do not generally constitute a major food source for large predators.

In spite of its considerable age and size (about 10 % of the world oceans), the Southern Ocean contains only 1 % of fish species (about 21,700 species; Nelson 1984) inhabiting Earth's waters. The perciform suborder Notothenioidei, the dominant group of Antarctic fish fauna, contains 97 % endemic

Fig. 28. Catch of krill during the last decade. (Data from CCAMLR 2003)

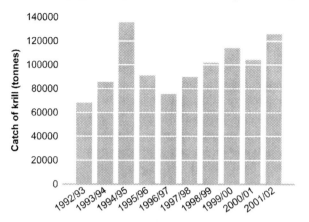

species, suggesting that it evolved in isolation (Eastman 1993). Few species of Antarctic fish live in the upper 200 m of the water column. The Antarctic cod (*Notothenia rossii*) is a secondarily pelagic species which feeds on krill and has been overfished in recent years. The silver fish (*Pleuragramma antarcticum*) is one of the few neutrally buoyant species among notothenioids; adapted to temporary or constant life in pelagic water, it is one of the major Antarctic fish species in number and biomass, and is the prey of most of the larger Antarctic carnivores. Small myctophids (lantern fish) are among the most abundant mesopelagic fish in the Southern Ocean. They show diel migratory patterns and are among the most important krill predators. Their biomass is thought to exceed that of *E. superba* (Eastman 1993).

Of the 43 species of birds (mainly procellariforms and penguins) which breed south of the APF, only 12 breed on the Antarctic continent and, after a short breeding season, most of them disperse northwards (Knox 1994). As a result, the distribution of birds in the Southern Ocean shows a strong bias towards the sub-Antarctic islands. In general, birds feed on zooplankton (mainly krill) and larval fish; some species are tertiary feeders, also preying on squid and fish, and others, such as skuas and giant petrels, are predators on smaller seabirds and penguin chicks or on carcasses of larger birds and marine mammals. Like seabirds, several species of marine mammals in the Southern Ocean feed on krill and, as populations of previously hunted species of seals and cetaceans recover to pre-exploitation levels, the competition for krill increases. The number of crabeater seals (curiously named, given that their major prey is krill, and not crab) increased in the 1960s and 1970s, so that they became the most numerous pinniped species in the world; their number is probably decreasing at present. Baleen whales move into Antarctic waters in early summer and, after 3–4 months of intensive feeding on krill, they return to warm water to breed during the Antarctic winter. The minke whale is the most common cetacean in the Southern Ocean because it was

exploited only at the end of commercial whaling; they were therefore not dec-
imated like greater whales (e.g. humpback and blue whales were reduced to 3
and 5% respectively of their pre-exploitation numbers; Croxall 1992).
Although whaling greatly perturbed the Southern Ocean ecosystem, it
allowed food to distribute through the marine ecosystem to the benefit of
competitors. Moreover, changes in the life cycle of animals in the krill system
were probably determined by the greater availability of food. The growth rate
and pregnancy rate of some species of baleen whales have increased, and the
mean age at sexual maturity of crabeater seals and whales is decreasing (Knox
1994). This reveals the rather complex response of marine mammals to
removal of competitors, and indicates that commercial fishing of krill and
cephalopods may have unpredictable effects on krill-feeding organisms and
top predators such as leopard seals and killer whales.

3.4 Benthic and Epibenthic Organisms

In Antarctica, ice is the main factor affecting the distribution of benthic
organisms on the continental shelf. Glaciers entering the sea carry coarse
material interspersed with boulders and gravel, which covers large areas of
the shelf. The littoral and sublittoral zone is scoured to a depth of 10–30 m
through the abrasion of shorelines by fast ice, pack ice and brash ice or by
floes driven ashore and piled upon each other during storms (push ice), and
through plucking by anchor ice. Icebergs calved from ice shelves, glaciers, and
grounded ice walls and floating on the continental shelf plough furrows in the
bottom down to depths of several hundred metres, with ruinous effects on
benthic communities (Gutt 2001). However, it has been found that in spite of
ice disturbance and scarce light penetration, Antarctica has very rich benthic
communities. Over 3,000 species of benthic invertebrates have been recorded
in the Southern Ocean (Arntz et al. 1994), and their characteristics are summarised in several reviews (e.g. Dell 1972; Arnaud 1977; White 1984; Dayton
1990; Knox 1994; Gambi and Bussotti 1999; Starmans et al. 1999). In general,
the most commonly reported features are abundance, high levels of
endemism, gigantism, longevity, slow growth rates, delayed maturation,
absence of some invertebrate groups and pelagic larval stages.

   The presence of ice, the unusual depth of the shelf (up to 500 m), the extensive area of deep water around the continent, and the lack of connection with
temperate shelves made the Antarctic shelf an insular evolutionary site
roughly equivalent to Lake Baikal or the Galapagos (Eastman and Clarke
1998). Moreover, the Southern Ocean oceanographic fronts constitute a major
zoogeographic boundary for epi- and mesopelagic organisms (in the upper
1,000 m). In 1913 C.T. Regan identified the zone between the 6 and 12 °C surface isotherms as the boundary between Antarctic and sub-Antarctic regions;

these zones have since been found to approximate the position of the APF and Subtropical Convergence respectively. As stated by Ekman (1953), "no other large faunal region in the world can match the Antarctic in the sharpness of its boundaries". In general, zoogeographic faunal units for benthic organisms in the Antarctic Region distinguish a Continental (or East) Antarctic Region (or Province; most continental coasts except for the Antarctic Peninsula), a West Antarctic Region (or Province; the Antarctic Peninsula and adjacent islands), and the South Georgia District or Province. Covering most of the Antarctic Region, the Food and Agriculture Organization of the United Nations established fishing areas 48, 58 and 88, which coincide with the limits defined by the Convention for the Conservation of Antarctic Marine Living Resources (CCAMLR). In the Indian Ocean sector (between 30 and 80° E), the northern limit of the Convention Area extends to 45° S, to ensure the inclusion of productive shelves around the sub-Antarctic islands (Marion, Prince Edward, Crozet and Kerguelen). Although these islands lie north of the APF, their fish fauna has an Antarctic origin and character (Eastman 1993).

The Liparididae and Zoarcidae, two North Pacific fish families, are well represented in the Southern Ocean, but the suborder Notothenioidei is the dominant group in coastal Antarctic waters. After the establishment of the ACC some 25×10^6 years ago, the notothenioids, originally a benthic nearshore group, probably survived or were the most successful in invading across the APF. With its six families (Harpagiferidae, Bovichtidae, Notothenioidae, Artedidraconidae, Bathydraconidae and Channichthydae), this group spread into water-column or ice-associated habitats of the Antarctic shelf and the upper continental slope to fill ecological roles usually filled by a variety of fish in other seas. Most of these bottom or coastal dwellers (they lack swim bladders) are endemic (97 % endemism for species and 85 % for genera; Gon and Heemstra 1990) and are excluded from the peripheral parts of the Antarctic Region. Through the reduction of skeletal material and the accumulation of fat, some notothenioid species are neutrally buoyant and have adapted to become pelagic species. *Pleuragramma antarcticum* (Antarctic herring) ranges in depth from 0 to 900 m and constitutes a major species (in number and biomass) in the water column of most Antarctic shelf areas. It feeds on krill, copepods and chaetognaths and, owing to its abundance and wide distribution, *P. antarcticum* plays a key ecological role as food for fish, penguins and other Antarctic marine vertebrates, especially in areas where krill is scarce. Other abundant pelagic fish in the Southern Ocean are the small lantern fish (Myctophidae). *Dissostichus mawsoni* and *D. eleginoides* (a sister species found largely north of the APF), the largest notothenioids (up to 160 cm and about 70 kg), are typical mesopelagic predators. They feed on pelagic fish, mysid shrimp and squid, and the main difference between the two species is the lack of antifreeze in *D. eleginoides*, which does not inhabit subzero waters. Through morphological and physiological adaptations some notothenioid fish, such as *Pagothenia borchgrevinki* and *P. brachysoma*, have specialised for life on the undersurface of sea ice (cryopelagic species; Andriashev 1970). As cryopelagic fish come into contact with minute ice crystals, their blood contains greater quantities of glycoprotein antifreeze compounds than that of benthic species (DeVries 1971). Although most Antarctic fish are benthic, some species such as *Cryothenia peninsulae* are semipelagic fish (i.e. they exhibit features of both pelagic and benthic notothenioids) and others, such as *Trematomus loennbergii* and *T. eulopidotus*, are typical epibenthic species. Only a brief description of benthic habitats and organisms will be given here. Readers interested in the biology and ecology of Antarctic fish can refer to specific books (e.g. Gon and Heemstra 1990; Di Prisco et al. 1991; Kock 1992; Eastman 1993; Nelson 1984; Di Prisco et al. 1998).

In the region of ice scour and disruption by anchor ice, sessile epibenthos is generally restricted to transitory groups of organisms, except in more sheltered habitats such as rock crevices. However, about 700 species of benthic macroalgae have been recorded in the upper infralittoral zone of Antarctica and in the sub-Antarctic islands (Fischer and Hureau 1985). Providing that there is some protection from ice abrasion, Rhodophyta *Iridaea cordata* is one of the most widespread species down to depths of about 15 m. In the Antarctic Peninsula and islands of the Scotia Arc, Phaeophyta *Ascoseira mirabilis*, with a thallus up to 4 m in length, is another common species inhabiting depths of up to 13 m. The endemic and common seaweed *Himantothallus grandifolius*, the largest Antarctic kelp with a thallus up to 10 m long and 1 m wide, together with other species such as *Leptophyllum coulmanicum*, may produce dense stands throughout the Antarctic inner continental shelf to depths well below 30 m. However, in terms of biomass, benthic microalgae may be more important than macroalgae. Shade-adapted benthic diatoms grow on the topmost few mm of sediments, rock surfaces, sponge spicules or attached to macroalgae and epibenthos. In the east McMurdo Sound, their standing stock has been estimated to range from 20 to 900 mg chl a m^{-2}, depending on depth and season (Dayton et al. 1986). Vagile organisms such as echinoderms, molluscs, polychaetes, peracarid crustaceans, and fish (e.g. *Trematomus bernacchii* and *T. hansoni*) are ubiquitous in shallow waters.

There is an abundant benthic fauna below the depth of sea-ice scour. Adaptive radiation within families provides reasonable diversity at the species level, although benthic fauna at higher taxonomic levels is less diverse in Antarctica than elsewhere in the world. Some groups such as decapods or gastropods, for instance, are poorly represented. Many invertebrates are particle-feeding or scavengers, inedible or of little value as food (e.g. sponges, starfish, sea spiders or brittle stars), and several species are toxic (cf. chemical defence against predation; McClintock 1989). The more abundant sessile particle-feeding invertebrates include sponges, hydroids, tunicates, bryozoans, sedentary polychaetes, actinarians, scleractinian corals and holothurians. Many types of sponges can cover more than 50% of the seafloor, and their spicules may form dense mats up to 1–2 m thick. Spicules and soft bottoms provide a suitable habitat for many infaunal organisms such as peracarid crustaceans, burrowing polychaetes, oligochaetes and bivalve molluscs. Motile organisms include echinoderms (particularly the echinoid *Sterechinus neumayeri* and the starfish *Odontaster validus*), bivalves (the Antarctic scallop *Adamussium colbecki* and the soft shell clam *Laternula elliptica*), gastropods (the Antarctic whelk *Neobuccinum eatoni*), nudibranchs (such as *Austrodoris mcmurdensis*, which feeds on sponges); the large isopod *Glyptonotus antarcticus*, the nemertine *Paraborlasia corrugatus*, and pycnogonids.

Antarctic benthic fish (mostly of the genus *Trematomus*) have heavier skeletons and are less buoyant than cryopelagic and epibenthic species. They are dorsoventrally depressed, with pelvic and anal fins supporting the body when resting on the substrate. In general, different species occupy different subhabitats: *Trematomus nicolai* lives in shallow waters (30–50 m deep), sometimes near masses of anchor ice; *T. pennellii* is usually found in waters less than 200 m deep, while *T. bernacchii* and *T. hansoni* can be found from shallow waters up to depths of 550 m. As a rule, benthic trematomids are feeding generalists, and their diet varies with location, according to the availability of prey organisms (i.e. polychaetes, amphipods, molluscs, crustaceans, small fish or fish eggs; Vacchi et al. 1994). Some species such as *Notothenia rossii* and *N. coriiceps* change habitat and diet at different life history stages. Larvae are pelagic, nearshore fingerlings and juveniles are benthic in beds of macroalgae, while adults are offshore semipelagic and feed heavily on krill (Duhamel and Hureau 1990).

The distribution of abyssal Antarctic benthos has received little attention. There is evidence that the distribution of main communities is strongly affected by sediment features, local hydrography and bottom dynamics (Gambi and Bussotti 1999). In general, the most represented groups of epifaunal organisms on fine sediments are foraminifers, bryozoans, ophiuroids, asteroids, crinoids, molluscs and crustaceans (mainly amphipods, tanaids and isopods). Infaunal assemblages are often dominated by polychaetes, although on the Ross Sea continental slope, bivalve molluscs such as *Yoldiella ecaudata* and *Genaxinus debilis* may be abundant in muddy and hard substrates respectively (Cattaneo-Vietti et al. 2000). Different sampling procedures make it hard to compare qualitative and quantitative data on fish fauna; however, Eastman and Hubold (1999) found a similar size and taxonomic composition between pelagic and benthic fish in the Ross and Weddell seas (i.e. in the East Antarctic Province). As the four dominant benthic species in the southern Ross Sea are also found in the Southern Scotia Sea, this suggests that the latter region served as a point of entry of sub-Antarctic organisms into the Antarctic region.

3.5 Antarctic Marine Food Webs and the Impact of Human Activity

Although the krill system is unique in the world oceans, and although it is undoubtedly a very important component of the Antarctic marine ecosystem, the seasonal pack-ice zone covers some 19×10^6 km^2 (Knox 1994), and food webs in the larger part of the Southern Ocean contain little krill. Human activity, such as the exploitation of marine resources or the bioaccumulation and/or biomagnification of persistent anthropogenic pollutants, will affect the various food webs in the Southern Ocean in different ways.

The area of the ACC (about 27×10^6 km^2) is free of ice throughout the year; it is rich in nutrients but relatively poor in phytoplankton (mainly nanoplankton) and usually has two production peaks each year. Zooplankton is dominated by herbivorous copepods, salps and small euphausiids, with higher biomasses usually concentrating at depths between 500 and 1,000 m. Myctophid fish, juveniles of benthic fish species and, above all, cephalopods are the main consumers in the productive mesopelagic layer. In this zone of the Southern Ocean, sperm whales, elephant seals and several species of albatross are among the main consumers of cephalopods and, probably, the main bioaccumulators of persistent pollutants. It has been suggested that Antarctic stocks of cephalopods have increased in the past century as a result of the overexploitation of sperm and baleen whales (i.e. the principal consumers of cephalopods and krill; Laws 1985). However, some major cephalopod fisheries are located in cool temperate waters, just north of the CCAMLR boundaries, and one of the target species (*Martialia hyadesi*) is also found in the CCAMLR Convention Area. Fishing poses special problems for stock management (Rodhouse 1990), because commercially fished species are fast-growing and short-lived (approximately one year). As recruitment depends on the breeding success of a single generation, cephalopod populations are prone to extreme inter-annual fluctuations and are therefore highly susceptible to overfishing, which also adversely affects seabirds. Populations of wandering albatrosses and petrels are also decreasing because they are caught on hooks deployed by vessels which tow huge, heavily baited lines to longline to catch various fish species, particularly the Patagonian toothfish *D. eleginoides*.

In the marginal ice-edge zone, spring and summer phytoplankton blooms allow the development of zooplankton communities, and the large stocks of krill directly or indirectly support vast populations of vertebrate consumers. Owing to high primary and secondary productivity, a vertical flux of organic matter, and probably of ad/absorbed chemical compounds, establishes in deep waters and sediments. Along with the relatively short krill food chain, this sedimentation probably helps reduce the potential bioaccumulation of persistent pollutants by seabirds and baleen whales. In this area, however, at first sealing, then whaling and, more recently, the exploitation of krill have determined dramatic perturbations in mammal and bird populations. While the harvesting of species such as baleen whales, at the higher level of the food web, increased the availability of food and enhanced the productivity of competitor species, the long-term effects of harvesting species such as krill, at lower levels in the food web, are unpredictable. Phytoplankton in the seasonal pack-ice zone is the most exposed to the effects of UV-B radiation, and the biological cycle of krill is closely linked to that of sea ice. Thus, climate and environmental change will also affect biotic communities and food chains in the marginal ice edge.

Near the continent, in the permanent pack ice or fast ice zone, phytoplankton production is restricted to a brief, intense summer period, and the zooplankton biomass is usually low. Therefore, a large proportion of algae inside or below fast ice, phytoplankton, and zooplankton organisms falling from above become an important source of food for benthic invertebrates, on which notothenioid fish feed in turn. Given the low biomass of euphausiids (*E. superba* is often replaced by the smaller *E. crystallorophias*), most seabirds breeding along Antarctic coasts, such as the Emperor and Adélie penguins, and marine mammals (Weddell seals) feed on crustaceans and pelagic, cryopelagic or epibenthic fish.

Regeneration of nutrients takes place on the bottom, and they are returned to the water column together with persistent pollutants, determining interexchanges between pelagic and benthic environments of the Antarctic continental shelf.

Most scientific stations in Antarctica are located at coastal sites, and marine ecosystems seem particularly at risk because the involvement of benthic invertebrates in the transfer of energy and persistent pollutants from phytoplankton, other autotrophic organisms and sediments to fish, nesting seabirds, and seals. The lengthening of the food chain enhances the bioaccumulation of pollutants (Bargagli et al. 2000). At Terra Nova Bay, for instance (Bargagli et al. 1998 c), concentrations of Hg in seawater and marine sediments are very low, and there is no evidence of metal inputs from the Italian Scientific Station. However, the total body content of Hg progressively increases from primary consumers (zooplankton and sponges) to benthic organisms feeding on algae and/or detritus (e.g. sea urchins) and more opportunistic feeders such as starfish and gastropod molluscs. Metal (in the form of methylmercury) is transferred from benthic invertebrates to demersal fish and higher vertebrates. Concentrations in bird feathers increase in the order snow petrel (zooplankton feeder)<Adélie penguin (zooplankton and fish feeder)<Emperor penguin (fish feeder)<Antarctic skua. The skua has an omnivorous diet consisting of marine organisms, eggs and chicks of penguin and skuas, adult snow petrels, and human refuse from the scientific stations (Court et al. 1997).

If seabirds are now recognised worldwide as important components of marine ecosystems, it is above all thanks to research in polar regions. Antarctic seabirds are considered among the most reliable indicators of environmental change in the Southern Ocean, and several international research programmes are promoted by CCAMLR to monitor changes in the size and distribution of populations breeding in Antarctica.

3.6 Summary

The unrestricted flow of water masses around Antarctica, together with the seasonal formation of a sea-ice cover and its melting in spring and summer, are among factors controlling the distribution and behaviour of Antarctic marine ecosystems. The presence of ice shelves and polynyas, and the winter freezing of water promote the formation of dense water masses which flow equatorwards to the cool oceans and atmosphere in low-latitude regions. This flow is counterbalanced by the inflow of warm air and seawater from lower latitudes. The massive eastward flow of the ACC, the only oceanic current moving completely around the Earth, allows the exchange of mass, heat and solutes with other oceans. The impact of climate change in the Southern Ocean is therefore likely to trigger large-scale changes in the world's oceans and climate.

In October, Antarctic sea ice covers an area larger than that of the continent; although satellite systems and in-situ observations are providing a great deal of data on broad-scale sea-ice extent and variability on daily to decadal scales, a number of processes of interest for climate-change studies need to be further investigated. There is widespread conviction that Antarctic sea ice is very sensitive to climate change, and most coupled atmosphere–sea ice models predict a reduction of sea-ice extent. However, at present no overall trends have been detected, and most models are still limited in their ability to reproduce detailed aspects of sea-ice distribution and timing.

For its physico-chemical and biological features, the Southern Ocean plays a key role in oceanic deep-water ventilation and in the global cycles of C, Si, Ca, S and other elements. Besides Si and other biolimiting elements such as Fe and Mn, many other factors determine large spatio-temporal variations in primary productivity. The complex interaction between biotic and abiotic factors makes it very difficult to interpret changes which have occurred in the Southern Ocean in the recent geologic past or to foresee possible effects of global climate change. Warming will probably reduce the density of surface waters which are now exposed to the atmosphere, with a consequent decrease in CO_2 uptake. However, increased stratification in the Southern Ocean will possibly affect overall deep ocean circulation and the global C cycle on century timescales.

The unusual depth of the Antarctic continental shelf, which is surrounded by deep waters and lacks connections with temperate shelves, the different water masses and the oceanic fronts form semi-closed systems where many species of benthic and pelagic organisms evolved for more than 20×10^6 years. Thus, many species of Antarctic marine organisms are endemic, have a circumpolar distribution and are characterised by late reproduction, long lifespans and large body sizes. Three main circumpolar biogeographical zones are usually distinguished: the zone of ice-free open waters (relatively rich in nutrients but poor in phytoplankton), the seasonal pack-ice zone (where seeding by ice algae contributes to high primary and secondary production), and the permanent sea-ice zone (with intense but brief primary production, mostly consumed by rich benthic communities). Each biogeographical zone shows distinct biotic communities and food webs: small euphausiids, myctophid fish and cephalopods in the zone of ice-free open waters, the typical short food chain algae–krill–whales (or seals or seabirds) in the seasonal pack-ice zone, and rather complex trophic relations in the permanent sea-ice zone (phytoplankton and benthic algae–benthic invertebrates–notothenioid fish–penguins or seals).

Contrary to popular belief, the Southern Ocean is far from being a pristine marine environment. In spite of its remoteness, during the last two centuries it has been affected by dramatic anthropogenic perturbations. Following Cook's visit to South Georgia in 1775 and the publication of his account of great numbers of seals, fur seals were slaughtered in the late 18th century until they became commercially extinct. The sealing industry was followed by whaling during the early–mid 20th century, and soon this industry also collapsed due to overexploitation; finfish, krill and cephalopod fisheries began to operate in the late 20th century. Research on the history of sealing and whaling in the Southern Ocean provides important indications not only on overexploited species, but also on the possible effects on competitor species. Populations of fur seals have recovered significantly, probably assisted by reduced whale numbers. However, this may not be the case for the harvesting of species such as krill, at lower levels in marine food webs. Enhanced UV-B radiation and changing climate and environmental conditions may also contribute to the reduction of ecologically important species of primary producers and consumers, with unpredictable impacts on competitor species, secondary consumers and their dependants. The future of Antarctic marine ecosystems thus depends on the ability of organisms to adjust to the impact of human activity (living resource exploitation and environmental pollution) and to changing environmental conditions. A better understanding of organism ecophysiology, the dynamics of different food webs, and the possible long-term effects of UV-B radiation and climate change on key communities is indispensable for the management of Antarctic marine ecosystems.

4 Persistent Contaminants in the Antarctic Atmosphere

4.1 Introduction

On a small scale, man has been altering the chemistry of the atmosphere ever since he learnt to make fire and cultivate fields. Between 1800 and the mid-1990s the world population increased six-fold, and the burning of fossil fuels in this period caused about an 800-fold increase in the global emission of carbon dioxide. In addition to the emission of natural elements and compounds, about 100,000 synthetic compounds were produced during the 20th century, and many of them were used to such an extent and were so persistent that they became widely dispersed.

The problem of pollution due to toxic and bioaccumulative pesticides, radionuclides and trace metals first emerged in the 1960s in several regions of the Northern Hemisphere. The development of spectrophotometric and chromatographic techniques and their ability to measure inorganic and organic contaminants in environmental media at concentrations of a few µg kg^{-1} or part per billion (ppb) allowed the detection of relatively high concentrations in Arctic and Antarctic environments, previously believed to be pristine. Once global pollution became of concern, further problems emerged. Pesticides and polychlorinated biphenyls (PCBs), collectively known as persistent organic pollutants (POPs), through a process known as "global distillation", evaporate into the air in tropical and temperate regions and are transferred by atmospheric circulation to higher latitudes, where low temperatures reduce or block further evaporation; some POPs condense and settle out, producing relatively high environmental levels. During the last decade, it has been found that some POPs are endocrine disrupters and threaten the health of both wildlife and humans. These pollutants are currently the subject of negotiations intended to bring them under global agreement, with some being phased out and others tightly controlled. Several POPs are still used in tropical countries. Moreover, the impact of the burning of forests, grasslands, and agricultural lands in these countries is emerging as much more extensive than previously thought. Fires probably constitute one of the most significant sources of atmospheric pollutants in the Southern Hemisphere. Besides chemically active gases such as nitrogen oxides and hydrocarbons, which can lead to the production of tropospheric ozone (harmful to organisms, including humans), aerosols in dense smoke plumes may affect the global climate by changing the planet's radiation budget, cloud properties and atmospheric precipitation. Biomass burning also releases many organic compounds and trace elements such as mercury. Elemental mercury, the dominant component of total atmospheric mercury, undergoes long-range atmospheric transport. Recent research in some polar areas shows that in spring, after sunrise, there is a remarkable deposition of elemental mercury from the lower tropospheric boundary layer. Polar regions may be "mercury cold traps" which collect the metal released by anthropogenic and natural sources at lower latitudes. Anthropogenic activity may also affect the chemistry of the stratosphere through the emission of ozone-depleting halons and other chlorine and bromine compounds (CFCs), nitrogen oxides and other contaminants from high-altitude aircraft. Although the Montreal Protocol called for CFC production phase-out in the developed world by 1996, in developing nations, the phase-out is more gradual. Besides CFCs and long-lived trace gases such as carbon dioxide and methane, the deposition of long-range transported radionuclides, trace metals and DDT-related compounds in continental Antarctica was first detected in the 1960s. In the following decades, with increasing global and environmental awareness, it was realised that Antarctica provided a unique opportunity to study the transport, behaviour and effects of persistent atmospheric contaminants. Following the recommendations of the Intergovernmental Forum on Chemical Safety, for protecting human health and the environment, in 1997 the United Nations Environment Programme (UNEP) promoted international action on some persistent organic pollutants. Negotiations resulted in the adoption of the 2001 Stockholm Convention and the selection of 12 initial substances on which to focus: aldrin, endrin, dieldrin, chlordane, DDT, toxaphene, mirex, heptaclor, hexachlorobenzene, PCBs, dioxins and furans. The project entails evaluation of sources, transport over a range of distances, levels in the environment, biological effects, existing alternatives to their use, and possible remedial action. To achieve these results, the Earth was divided into 12 regions, and Antarctica was designated by UNEP as a region for the global assessment of persistent toxic substances. Antarctic Regional Assessment was delegated to the SCAR, and the recently published report (UNEP 2002a) constitutes the first overview of sources, transport and impact of 26 organic compounds in Antarctica (including the 12 POPs of the Stockholm Convention). Apart from this report on specific contaminants, another important source of data and references is the summary of environmental monitoring activities produced by a group of environmental officers (AEON) under the aegis of the Council of Managers of National Antarctic Programmes (COMNAP; COMNAP-AEON 2001). The New

Zealand Antarctic Institute published a report (Waterhouse et al. 2001) on the state of the environment in the Ross Sea region, and other reviews of trace metal contamination and biomonitoring in Antarctica have been published recently (Bargagli 2000, 2001). However, most available publications refer to specific groups of chemicals or specific regions, and much data on persistent contaminants in the Antarctic environment are still scattered across a wide range of journals. Antarctic research and logistic organisations have established committees for the development of internationally coordinated monitoring networks; however, unlike the Arctic, where a coordinated assessment of environmental pollution is in progress (e.g. AMAP 1997, 1998), an Antarctic Monitoring and Assessment Programme does not yet exist. This and the following chapters will attempt to fill, at least in part, the gap by assembling and organising available data into a single, wide-reaching survey. Literature data on persistent atmospheric contaminants in Antarctica and the Southern Hemisphere are scarce in comparison to those on the Northern Hemisphere, and refer to a rather long period of time. The reliability of analytical determinations of atmospheric contaminants is constantly improving and, although this and the following chapters are based on data published in peer-reviewed literature, it is impossible to assure a uniform quality level of data. Whenever possible, preference will be given to more recent papers, although comparisons with older data are essential.

Environmental contamination and pollution have been defined in many different ways; most definitions (e.g. Moriarty 1983; Bacci 1994) state that environmental pollution occurs when there is impairment of a biological system (organism, population, or community). Thus, the use of the terms pollution or pollutants for chemical elements or compounds occurring in the Antarctic atmosphere in very low concentrations seems inappropriate. The terms environmental contamination and contaminant will be used instead, except in a few localised areas where measurable damage to living organisms cannot be excluded.

4.2 The Atmosphere of the Southern Hemisphere

The atmosphere is composed of a mixture of gases, vapours and minute particles which are tied to the Earth by gravity. This gaseous envelope decreases in density with increasing altitude, passing gradually into space some hundreds of kilometres away from the Earth. The troposphere (from the Earth's surface to the tropopause, at an altitude of 9–15 km, depending on latitude and season) holds about 80 % of the atmospheric mass, and more than half of this mass is concentrated in the first 5 km.

The composition of the atmosphere has changed considerably with the evolution of the Earth, especially when cyanobacteria began to release O_2 in the ancient seas. As the atmosphere has evolved under the influence of biota, the greenhouse effect of the atmosphere in turn stabilised the Earth's temperature, while stratospheric O_3 provided an effective shield from much of the Sun's ultraviolet radiation, thereby allowing the evolution of terrestrial organisms. The atmosphere sustains life by transporting water from the oceans to land, and by providing essential elements such as oxygen and nitrogen. Despite its importance to living organisms, the atmosphere has become a dumping ground for elements and compounds emitted by human activity. A small proportion of these chemicals, depending on sources and their physicochemical properties (e.g. vapour pressure, volatility, solubility and susceptibility to be scavenged by precipitation), may reach the Antarctic environment. Indeed, in spite of the remoteness of Antarctica, its atmosphere and the Southern Ocean are inextricably linked to atmospheric and oceanic circulation at lower latitudes, and the large equator–pole temperature gradient drives the poleward transport of chemicals.

As discussed in previous chapters, the Antarctic atmosphere loses more heat by radiative cooling than it gains by surface energy exchange; the deficit is balanced by atmospheric transport of heat, moisture (including persistent contaminants), from lower latitudes. Although this energy balance constrains large-scale circulation of the atmosphere, Antarctica does not simply play a passive role in this transport, but it is involved in complex interactions with long waves. The high topography of the continent, strong cooling over the surface of ice sheets, and the presence of a persistent surface-temperature inversion over the interior of the continent generate the katabatic wind regime and limit the poleward propagation of cyclones. The katabatic drainage flow is compensated by subsiding air from the mid-troposphere which, in turn, determines horizontal convergence and cyclonic inflow in the upper troposphere (James 1989). Cold air over the continent is thus exported at low levels, while warmer air from mid-latitudes is advected at upper tropospheric levels. This poleward flow in the high troposphere significantly decreases during winter, when the circumpolar vortex is well established (Mroz et al. 1989). For instance, there is evidence that concentrations of aerosols, ^{210}Pb and its terrestrial precursor [222]Rn at Mawson and other Antarctic stations are higher during the austral summer (Lambert et al. 1990; Wagenbach 1996).

Studies on the hemispheric water vapour budget show that, as in the case of heat, in the tropics and subtropics there is an excess of evaporation with respect to precipitation, which is balanced by a water vapour sink at high latitudes. However, this does not necessarily mean that all atmospheric precipitation in Antarctica originates in the tropics or subtropics. North of the Antarctic coast, atmospheric circulation is dominated by the circumpolar low-pressure trough (at an annual average latitude of about 66° S, and with three climatological low-pressure centres located approximately at 20° E, 90° E and 150° W; King and Turner 1997). As moist air moves southwards, water vapour and a large amount of chemical substances are lost by precipitation;

they are then replaced by evaporation and the bursting of entrained air bubbles in the Southern Ocean. The isotopic composition of snow in Antarctic coastal areas reflects that of seawater at the northern edge of the pack-ice zone (about 65° S; Bromwich 1988). On the contrary, model studies (Ciais et al. 1995) indicate that the main source of water vapour in the continental interior is located between 20 and 40° S.

4.2.1 Trace Gases

The main atmospheric gases (N_2, O_2 and Ar) account for more than 99.9 % of the total volume. They have limited interaction with incoming solar radiation and do not interact with infrared radiation emitted by the Earth; thus, their concentrations are nearly invariant. In contrast, water vapour (with a highly variable volume, but typically in the order of 1 %) and many trace gases, such as CO_2, CH_4, O_3 and N_2O, absorb and emit infrared radiation. In spite of their small volume, the so-called greenhouse gases therefore play an essential role in the Earth's energy budget. The atmospheric distribution of O_3 and its role in this budget is unique – O_3 in the troposphere and lower stratosphere acts as a greenhouse gas, while in the stratosphere it absorbs solar UV radiation. Tropospheric O_3 is the third most important greenhouse gas after CO_2 and CH_4; it is a photochemical product which develops from emissions of nitrogen oxides (NO_x), CO, CH_4 and volatile organic compounds (VOCs).

In general, more reactive trace gases containing one or more H atoms are removed in the troposphere by reactions with hydroxyl radicals (OH), while N_2O, CFCs, perfluomethane (CF_4) and perfluorethane (CF_6) are only destroyed by solar ultraviolet radiation in the stratosphere. Given the time required to transport these gases at very high altitudes, their lifetime in the atmosphere is more than 20 years. The mean residence time of natural or anthropogenic compounds in the atmosphere is a very important parameter because it relates emissions to the atmospheric burden. As a rule, gases with a short atmospheric lifetime, such as CO, tropospheric O_3 and NO_x, show much more variable concentrations in space and time (Table 1), while those with long mean residence times are well mixed. As the exchange between northern and southern tropospheric air masses takes about one year, greenhouse gases such as N_2O, CO_2 and CH_4 are rather uniformly mixed throughout the global troposphere.

Exchange between the troposphere and stratosphere accounts for a large proportion of the stratospheric mass each year, and it has been estimated that stratospheric gases have a mean residence time of about 1.3 years (Warneck 1988). This exchange time is rather short compared to the residence time of major gases and many anthropogenic greenhouse gases. For these molecules the proportional composition of the atmosphere hardly varies with altitude. Methane and N_2O, for instance, have a negligible vertical gradient in the tro-

Table 1. Estimated atmospheric lifetime, variability and annual increase in global mean concentrations (ppb) of atmospheric trace gases and their major anthropogenic sources (period 1996–1998; IPCC 2001)

Origin	Gas	Lifetime	Concentration variability	Global mean concentration (1998)	Global annual increase	Main anthropogenic sources
Tropospheric	CO_2	5–200 years	10^{-3}	365,000	+1,500	Fossil fuel (about 75%), deforestation
	CH_4	9–12 years	10^{-2}	1,745	+5.0	Fossil fuel, agriculture
	N_2O	114–120 years	10^{-3}	314	+0.8	Agriculture, chemical industry
	CO	30–90 days	10^{2}	80	+0.006	Transportation (about 50%), biomass burning
	O_3	4–20 days	10^{1}	Remote, <10; urban, >100	?	Industries, biomass burning, aircraft
	NO_x	<4–12 days	10^{4}	Remote, <0.001; urban, <100	?	Fossil fuel (transportation, about 40%); biomass burning
Stratospheric	H_2O	1–6 years	?	3,000–6,000	+40	CH_4 oxidation

posphere and, although their emissions occur mainly in the Northern Hemisphere, their average atmospheric surface concentrations in the Southern Hemisphere are only slightly lower (about 5 % for CH_4 and about 0.8 ppb for N_2O; IPCC 2001).

In contrast to well-mixed gases, reactive gases such as CO, NO_x and VOCs, and tropospheric O_3 have short lifetimes, their atmospheric abundance shows large gradients, and the global burden is uncertain. In general, any chemically reactive gas, whether a greenhouse gas or not, can produce an indirect greenhouse effect through its impact on atmospheric chemistry. For instance, although carbon monoxide (CO) is not a greenhouse gas, it determines tropospheric levels of OH, indirectly affecting atmospheric concentrations of CH_4 and leading to the formation of tropospheric O_3. More than 50 % of CO is released by anthropogenic activity, especially in the Northern Hemisphere where atmospheric concentrations are about twice those in the Southern Hemisphere. CO concentrations in high northern latitudes vary from about 60 ppb during summer to 200 ppb during winter; at the South Pole these value are about 30 and 65 ppb respectively (IPCC 2001). Molecular hydrogen (H_2) can reduce OH, with an indirect increase of atmospheric concentrations of CH_4 and HFCs. About half of H_2 emissions are anthropogenic, and the gas is removed from the atmosphere by reaction with OH and especially by microbial uptake in soils. Thus, although 70 % of H_2 emission occurs in the Northern Hemisphere, average concentrations are higher in the Southern Hemisphere, which has a much smaller landmass (Novelli et al. 1999; Simmonds et al. 2000). Nitrogen oxides (NO_x) catalyse tropospheric ozone formation and indirectly reduce the atmospheric burden of CO, CH_4 and HFCs. NO_x emissions are increasing in East Asia and several countries of the Southern Hemisphere; these gases are also released in the free troposphere by aircraft, ammonia oxidation and lightning (Lee et al. 1997). The dominant NO_x sink is oxidation of NO_2 by OH to form nitric acid, which then collects on aerosol, dissolves in precipitation or deposits directly. Volatile organic compounds include non-methane hydrocarbons NMHCs and oxygenated NMHCs (aldehydes, organic acids, etc.), and may produce O_3 in the presence of NO_x and light. Vegetation is the main source (e.g. isoprene and monoterpene, which also play a role in aerosol formation) of VOCs, and atmospheric concentrations are therefore high near the tropics (IPCC 2001). However, there are also anthropogenic sources of VOCs such as motor vehicles and biomass burning. Tropospheric O_3 is a direct greenhouse gas; through its chemical impact on OH, it modifies the lifetime of other greenhouse gases such as methane. Concentrations are extremely variable in space and time (from less than 10 ppb over remote tropical oceans, up to 100 ppb in the upper troposphere or in downwind metropolitan areas; IPCC 2001). Tropical industrialised and biomass-burning regions are among the main producers of O_3. Although global trends in stratospheric O_3 are extremely difficult to infer, evidence suggests that during the last decades there has been little increase in global tropospheric O_3 at the few remote locations where it is regularly recorded. At the South Pole, long-term records of surface O_3 since the early 1960s show the largest negative trend, particularly in summer values (Oltmans and Levy 1994). The decrease was attributed to the greater penetration of UV radiation to the surface, and enhanced transport of ozone-depleted air masses from the coast.

The hydroxyl radical OH is the atmosphere's primary cleansing agent, and its local abundance is therefore controlled by concentrations of NO_x, CO, VOC, CH_4 and O_3, and by the intensity of solar radiation. Its concentrations thus vary greatly according to the time of day, season and geographic location. Tropical regions play a critical role in determining the global oxidising power of the atmosphere, because high levels of UV radiation and humidity promote the formation of OH through reactions which follow the photolysis of O_3.

4.2.2 The Impact of Biomass Burning

Biomass burning is a global phenomenon, largely anthropogenic in origin, which serves for a variety of purposes: food preparation, clearing land for agriculture, weed and brush control on cultivated lands, production of charcoal. Most biomass burning occurs during the dry season in the tropics, with about 80 % occurring between 25° N and 25° S (Dignon and Penner 1991). As this section only considers few recent studies on the Southern Hemisphere, for a more thorough discussion of environmental and climatic effects of biomass burning the reader can refer to specific publications such as Remmert et al. (1990), Levine (1991, 1996), and Clark et al. (1997).

Satellite imagery and data (Husar et al. 1997; Ross et al. 1998) show that dense smoke plumes occur annually downwind of fires in South America (August–October), southern Africa (July–September), and Central America (April and May). Combustion products of biomass burning include water vapour, CO_2, CO, CH_4, NMHC, NO_x, N_2O, NH_4 and smoke particles. The amount and proportion of different gaseous and particulate compounds in smoke depend on the characteristics of the biomass, conditions and the duration of smouldering and flaming, which affect chemical reactions such as pyrolysis, oxidation and depolymerisation, char formation and steam stripping. In forest fires, for instance, the flaming stage is usually followed by a long, cooler smouldering stage, in which the incompletely consumed wood emits smoke, mainly composed of organic particles without black carbon. On the contrary, grasses burn quickly in strong flaming fires, with no smouldering and the release of large quantities of black carbon (Simoneit et al. 1999; Kaufman et al. 2002). Convective transport of trace gases from biomass burning affect O_3 concentrations in the middle and upper troposphere. Aldehydes, ketones and other oxygenated hydrocarbons produced in fire are subsequently oxidised, in part forming formaldehyde (HCHO). The photolysis of

HCHO in the troposphere is the source of HO_2 which, coupled with NO_x and hydrocarbons produced by biomass burning, takes part in the catalytic cycle which produces tropospheric O_3.

Emissions by tropical fires in South America and southern Africa play a significant role in radiative forcing over large spatial scales. Ozone concentrations comparable to those occurring downwind of major urban/industrial areas in the Northern Hemisphere have been measured in some areas of the South Atlantic (Thompson et al. 1996). The effects of biomass burning have also been detected in the atmosphere over the remote South Pacific (Schultz et al. 1999; Talbot et al. 1999). Simoneit and Elias (2000) used levoglucosan (a major compound released during thermal degradation of cellulose and hemicellulose) as a tracer of biomass-burning emissions, and found it in all air samples collected over the Atlantic, up to 62° 21'S.

Biomass burning also affects the climate through aerosol, which cools the Earth's surface by reflecting and absorbing solar radiation and modifying cloud properties. Aerosols containing black graphitic and tarry carbon are dark and strongly absorb incoming sunlight, warming the atmosphere and cooling the surface. Black carbon over the Indian Ocean (Satheesh and Ramanathan 2000) warmed the lowest 2–4 km of atmosphere and reduced the amount of sunlight reaching the surface by about 15 %. The reduction of the vertical temperature gradient in the troposphere may reduce evaporation and cloud formation, with a possible reduction in precipitation. Moreover, the large amount of smoke from vegetation fires is dominated by fine organic particles with polar functional groups, such as carboxylic and dicarboxylic acids. These particles are water-soluble and form numerous, small nucleation droplets in clouds. These droplets have a large surface area and determine a corresponding increase in reflectivity of up to 30 % (Bréon et al. 2002; Kaufman et al. 2002). Satellite data in a smoke plume over Indonesia (Rosenfeld 1999) revealed small droplets at the cloud base (a radius of 5–8 µm compared to one of about 15–25 µm in clean atmosphere), and a lack of increase in droplet size as the cloud developed. Under these conditions precipitation does not occur or is delayed.

4.2.3 Aerosols

Aerosols are liquid or solid particles suspended in the air and, in contrast to greenhouse gases, their sources are more difficult to identify because some species are not directly emitted but are formed in the atmosphere by gaseous precursors. Their atmospheric lifetime and radiative effects cannot be reliably predicted because some aerosol types have a wide range of physical properties and often combine to form mixed particles. In general, aerosols with dry diameters between 0.1 and 1 µm (i.e. those in the accumulation mode) can hydrate to diameters between 0.1 and 2 µm; they have high scattering efficiency, the longest atmospheric lifetime, and form the majority of cloud condensation nuclei. Particles with a diameter<0.1 μm coagulate more quickly, and those with a diameter>2 μm are efficiently removed by nucleation to cloud drops or by impaction onto the surface. Owing to their relatively short lifetime, most aerosols are regional in nature. However, how far they are transported from sources or vertically distributed through the atmosphere also depends on meteorological conditions. Elevated aerosol layers can be picked up by strong winds and transported for several thousand kilometres (Prospero 1996). During transport, particles may change their physico-chemical properties (through atmospheric chemical reactions, in-cloud processes, wet and dry deposition) and their cooling influence on climate (directly through the reflection of sunlight to space, indirectly through changes in cloud properties). These processes have been widely reviewed (e.g. D'Almeida 1991; Hobbs 1993; Prinn 1994; Charlson and Heintzenberg 1995; Kulmala and Wagner 1996; Pruppacher and Klett 1997; Seinfeld and Pandis 1998) and are currently the subject of much interest because they are still largely unknown, despite their importance in regional pollution and global climate. In recent years there has been a rapid development of instruments for satellite observation of aerosol properties (e.g. MODIS, POLDER or TOMS; Hsu et al. 1999; Boucher and Tanré 2000), of monitoring networks such as AERONET (Holben et al. 1998), and numerous field campaigns were performed within the framework of specific research projects on tropospheric aerosols (e.g. TARFOX, ACE-1, ACE-2, INDOEX; Bates et al. 1998; Satheesh et al. 1999; Johnson et al. 2000).

Figure 29 shows the estimated relative contribution (%) of present-day emissions of primary particles and aerosol precursors in the Southern Hemisphere, with respect to global-scale emissions (IPCC 2001). Although these estimates are affected by large uncertainties, in a global context the main sources of aerosols in the Southern Hemisphere are oceans, biomass burning, soils, and land biota.

Given the larger expanse of oceans with respect to landmasses, sea-salt particles are the most widespread type of primary particles in the atmosphere of the Southern Hemisphere. Their formation is due to the bursting of entrained air bubbles during whitecap and is largely dependent on wind speed. Given their wide size range (from about 0.05 to 15 μm in diameter), sea-salt particles may have very different atmospheric lifetimes. Soil and rock dust particles are other major contributors to aerosol. Their diameter ranges from <1 to about 20 μm; the largest particles are quickly removed from the atmosphere by gravitational settling, while those in the sub-micron size range may have an atmospheric lifetime of several weeks. The deflation of dust is mainly due to wind speed, surface roughness, grain size and the moisture content of soils. Major dust sources are therefore in desert regions, dry lakebeds and semi-arid desert fringes. The plumes originating in North Africa, especially during El Niño years, are the largest and most persistent to be found

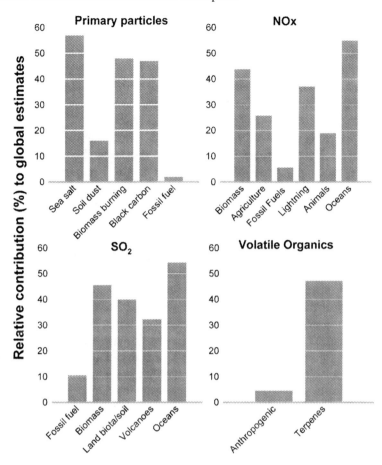

Fig. 29. Estimated emissions of primary particles and aerosol precursors in the Southern Hemisphere with respect to global estimated emissions for the year 2000. (Data from IPCC 2001)

over any ocean region (Prospero 1996). The disturbance of soil surfaces by human activity enhances dust mobilisation, and it has been estimated that up to 50 % of the atmospheric dust load should be considered of anthropogenic origin (Tegen et al. 1996). On the contrary, crusting of soil surfaces and scarce disturbance by human activity reduce the release of particles. In the Southern Hemisphere, for instance, almost no dust is observed from arid regions of Australia, which are old, highly weathered and have an almost flat topography (Husar et al. 1997).

The burning of biomass and fossil fuels, and the atmospheric oxidation of biogenic and anthropogenic volatile organic carbon (VOC) generate carbonaceous aerosols, i.e. submicron-sized particles consisting of organic substances and various forms of black carbon such as soot, charcoal and light-absorbing refractory organic matter. Although highly variable in space and time, carbonaceous aerosol constitutes an important fraction of Southern Ocean atmospheric aerosol; particles injected in the upper troposphere have a long lifetime and can be transported from southern mid-latitudes to Antarctica. In recent years substantial progress has been made in the measurement of carbonaceous aerosols, although a reliable assessment of different organic carbon species, especially of black carbon concentrations, still poses complex and partly unresolved analytical problems (e.g. Heintzenberg et al. 1997; Martins et al. 1998). However, there is evidence that the burning of grasslands and emission of large quantities of black carbon in the Southern Hemisphere determine a significantly higher percentage of aerosol optical thickness (i.e. attenuation of sunlight) than fires in boreal regions (Ross et al. 1998; Dubovik et al. 2002).

Besides the burning of biomass and fossil fuels, a major carbonaceous fraction in primary aerosols consists of biogenic particles such as plant debris, humic and fulvic substances, bacteria, fungi, viruses, algae, pollen and spores. Biogenic aerosols were not included in Fig. 29 because estimates of their atmospheric abundance are not available. However, since the earliest studies on Antarctic terrestrial biogeography (Ridley 1930), it was recognised that long-range transport of small organisms and diaspores (particles of any kind with the potential to reproduce a living organism) was the leading process responsible for circumpolar distribution patterns and the establishment of species in remote and isolated continental habitats (van Zanten 1983; Benninghoff and Benninghoff 1985; Broady et al. 1987; Linskens at al. 1993; Bargagli et al. 1996a).

Very fine particles of secondary aerosol are formed in the atmosphere by oxidation of biogenic hydrocarbons such as monoterpenes and other VOCs. Ozone and NO_3 are the main oxidants in polluted regions, while in remote regions and on a global scale most VOC oxidation occurs through reaction with OH. Although most precursors of these aerosols are of natural origin, anthropogenic activity may affect oxidation mechanisms and the amount of aerosol produced. NO_x emissions, for instance, increase O_3 and NO_3 concentrations in the troposphere; in regions with atmospheric pollution these gases may have determined a three- to four-fold increase in biogenic aerosol with respect to pre-industrial times (Kanakidou et al. 2000). Dimethylsulphide (DMS) is another important biogenic precursor of aerosol in the Southern Hemisphere. As discussed in the previous chapter, marine phytoplankton, especially that in the Southern Ocean, is a very important source of DMS. In the Southern Hemisphere (Fig. 30), biogenic sources probably prevail with respect to other sources of sulphate precursors, such as SO_2 from combustion of fossil fuel burning and volcanoes. In aerosol particles, sulphate occurs as sulphuric acid, ammonium sulphate and intermediate compounds, the relative proportions depending on the amount of gaseous ammonia available for the neutralisation of SO_4^{2-} formed from SO_2. If ammonium is in excess with

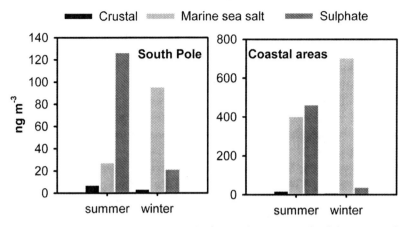

Fig. 30. Average concentrations (ng m⁻³) of crustal, marine and sulphate aerosol at the South Pole and at coastal Antarctic stations. (Data from Bodhaine 1996; Wagenbach 1996)

respect to the amount required to neutralise SO_4^{2-}, nitrate can form small particles of aerosol. According to van Dorland et al. (1997), radiative forcing due to ammonium nitrate accounts for about one tenth of sulphate forcing. However, assuming increasing emissions of ammonia from agriculture, direct forcing by ammonium nitrate could increase in the future, becoming comparable to that by sulphate (Adams et al. 2001).

4.2.4 Volcanic Emissions

About 380 volcanoes were active during the last century, and around 50 of them were active per year (Andres and Kasgnoc 1998). The distribution of active volcanoes is linked to the active zones of plate tectonics, and more than two-thirds of the world's volcanoes are in tropical regions of the Northern Hemisphere. Emissions of gases depend on thermodynamic factors such as temperature and pressure, and on the magma type (basaltic, felsic or andesitic), which in turn depends on the tectonic environment. In general, most basaltic volcanoes (with magmas rich in Mg and Fe and poor in Si) occur along mid-ocean ridges; sub-aerial eruptions (into the atmosphere) may occur in areas such as the Azores and Iceland or in intra-plate volcanoes such as the island chain of Hawaii. These volcanoes mostly show effusive eruptions with a low gas content (mostly CO_2 and sulphur), and only in rare cases do volcanic aerosols reach the stratosphere. Felsic (rich in silicate and alkali) and andesitic (intermediate silicate content) magmas are typical of volcanoes along converging oceanic plates (e.g. Indonesia) or where a continental plate overrides an oceanic plate (e.g. in the Andes). These volcanoes erupt less frequently than basaltic volcanoes, but they play a major role in global climate change and in the composition of the atmosphere in the Southern Hemisphere; because their eruptions are often more explosive, they can inject large amounts of ashes and gases directly into the stratosphere. These volcanoes emit gases during non-explosive phases, contributing a large amount of sulphur to global volcanic emissions. Although the Northern Hemisphere is usually a negligible source of aerosol to high latitudes in the Southern Hemisphere, large explosive volcanic eruptions may affect the composition of the Antarctic atmosphere.

The main compounds (by volume) of volcanic gases at the vent are 50–90 % water vapour (H_2O), 1–40 % CO_2, 2–35 % sulphur gases (SO_2, H_2S and SO_4^{2-}) and 1–10 % HCl (Symonds et al. 1994). The amount of sulphur gases is important to the global sulphate aerosol burden and represents by far the most relevant species as far as the climate impact of active volcanoes is concerned. Although the emission of sulphur from a single volcano may vary, according to its state of activity, the total amount released into the atmosphere from quiescent degassing and eruptions has been estimated to be in the range of 7.2–14.0 tg S year^{-1} (Spiro et al. 1992; Andres and Kasgnoc 1998). The concentration of anthropogenic sources of SO_2 is higher than that of volcanic sulphate but the latter, with a contribution of about 35 % to the tropospheric sulphur burden, has only a slightly smaller radiative effect (Graf et al. 1997). Indeed, volcanic sulphate aerosols in the upper troposphere contribute to the formation of ice particles and consequently, to an indirect radiative effect. There is evidence (Sassen 1992) that volcanic aerosols are involved in cirrus cloud formation, and years with high-level clouds are usually associated with intense explosive volcanic activity.

Cataclysmic volcanic eruptions which inject ash and gas into the stratosphere are sporadic and unpredictable, usually occurring a few times per century. However, there is evidence (Graf et al. 1998) that, besides explosive eruptions, many felsic and andesitic volcanoes emit gases during non-explosive phases and may contribute a large amount of sulphur to total global volcanic emissions. Sulphate has a residence time of a few years in the stratosphere, and it can affect global climate in this period through a transient cooling of surface temperatures (for a review, see Robock 2000). The stratospheric sulphate aerosol mass and optical thickness reached peak values about 3 months after the Mt. Pinatubo eruption; materials stored in the stratosphere then gradually returned to the troposphere and it took about four years for radiative forcing (about –4 W m^{-2}) to decay exponentially to background values (McCormick et al. 1995).

Since the discovery (Hammer 1977) that major volcanic events are recorded in polar ice sheets as sulphuric acid layers, ice cores from Antarctica have been widely used to study the link between climate and volcanism (e.g. Langway et al. 1988; Moore et al. 1991; Stenni et al. 1999; Zhang et al. 2002). Delmas et al. (1992) detected 23 major volcanic eruptions by applying physical and chemical analytical techniques to a 1,000-year ice core drilled at the South Pole. The 19th century was the period most affected by global explosive volcanic activity, and several eruptions were tentatively identified by comparing similar Antarctic and Greenland records. However, ice core records are noisy, and small volcanic events of regional significance may produce the same signal as distant large eruptions. In continental Antarctica, besides an active volcano (Mt. Erebus on Ross Island, southern Victoria Land), there are sites with fumaroles in Marie Byrd Land (LeMasurier and Rex 1982) and northern Victoria Land (Mt. Melbourne and Mt. Rittmann; Bargagli et al. 1996a) which indicate quite recent volcanic eruptions. Moreover, the amount of volcanic sulphate (non-sea salt sulphate; $nssSO_4^{2-}$) deposited on Antarctic snow is a small fraction of that arising from sea-salt aerosols. Robock and Free (1996) used data from both hemispheres to produce a new Ice core Volcanic Index (IVI); Robertson et al. (2001) introduced the volcanic aerosol index (VAI) by combining historical observations, ice core data from both hemispheres and satellite data to estimate the stratospheric optical depth for the past 500 years. Figure 31 shows the explosive volcanic eruptions south of 20° N, most commonly detected in Antarctic ice cores.

Major volcanic eruptions also play a very important role in the global atmospheric cycle of Hg and of many other trace metals. Based on the emission of metals from Kilauea, Hinkley et al. (1999) suggested a revision of the

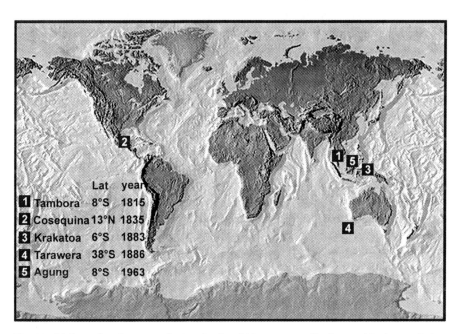

Fig. 31. Major volcanic events during the last 200 years usually detected in Antarctic ice cores (for references, see text)

estimated worldwide output by quiescent (non-explosive) volcanoes. More recently, Matsumoto and Hinkley (2001) found that the mass and proportion of Cd, Pb and other metals in pre-industrial Antarctic ice match the output rate to the atmosphere by quiescent degassing of volcanoes, and that the isotopic composition of Pb in ice is similar to that in emissions of a suite of ocean island volcanoes, mostly located in the Southern Hemisphere.

4.3 Persistent Contaminants in the Antarctic Atmosphere

By virtue of their remoteness and lack of an indigenous human population, Antarctica and the Southern Ocean are the cleanest regions on Earth. However, minimal and localised emissions of anthropogenic contaminants occur in areas affected by scientific and logistic operations, tourism and fishing. Sulphate aerosols and trace metals are released by active volcanoes and fumaroles, and a significant amount of persistent atmospheric contaminants is imported from continents at lower latitudes. In recent decades the Antarctic atmosphere has changed due to global transport of radiatively important trace gases such as CO_2, CH_4 and especially CFCs, also responsible for Antarctic ozone depletion. The atmospheric transport of trace metals, radioactive debris and POPs from other continents is revealed by similarities between contamination patterns in Antarctica and those observed in the remainder of the Southern Hemisphere (UNEP 1996). Air masses reaching the Southern Ocean carry moisture, aerosols and contaminants lifted into the free troposphere via deep convection in tropical and subtropical regions. However, it is difficult to identify individual air mass trajectories and the main sources of contaminants to different Antarctic regions. Atmospheric circulation at mid-latitudes is dominated by westerly flows from subtropical highs and sub-polar lows. The "westerlies" constitute an almost-permanent sequence of cyclonic waves, which slowly migrate eastwards. Each cyclonic cell has a diameter of 600–1,000 km and the axis of the belt is located between 61 and 64° S (Hanna 1996). Cyclonic storms in this belt act as a filter, removing particles and reactive gases from the atmosphere and depositing them in the Southern Ocean (Shaw 1988). Prevailing katabatic surface winds flow radially out of Antarctica. Studies on atmospheric concentrations of persistent atmospheric contaminants such as POPs, along north–south transects, usually reveal much higher values in subtropical areas or the belt of westerlies than at Antarctic coastal sites (e.g. Tanabe et al. 1983a; Kallenborn et al. 1998; Ockenden et al. 2001).

Cyclones may also raise air masses and aerosols to the upper troposphere, where the mechanism of mass compensation favours the penetration of particles in the polar anticyclone and their subsequent deposition over Antarctica. Based on isotopic relations, Basile et al. (1997) demonstrated that most of the atmospheric dust trapped in East Antarctica ice has a Patagonian origin. According to Iriondo (2000), Australia and southern Africa have difficulty in providing aerosols to Antarctica because they are located in the subtropical high-pressure belt, while Patagonia is the only sizeable landmass in the belt of westerlies.

The poleward transport of air masses at higher tropospheric levels is considerably reduced between autumn and late winter by the circumpolar vortex which establishes over the continent (Mroz et al. 1989). Long-term records of mineral dust, black carbon, ^{210}Pb and its terrestrial precursor ^{222}Rn, at some coastal Antarctic stations (e.g. Lambert et al. 1990; Wolff and Cachier 1998; Wagenbach et al. 1998) and at the South Pole (Tuncel et al. 1989), usually show higher concentrations during the austral summer. Near-surface dry aerosol in winter consists mostly (>90%) of sea salt, while in summer it also contains a large proportion of biogenic sulphur (produced by photochemical oxidation of DMS) and significant amounts (about 10–20 ng m^{-3}) of mineral dust (Wagenbach 1996; Fig. 30). Sea-salt aerosol concentrations at coastal sites are higher in winter than in summer, and the analysis of major ions in bulk aerosol and concurrently sampled fresh snow usually reveals a depletion of SO_4^{2-} to Na^+ or Cl^- ratios with respect to bulk seawater. It has been suggested (Wagenbach et al. 1988; Wolff et al. 1998) that, in addition to marine areas with free water, another winter sea-salt source is the surface skim of highly saline brine on freshly formed marine ice. During summer fractionation is expected to be low and sea-salt particles provide a medium for various trace gases to react with. Chloride is usually strongly depleted with respect to bulk seawater values. Sulphate, methanesulphonate (MSA) and nitrate are among the ions responsible for chloride loss. Kerminen et al. (2000) found that nitrate was strongly associated with sea salt, and hypothesised that the most likely formation pathway is the interaction between gaseous nitrogen compounds and sea-salt particles in the Antarctic atmosphere.

Black carbon particles, often described as soot, are probably one of the most reliable tropospheric tracers of the transport of contaminants from biomass-burning areas. These particles are emitted as sub-micron aerosols during combustion, they are not formed from precursors in the atmosphere by secondary mechanisms, or transformed by atmospheric reactions, and they may have a long residence time before deposition. After removing the local contamination, concentrations of black carbon at the South Pole (Bodhaine 1996) and at coastal Halley Station (Wolff and Cachier 1998) are generally low (several pg m^{-3}); the highest values (about 2 ng m^{-3}) occur at coastal stations in summer. Although concentration patterns are modulated by the efficiency of transport to Antarctica, they seem to be controlled by the timing of biomass burning in the tropics (Wolff and Cachier 1998).

4.3.1 The Mercury Cold Trap

Like other trace metals, Hg occurs naturally in the environment and is emitted into the atmosphere by both anthropogenic and natural sources. However, in contrast to other trace metals which are inherently associated with atmospheric aerosols, Hg in ambient air mainly exists as gaseous elemental Hg (Hg°). This form has a long atmospheric residence time (from months to more than one year) and is removed by wet or dry deposition, after transformation into ionic or particulate forms (Vandal et al. 1993; Fitzgerald et al. 1998; Bargagli 1999). Once deposited in terrestrial and aquatic ecosystems, the metal can be re-emitted in air and the atmosphere is consequently one of the main pathways for the distribution of Hg on the Earth's surface. In oceans, for instance, atmospheric inputs of Hg greatly exceed fluvial inputs.

In terrestrial and aquatic ecosystems, microorganisms and abiotic processes can produce methylmercury (MeHg), a compound of particular concern because it accumulates in organisms and moves up food chains (biomagnification). In Antarctic marine organisms at the higher level of the food web, MeHg concentrations are thousands of times greater than those in seawater and roughly correspond to those in related species from other seas (Bargagli et al. 1998a).

Numerous attempts have been made to compile regional and global inventories of Hg emission (e.g. Pirrone et al. 1996; AMAP 2000; Pirrone et al. 2001), but natural contributions to the total atmospheric burden from volcanoes, geothermal fields, forest fires, out-gassing from soil and water surfaces, and biovolatilisation are difficult to estimate. Moreover, fluxes of Hg from natural sources may also include previous deposition of Hg from anthropogenic sources. It is therefore impossible to quantitatively assess the relationship between anthropogenic releases of Hg to the atmosphere and the potential impact on ecosystems.

Early measurements of baseline concentrations of total atmospheric Hg in Antarctica and New Zealand gave values ranging from 0.52 to 0.84 ng m^{-3} (Bibby et al. 1988; de Mora et al. 1993). More recent data from coastal Antarctica and other remote areas show that background concentrations of total Hg in the lower troposphere of the Southern Hemisphere are slightly higher (0.9–1.3 ng m^{-3}; Slemr 1996; Ebinghaus et al. 2002; Sprovieri et al. 2002; Fig. 32). Long-term latitudinal monitoring over the Atlantic Ocean shows a pronounced gradient between the two hemispheres, with 25–30 % lower values in the Southern Hemisphere (Slemr 1996). According to the same author, values in both hemispheres increased in the 1977–1990 period (up to 1.50 ng m^{-3} in the Southern Hemisphere), and then decreased (by about 20 %) in the 1990–1994 period.

The tropospheric chemistry of Hg involves gas-phase reactions, aqueous reactions (in cloud and fog droplets, and deliquesced aerosol particles), and partitioning between gaseous, solid and aqueous phases. This chemistry has

Fig. 32. Temporal trends of total gaseous Hg concentrations (ng m^{-3}) in the lower atmosphere over the southern Atlantic Ocean, and at Terra Nova Bay (*dark shading*) and Neumayer stations (*light shading*; for references, see text)

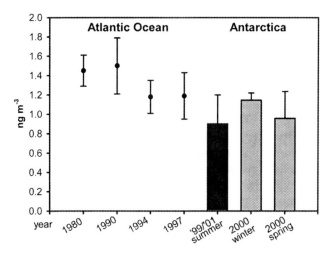

been widely discussed in recent years because highly time-resolved measurements of Hg° at Alert (82.5° N, Canadian North West Territories; Schroeder et al. 1998) showed frequent episodic depletions in atmospheric Hg concentrations during the 3-month period following polar sunrise. The springtime depletion of Hg° was correlated with ground-level ozone-depletion events. More recent studies (Lu et al. 2001; Lindberg et al. 2002) in the Arctic confirm that the lower tropospheric levels of Hg during spring are at times unusually depleted, with a simultaneous increase in Hg concentrations in snow and Arctic ecosystems. The decrease in Hg° concentrations is probably due its oxidation by halogen atoms or halogen-containing radicals such as BrO, resulting in either gaseous reactive Hg and/or particle-associated Hg species, which are much more readily deposited than Hg°. Several reactive halogen compounds exhibit a strong diel pattern, which indicates the importance of sunlight and photochemical reactions. The same compounds and reactions are probably involved in surface O$_3$ destruction. The production of reactive gaseous Hg and depletion of Hg° generally peaks at midday under maximum UV radiation, and the process occurs as long as O$_3$ and reactive halogens are present (Lindberg et al. 2002). At Barrow (Alaska, 71° N), the highest concentrations of reactive gaseous Hg (up to 0.95 ng m^{-3}) coincided with periods of increased levels of BrO, which followed elevated wave activity and sea-salt aerosol generation in the Beaufort Sea. During the year 2000, prior to melting, Hg concentrations in snow increased steadily from <1 to >90 ng l^{-1}, with an estimated average air–snow flux of about 50–60 µg m^{-2} (Lindberg et al. 2002). As there is evidence that the percentage of bioavailable Hg in melting snow increases in spring, when organisms resume their metabolic activity, this has important implications for polar ecosystems. Based on a recent increase in Hg levels in Arctic biota and on changes in the Arctic climate during the last decades (e.g.

increase in temperature, transport of photooxidants, production of reactive halogens, surface UV-B exposure), Lindberg et al. (2002) hypothesise that the enhanced springtime deposition of Hg is a recent phenomenon.

Sea-salt aerosols contain large amounts of chloride ions, with which Hg can form aqueous-phase complexes, and high concentrations of reactive gaseous Hg have been measured in the air directly above the sea surface (Pirrone et al. 2000). Concentrations of oxidised Hg compounds in the marine boundary layer are higher during the day and there is evidence of daily ozone destruction at sunrise in this layer (Nagao et al. 1999). It seems likely that processes involved in the depletion of polar atmospheric Hg are similar to those occurring daily at the marine boundary layer (UNEP 2002b).

Through highly time-resolved measurements of total gaseous Hg performed for 12 months (January 2000–January 2001) at the German Antarctic Station of Neumayer (70° 39'S, 8° 15'E), Ebinghaus et al. (2002) found that events of tropospheric Hg depletion during and after spring sunrise may also occur in coastal Antarctica. Between March and July the total concentrations of gaseous Hg were rather constant (1.146±0.075 ng m^{-3}, roughly corresponding to the average background concentration in the Southern Hemisphere; Fig. 32), while between August and November several simultaneous events of Hg and O_3 depletion, each lasting a few days, were detected, with minimum daily average Hg concentrations of about 0.1 ng m^{-3} and a mean value of 0.958±0.278 ng m^{-3} for this period. Like in the Arctic, ground-level air masses from sea ice and carrying BrO were deemed to be a prerequisite for gaseous Hg depletion. By comparing data collected at Neumayer with mean daily concentrations of gaseous Hg measured at Alert (from 1995 to 1999), Ebinghaus et al. (2002) found that Hg-depletion events at the Antarctic station were less frequent and of shorter duration than those at the Arctic site; moreover, they occurred 1–2 months earlier (August–November in Antarctica, March–June at Alert). The different latitudes of the two sampling sites and the earlier beginning of sunrise during springtime at Neumayer Station and in Antarctic sea-ice regions, where photochemical reactions between O_3, Br atoms (and/or BrO radicals) and Hg° occur, was supposedly responsible for the relatively earlier Hg° depletion in Antarctica. Using sea-ice charts and air mass trajectory calculations, Temme et al. (2003) found that the first strong Hg° depletion at Neumayer occurred at the beginning of August, in coincidence with the arrival of air masses which resided most of the time over the ocean covered more than 40 % by sea ice, before reaching the station. These results support the theory that reactive Br, which destroys O_3 and can oxidise Hg° in a subsequent reaction, is released from sea salt, associated with either the sea-ice surface or sea-salt aerosols. Hg depletion suddenly ended at the beginning of November, when sea ice in the Weddell Sea around Neumayer melted and the duration of air mass contact with sea ice sharply decreased.

During the Antarctic summer, peak concentrations of up to 0.33 ng m^{-3} of reactive gaseous Hg and/or total particulate phase were measured at Neumayer (Temme et al. 2003) and Baia Terra Nova Station (Sprovieri et al. 2002). This value corresponds to those measured near strong anthropogenic sources such as coal-fired power plants, or to average daytime concentrations during springtime depletion at Barrow (Lindberg et al. 2002). It has been hypothesised (Temme et al. 2003) that high concentrations of reactive Hg at coastal Antarctic sites during summer are the result of gas-phase oxidation of Hg° by several potential oxidants such as OH, O_3, H_2O_2, HCl or NO_3, resulting from photodenitrification processes in the snow pack. However, the re-emission of Hg from snow pack and seawater following the springtime depletion period and the possible anthropogenic contribution of condensation nuclei (the number of scientists in Antarctic stations increases during summer) may also enhance concentrations of reactive gaseous Hg.

Springtime depletion of atmospheric Hg° has been reported at some widely dispersed Arctic coastal sites (Bottenheim et al. 2002) and at only one Antarctic coastal station. In spite of the scarce availability of data and our limited understanding of complex photochemical reactions, evidence suggests that tropospheric elemental Hg in polar regions is very sensitive to photochemically driven oxidation processes during the sunlit time of year. During spring and summer, coastal Antarctic ecosystems seem affected by higher Hg deposition rates than many other regions of the Earth. As the re-emission of Hg from terrestrial and aquatic ecosystems depends especially on temperature and microbial activity (Klusman and Jaacks 1987; Ferrara et al. 1988; Schroeder et al. 1998), re-emission is probably negligible during the dark, cold Antarctic winter. Lindberg et al. (2002), for instance, estimated that melt-related re-emission at Barrow represented only 10–20% of the deposited Hg; through measurements in runoff, they found that the metal was transported to the tundra during snowmelt. Polar coastal ecosystems, especially those unaffected by katabatic winds, thus seem destined to collect a disproportionate part of the global atmospheric Hg burden. As Hg° undergoes transport at least on a hemispherical scale, all local sources in the Southern Hemisphere may contribute to increase background concentrations of Hg in Antarctica.

While in North America and Europe there is evidence of a decrease in anthropogenic emissions of Hg and other trace metals, in Asia and many countries of the Southern Hemisphere the growing demand for energy, the burning of coal and biomass, gold mining, and poor emission control are probably increasing the atmospheric burden of metals (Pacyna and Pacyna 2001; UNEP 2002b). Even if this trend were to change in the near future, as in the case of other persistent atmospheric pollutants such as CFCs, the decrease in Hg° concentrations in the atmosphere would occur very slowly, and it seems probable that Antarctica would serve as a cold trap for Hg for several decades. Moreover, as suggested by Lindberg at al. (2002), climate change might actually have a greater impact than global emissions on the Hg cycle. Changes in sea-ice cover, and in the pattern and amount of atmospheric precipitation could increase the release and transport of halogens and other potential oxidants of Hg° , determining enhanced deposition in coastal ecosystems.

Recent findings on the atmospheric cycle of Hg in polar regions help explain why, a decade earlier, Bargagli et al. (1993) had found that epilithic macrolichens in northern Victoria Land had higher Hg concentrations than samples of related species collected in remote areas of Italy. At the time, this unexpected finding was mainly attributed to local natural emissions of Hg from Victoria Land fumaroles and Mt. Erebus (an active volcano), and/or to the slow growth rate of Antarctic lichens with respect to those in temperate regions. The results were questioned by the scientific community, which suspected Hg contamination of samples. Lichens are one of the most reliable biomonitors of atmospheric Hg (Bargagli 1990); in Victoria Land they reactivate their metabolism in spring and summer, receiving most nutrients from atmospheric deposition (Bargagli et al. 2003). It thus seems very likely that the enhanced deposition of Hg during the first few months of polar sunrise, and the high concentrations of reactive gaseous Hg recently measured at Baia Terra Nova (Sprovieri et al. 2002) are more acceptable explanations for the unexpected bioaccumulation of Hg in lichens growing in pristine terrestrial ecosystems.

4.3.2 Trace Elements in Antarctic Aerosol

Unlike Hg and Se, most trace elements are emitted as fine particles which can be transported in atmospheric air masses over distances ranging from hundreds to several thousand kilometres. Antarctica is thus the best region in which to follow global trends in background concentrations of airborne trace elements. Moreover, as discussed in the previous chapter, atmospheric transport and deposition of Fe, Mn and other trace elements play an important role in the chemistry and productivity of the Southern Ocean. Research on the elemental composition of Antarctic aerosol began in the 1970s (e.g. Zoller et al. 1974; Parungo et al. 1979) and, although new works were published more recently (Harvey et al. 1991; Artaxo et al. 1992; Hara et al. 1996; Zreda-Gostynska et al. 1997), it is difficult to compare the results of different studies. The amount and composition of airborne particles in Antarctica change with seasons and between coastal and inland sites; moreover, different sampling and analytical procedures (bulk analysis or microprobe analysis of single particles) were adopted. Artaxo et al. (1992) combined bulk analytical techniques with microanalytical techniques to study the elemental composition of aerosol in the Antarctic Peninsula. The aerosol was dominated by sea-salt particles and small amounts of soil dust and sulphate; some trace elements such as Cr, Ni, Cu, Pb and Zn showed high enrichment factors with respect to soil composition, possibly indicating local or regional contamination from

Fig. 33. Average concentrations of trace elements (ng m⁻³) in the coarse fraction (2–15 μm) of aerosol particles collected at King George Island (Artaxo et al. 1992) and the Victoria Land coast (Mittner et al. 1994) during summer

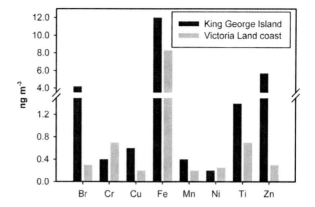

anthropogenic sources. A rough comparison (Fig. 33) between average element concentrations measured in coarse particles (2–15 μm in diameter) collected during summer (13 December–8 February) at King George Island (Artaxo et al. 1992) and at Terra Nova Bay (2.3–8.0 μm in diameter; Mittner et al. 1994) indicates a higher input of most elements in the Antarctic Peninsula region than at the northern Victoria Land coast.

Hara et al. (1996) analysed individual aerosol particles (1.6–5.4 μm size range) collected 3 m above ground at Syowa Station. They found remarkable seasonal variations in sulphates, MSA, nitrate, NaCl, KCl, silicates and trace element concentrations. Metallic elements were divided into two categories according to their behaviour: those which are components of soil particles (Ti, V, Fe, Mn, Ga, Zr, Sn and Pb) and those, including elements such as Cr, Co, Cu, Ba and Pb, which were detected in the form of chlorides, oxides or nitrates. The Pb detected in the latter chemical forms was ascribed to anthropogenic sources. Knowing to what extent metal concentrations measured in Antarctic aerosols are of anthropogenic origin is important for modelling their transport within air masses and preparing proper strategies for emission reduction. Unfortunately, it is difficult to evaluate trace metal inputs from human activity (in Antarctica and/or elsewhere in the Southern Hemisphere), due to the fact that these inputs are "masked" by much higher, natural local sources of elements such as sea salt, volcanic and biogenic emissions, soil and rock dust particles. Identification is difficult even for fly ash particles, released during the burning of fossil fuels, which appear under the scanning electron microscope as typical hollow spheres or as smooth surfaces. Many other airborne particles, such as cosmic dust ablated in the atmosphere and particles produced during explosive volcanic activity, may become spherical through melting processes and have long been recognised in sediments and snow from Antarctica and other remote regions of the Earth (Murray 1876; Thiel and Schmidt 1961; King and Wagstaff 1982).

According to a recent inventory of global anthropogenic emissions of trace elements (Pacyna and Pacyna 2001), in the mid-1990s the combustion of coal, oil and gasoline was the major source of airborne Cr, Hg, Mn, Ni, Sb, Se, Sn, Tl and V, while non-ferrous metal production was the largest source of atmospheric As, Cd, Cu, In and Zn. Estimates of global anthropogenic and natural emissions (Nriagu 1989; Pacyna and Pacyna 2001) suggest that anthropogenic emissions of Pb, V, Cd and Ni prevail on natural emissions, those of Cu, Hg, Mo, Sb and Zn are comparable, while global natural sources of As, Cr and Se are by far more significant than anthropogenic sources (Fig. 34). However, these should be considered rather tentative estimates, especially for natural sources.

It is well known that concentrations of several elements such as As, Br, Cd, Cu, In, Pb, Sb, Se, W and Zn are usually higher in the atmospheric load and in deposited materials than in rocks and soils which are the sources of the bulk of dusts (Bowen 1979). Since the 1970s (e.g. Duce et al. 1975; Weiss et al. 1978), it has been questioned whether the enrichment of trace elements in airborne

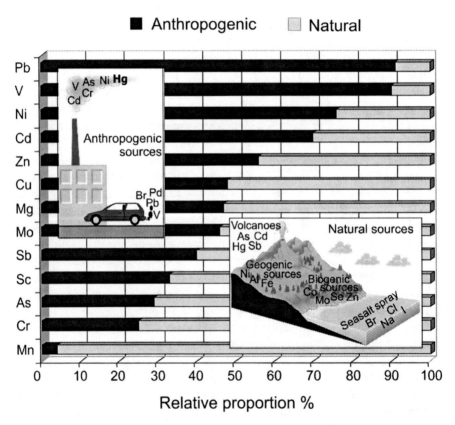

Fig. 34. Estimated relative contribution (%) of anthropogenic and natural sources to airborne trace elements

dust is a natural process or whether it is also affected by anthropogenic emissions. The elemental composition of modern dusts collected for several years in the south-western United States shows that concentrations of As, Bi, Cd, Cu, Pb, Se, Sb and Zn are much higher in at least finer-grained dusts than in the average terrestrial crust (Hinkley et al. 2002). Owing to the relatively long atmospheric residence time of fine particles, trace elements from natural or anthropogenic sources may have been acquired during their residence in air (Hinkley et al. 1997). An enrichment in metals such as Cd, Co, Cr, Cu, Mn, Pb, V and Zn has also been documented in marine aerosols with respect to sea-water, when the aerosol is formed by bubble bursting through the sea-surface microlayer (e.g. Arimoto et al. 1987; Nriagu 1989). Marine biogenic activity is an important sources of Cd in Antarctic aerosol. Surface waters in the Southern Ocean, like those in other marine areas of enhanced upwelling, have high Cd concentrations in spring, at the beginning of the algal bloom. The metal is ad/absorbed by phytoplankton, and very high concentrations occur in the liver (or digestive glands) and kidney of most Antarctic organisms (Bargagli et al. 1996b). Cd enrichment has been detected in aerosol collected at a coastal Antarctic site (Heuman 1993), and there is also evidence of the production of methylated Cd and Pb by marine bacteria in polar regions (Pongratz and Heuman 1999). However, Matsumoto and Hinkley (2001) found that, in Antarctic ice representing pre-industrial atmospheric deposition, dust and marine aerosols account for only a few percent of the Cd, I and Pb contents. The masses and proportions of these metals, and the proportion of Pb isotopes, indicate that deposition rates in pre-industrial ice match the output rate to the atmosphere by quiescent (non-explosive) degassing of ocean island volcanoes, mostly in the Southern Hemisphere.

Attempts to identify sources of trace elements to the atmosphere other than dust and salt have been hindered by uncertainties in estimates, especially of volcanic emissions. A significant amount of Hg and other trace elements are emitted as vapour species, which are usually present in greater concentrations in gas than in magma. After eruptions, trace elements condense on ash and other particles or form sublimates and agglomerates. Based on detailed records of atmospheric injection of metals from the quiescently degassing Kilauea volcano, and by combining their results with those from recent studies on other volcanoes, Hinkley et al. (1999) estimated that volcanoes probably emit a substantially smaller amount of metals (especially As, Cu and Se) than Nriagu's (1989) previous estimate. However, through fractionation processes at the melt–vapour interface, several commonly rare metals may become exceptionally abundant in the plumes of quiescent volcanoes (Hinkley et al. 1994), and this contribution may account for a significant portion of trace metal enrichment in dust deposited in Antarctic snow.

Estimates of the amount of elements emitted by individual volcanoes are usually based on the parallel collection of metal-bearing particles and SO_2, and results are then related to total sulphur emissions, which are monitored as

SO_2 at several volcanoes worldwide. However, the proportions of emitted metals and metal-to-sulphur ratios may vary by orders of magnitude in different types of volcanoes, and temporal variations have also been found for individual volcanoes (Hinkley et al. 1999). The summit crater of Mt. Erebus, the largest (3,794 m high) and most active volcano in Antarctica, contains a convecting lake of anorthoclase phonolite magma, which feeds a plume of acidic gases and aerosols and explosive strombolian eruptions (Kyle et al. 1990). The emission rate of sulphur is rather low, and the alkaline magma is characterised by high emissions of halogens (HCl and HF) and aerosols, with trace elements showing an unusually large metal-to-sulphur ratio. Several studies have focused on the potential impact of Mt. Erebus emissions on the Antarctic environment (e.g. Chuan 1994; Palais et al. 1994; Sheppard et al. 1994; Zreda-Gostynska et al. 1997; Harris et al. 1999). Since an early survey on particulate and vapour-phase aerosol emissions from the plume (Germani 1980), it was found that there was an enrichment in halogens (F, Cl, Br), volatile chalcophile elements (As, Cd, In, Sb, Se) and other elements such as Na, K, La, Ce, Sm and Th. Mt. Erebus was supposedly a source of As, Au, Br, Cs, Cu, S, Sb, Se and Zn in South Pole aerosol. By flying through the Mt. Erebus plume and measuring aerosol concentrations in 1983–1984, when small Strombolian eruptions were frequent, and during the less active period from 1985 through 1989, Chuan (1994) estimated an emission rate of up to 20 Mg day^{-1} for particles smaller than 50 μm. Crystalline elemental Au is a unique signature of aerosol emissions from Mt. Erebus (Meeker et al. 1991) and, during his sampling flights, Chuan (1994) found that these particles and total concentrations of aerosol at an altitude of 8 km up to 88.7° S (where sampling terminated prior to descent to the South Pole Station) were greater than background values. Zreda-Gostynska et al. (1997) found that many elements in the Erebus plume are common impurities in Antarctic snow, and that they can be detected over a wide area of the continent. In addition to Mt. Erebus and to continental sites with fumaroles, other active volcanoes occur on sub-Antarctic islands (Deception Island and South Sandwich Islands). Thus, it cannot be excluded that more reliable assessments of metal fluxes will show that volcanic emissions in Antarctica and elsewhere in the Southern Hemisphere account for a larger proportion of metals in ice and snow than presently supposed.

Although most trace elements in Antarctic aerosol have a natural origin, Pb and Cu contamination from anthropogenic sources has been widely documented in Antarctic snow (e.g. Boutron and Patterson 1987; Wolff and Suttie 1994; Barbante et al. 1998). A recent paper (Planchon et al. 2002a) indicates that in the second half of the 20th century there was an increase in Pb, Cd, Ag, Bi, Cr, Cu, U and Zn concentrations in snow from Coats Land (about 2,500 km from Mt. Erebus). Concentrations of several metals, especially Pb, Cr, Ag and U, in snow in recent years were somewhat lower than those observed in the 1970s–1980s, probably indicating a decline in atmospheric contamination by human activity. On the grounds of historical changes in ore and/or metal production and emission inventories, the increased deposition of metals on Antarctic snow was attributed to anthropogenic sources in the Southern Hemisphere, especially to non-ferrous metal mining and smelting in Chile, Peru, Zaire, Zambia and Australia.

Although the relative inputs from natural and anthropogenic sources cannot be easily discriminated, the results of research on the composition of recent snow suggest that atmospheric contamination by trace metals is becoming a global problem. Moreover, there is evidence that many elements which once occurred in extremely small concentrations in the environment are now increasingly released by human activity (Schüürmann and Markert 1997). Catalytic converters in motor vehicles are increasing global emissions of Pt and other companion elements such as Pd, Rh, Ru, Os and In. Increased concentrations of Rh, Pd and Pt with respect to ancient Greenland ice samples were measured by Barbante et al. (1999) in surface snow from the Alps, Greenland and Antarctica. Several lanthanoids, another group of scarcely investigated elements, are already employed as components of alloys for magnetic materials, for manufacturing laser crystals or as constituents of high-temperature superconductors, and their significance as environmental contaminants will probably increase in the future.

4.3.3 Radionuclides

Radioactive substances such as ^{14}C or tritium are natural components of the atmosphere, which are constantly formed as a result of cosmic radiation from outer space. Although life on Earth has evolved in a radioactive environment, since the first atomic explosion at Alamogordo (New Mexico, 16 July 1945) the biosphere has also been affected by increasing amounts of artificial radionuclides from ^{235}U fission produced by the detonation of two atomic bombs in 1945, and subsequent testing of nuclear weapons in the atmosphere until the 1980s. There have also been continuing discharges of radioactive wastes from weapons production facilities, fuel processing sites, nuclear power plants and research reactors. Accidental releases of radioactive materials can occur from land-based nuclear power reactors, ships or submarines with nuclear propulsion devices, space vehicles carrying nuclear facilities, and from mining, storage and transport of radioactive materials. With the exception of 12 atmospheric nuclear weapon tests by the United Kingdom at Woomera (31° S, 137° E) and Maralinga (30° S, 131° E), and 26 tests performed by France in Mururoa (Tuamotu Archipelago; 21° S, 137° W), the remaining 383 atmospheric weapon tests were all carried out in the Northern Hemisphere (Carter and Moghissi 1977). Initial atomic weapons had a modest impact on stratospheric transport routes but, with the development of more powerful thermonuclear weapons (hydrogen bombs) in the mid-1950s, radioactive materials were injected into the stratosphere. After a primary pulse in conjunction with a tropospheric component, these materials produced a succession of radioactive clouds and global-scale and long-term stratospheric fallout. Measurements performed in Antarctica since 1956 (Picciotto and Wilgain 1963) show the deposition of fission products released in the Northern Hemisphere. In contrast to weapon testing pathways, accidents such as that occurred at Chernobyl in 1986 usually determine a tropospheric injection of smoke and radionuclides, with a residence time in the atmosphere of a few days or weeks. The consequent fallout, which is maximised by rainfall and snowfall events, therefore has a hemispheric impact.

Radionuclides in the atmosphere may occur in different physico-chemical forms and show variable size (molecular mass), structure and charge properties. An understanding of the release scenario, composition and structure of radioactive materials is useful for dispersion and transport models; however, radionuclides in the atmosphere may change their physico-chemical characteristics, and volatiles can condense on particle surfaces or can form condensed particles. These processes affect the transport and deposition of radionuclides and their environmental fate. In general, radionuclides associated with colloids and particles are relatively inert in the environment, whereas low-molecular mass species are believed to be more mobile and bioavailable (Salbu et al. 1998). However, like airborne trace metals, radionuclides can be mobilised after deposition by weathering processes, depending on environmental conditions and particle characteristics.

Antarctic snow and ice give the clearest profiles of the deposition rate of radioactive debris from nuclear bomb tests, such as ^{90}Sr, ^{137}Cs, ^{239,240}Pu and ^{241}Am, partially cosmogenic radionuclides such as ^{3}H, or natural radionuclides (e.g. ^{7}Be, ^{210}Pb, ^{32}Si; e.g. Picciotto and Wilgain 1963; Cutter at al. 1979; Koide et al. 1979; Nijampurkar and Rao 1993; Pourchet et al. 1997). As the dates of arrival and deposition of radionuclides in polar regions are known, they provide a means to estimate snow-accumulation rates in different Antarctic areas or to describe air mass circulation patterns and stratospheric residence times. It has been estimated (Pourchet et al. 1997) that the total deposition of ^{137}Cs over Antarctica represents 0.09 % of the total deposition on Earth, and that the dry fallout of artificial radionuclides accounts for 60–80 % of the total and about 40 % of ^{210}Pb. On the whole, the radioactive fallout of artificial long-lived radionuclides has been ten times greater than that of natural radionuclides. Chronological profiles of tritium at the South Pole show a peak in 1966 (2,000 tritium units with respect to the pre-bomb value of about 32; Jouzel et al. 1979), while those of ß radioactivity (mainly due to ^{90}Sr and ^{137}Cs, with half-lives of around 30 years) increased in the 1950s (UK and US tests) and peaked in 1964–1965 (mainly from USSR tests in the Northern Hemisphere, with inter-hemispheric transport of about 2 years). After the partial ban on atmospheric nuclear tests, in the late 1960s there was a decline in radionuclide fallout, although French and especially Chinese atmospheric nuclear tests continued for many years. By integrating the results of many studies performed in the sector 90–180° E, Pourchet et al. (2003) traced a map of the deposition and distribution of ^{137}Cs in this sector of Antarctica, for the whole deposition period (1955–1980). According to the pattern of atmospheric deposition and distribution of the other persistent atmospheric contaminants, radionuclide fallout was found to be significantly higher over coastal areas.

A global network of aerosol sampling is operated by the Environmental Measurements Laboratory (Surface Air Sampling Program, NOAA/OAR/ CMDL), which can be used to track atmospheric release of artificial radionuclides due to nuclear weapons or nuclear accidents, revealing spatio-temporal trends in the worldwide distribution of artificial and natural radionuclides, and allowing the evaluation of transport and aerosol scavenging components of global climate model aerosols. Data from this network on monthly concentrations of radionuclides in surface air at the South Pole show that concentrations of artificial radionuclides in the 1990s were below detection limits in all analysed samples. Like soil dust, black carbon and other Antarctic aerosol, average concentrations of ^{7}Be and ^{210}Pb in surface air were significantly higher in summer, when the disappearance of the circumpolar vortex favours the penetration of air masses from lower latitudes.

4.3.4 Persistent Organic Pollutants (POPs)

The discovery and extensive use of organochlorine pesticides such as aldrin, chlordane, DDT, dieldrin, endosulfan, endrin, heptachlor, lindane and toxaphene began during and after World War II. Although chemists were aware that these compounds are very stable, there was little concern about possible long-term environmental effects. In 1962 the book *Silent Spring* by Rachel Carson raised public concern, drawing a link between the use of organochlorine insecticides and declining bird populations; as a result, in the following years intense research was carried out on the environmental fate and biological effects of what we now call POPs. Since then many books have been published on POPs (e.g. Edwards 1973; Hutzinger et al. 1974; Kurtz 1990; Mackay 1991; Howard 1991; Mackay et al. 1992; Beek 2000).

Many thousands of chemicals fall into POPs, and certain "families" such as PCBs include more than 200 compounds, which differ from each other by level of chlorination and substitution position. POPs are persistent and may have a half-life of years to decades in soil and sediments, and of several days in the atmosphere; they are therefore prone to long-range atmospheric transport. POPs are hydrophobic and lipophilic, and partition strongly in organic matter, particularly in fatty tissues. Because they are slowly metabolised, they accumulate in organisms and food chains and may have an adverse impact on human health and the environment (Jones and de Voogt 1999). As a complex array of POPs bioaccumulate simultaneously in all ecosystems, it is usually difficult to establish if a biological effect in a species is due to a particular chemical, family of chemicals, their metabolites or many chemicals acting synergistically. While early ecotoxicological studies concentrated mainly on eggshell thickness of birds of prey and reproductive potential of fish-eating birds and marine mammals (e.g. Ratcliffe 1970; Reijnder 1986), during the last decade there has been increasing concern over possible adverse effects of POPs on humans and wildlife. Evidence suggests that POPs, together with other chemicals, are immunotoxic, endocrine disrupters and tumour promoters (e.g. Vallack et al. 1998).

Because of vapour pressure under ambient temperature, POPs may volatilise from waters, soils and vegetation into the atmosphere, where they are unaffected by breakdown reactions and are transported for long distances before re-deposition. The cycle of volatilisation and deposition may be repeated many times, and POPs assume a global-scale redistribution according to the theory of global distillation and cold condensation in polar or mountainous regions (Wania and Mackay 1993). Concern about their potential toxicity, propensity for long-range atmospheric transport and global-scale redistribution has resulted in international measures to ban or restrict the production and use of POPs. Diplomats from 122 countries drew up a treaty in Johannesburg in December 2000 for the control of the production, import, export and use of toxic and persistent chemicals. The agreement was formally signed at a conference in Stockholm in May 2001, and this treaty will become international law after the legislatures of 50 countries ratify its terms (http://www.chem.unep.ch/pops/). Twelve chlorinated compounds or classes of compounds which were used or are still in use, such as chlorinated pesticides, hexachlorobenzene (HCB), polychlorinated biphenils (PCBs), polychlorinated dibenzo-p-dioxins (PCDDs) and polychlorinated dibenzofurans (PCDFs), have been selected and regulated by the Stockholm Convention. However, other persistent organic substances of known anthropogenic origin have been detected in remote regions of the Earth (e.g. polychlorinated paraffins, polychlorinated terphenyls, and polychlorinated naphthalenes; Beek 2000; Paasivirta 2000; Ballschmiter et al. 2002). Two semi-volatile hologenated compounds (a group of mixed bromochloro C_{10} compounds and a heptachlorinated C_9 compound), showing the same characteristics as anthropogenic POPs, are probably of natural marine origin (Vetter et al. 1997; Tittlemier et al. 1999; Vetter et al. 2000). Thus, as in the case of trace metals, it is rather difficult to establish accurate global emission inventories for POPs. Moreover, owing to the large number of substances with different physico-chemical properties and the temperature-dependent behaviour of individual compounds, their input and behaviour in the environment cannot be generalised. As a result, most available emission inventories (e.g. Li 1999; Bailey 2001; Breivik et al. 2002) or global-scale models on the long-range transport of POPs (e.g. Strand and Høw 1996; Wania and

Mackay 1999; Scheringer et al. 2000) usually refer to selected isomers/congeners.

Persistent organic pollutants are transported to polar regions mainly in the atmosphere, and the main forces driving their transport, temporary deposition and remobilisation are POP concentrations, the physical properties of each compound (i.e. Henry's constant, vapour pressure, and the octanol–water partition coefficient), temperature gradients and weather conditions. The range of transport increases with volatility. HCHs and HCB are usually the dominant compounds in the polar atmosphere, while less volatile compounds such as PCBs, DDTs and dieldrin tend to condense closer to their sources. Only a small fraction of less volatile POPs is transferred from tropical to polar ecosystems, and their inter-hemispheric exchanges are restricted by the presence of Hadley cells over the tropics, which produce a strong upward flux in the tropics and a subsiding flux in the subtropics (Levy 1990). Ocean currents are potentially important in inter-hemispheric mixing, but major currents circulate in an anticlockwise direction in the Southern Hemisphere and in a clockwise direction in the Northern Hemisphere. For instance, there is no evidence of inter-hemispheric mixing of PCBs, and it has been estimated that PCB concentrations in the Northern Hemisphere oceans are fourfold those in Southern Hemisphere oceans (Tanabe and Tatsukawa 1986).

The main sources of POPs in the Southern Hemisphere are urbanised areas and those with intensive agriculture, and tropical and subtropical regions where spraying is used for disease vector control. According to several authors (e.g. Mowbray 1986; Forget 1991), the demand and use of many POPs in the 1990s was increasing in tropical Asian countries and Southern Pacific islands, and considerable quantities of PCBs were used in older electrical devices and deposited as landfill in some developing countries (Iwata et al. 1994). Available data indicate that DDT has been widely used in the Southern Hemisphere, and South America has historically been the heaviest user of DDT, toxaphene and lindane. In contrast, the use of HCH (in contrast to its γ isomer, lindane) has been largely restricted to the Northern Hemisphere.

The distribution of POPs in eastern and southern Asia and Oceania (Iwata et al. 1994) shows higher levels of HCHs (sum of α, β and γ isomers) and DDTs (sum of p,p'-DDT, p,p'-DDD, p,p'-DDE, and p,p'-DDT) at lower latitudes than at higher ones. Concentrations of CHLs (sum of cis-chlordane, trans-chlordane, cis-nanochlor and trans-nanochlor) and PCBs (sum of isomers and congeners) show less prominent latitudinal variations. The highest HCH concentrations (about 11,000 ng m^{-3}) were detected in air from Calcutta (India) and Hue (Vietnam) and were attributed to their use in mosquito control programmes in tropical urban areas. DDT concentrations in some cities of India, Thailand, Vietnam and the Salomon Islands were in the same range as those reported for Brazzaville (Congo; Ngabe and Bidleman 1992), and 2–3 orders of magnitude higher than in Japan, Australia, European countries and the

USA. Atmospheric concentrations of PCBs in tropical urban air were comparable to those in European and North American urban air, and were much higher than values measured in the open ocean. In general, HCB and HCHs show a remarkable tendency to distribute globally, while DDTs, CHLs and PCBs show a lower potential for long-range transport from point sources at lower latitudes.

Data on ambient air levels of POPs in the Southern Ocean and Antarctica originate from cruises close to the continents (Tanabe et al. 1982; Kawano et al. 1985; Weber and Montone 1990; Bidleman et al. 1993), and from some sampling sites in the continent and sub-Antarctic islands (Risebrough et al. 1990; Larsson et al. 1992; Kallenborn et al. 1998; Montone et al. 2003). This section briefly reports on available data; for a more comprehensive review the reader can refer to the Antarctica Regional Report on persistent toxic substances (UNEP 2002a). Most shipboard sampling along transects to the Antarctic coasts measured total DDT concentrations: values were usually lower near the Antarctic coast and there was a remarkable decrease in average values from the early 1980s to the early 1990s. Very low concentrations (0.07–0.4 pg m^{-3}), for instance, were measured in the period December 1994–April 1995 at Signy Island (Kallenborn et al. 1998), a site which is closer than Antarctica to potential sources of DDT. Atmospheric concentrations of chlordane, heptachlor, HCB and HCHs showed a distribution pattern similar to that of DDT, with lowest values at the Antarctic end of the transect. In general, their concentrations decreased during the 1980s and early 1990s (Fig. 35), but there are no recent measurements to ascertain whether this decline has progressed further.

In contrast to pesticides, some industrial POPs have been used in Antarctica; Risebrough et al. (1990), for instance, reported high PCB contamination at McMurdo Station on Ross Island. In the 1980–1982 period, Tanabe et al. (1983a) performed extensive sampling during voyages in the Indian and

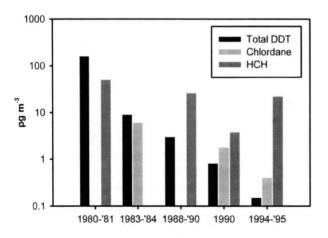

Fig. 35. Indicative trend (1982–1995) of pesticide concentrations in the Southern Ocean and Antarctica (see text for sources of data and locations)

Southwest Pacific sectors of the Southern Ocean, and found a marked decline in total air PCBs from Mauritius to Syowa Station. Year-round sampling at this station revealed higher concentrations in summer than in winter. Throughout 1999, Ockenden et al. (2001) measured PCB concentrations in air from the Falkland Islands and Halley Station (eastern coast of Weddell Sea) and found that vapour-phase concentrations were lower at the latter site, although the difference was less for more chlorinated congeners. By comparing literature data on mean annual 1994 air concentrations of PCB congeners at Alert (Canadian Arctic) with those recorded at Halley Station in 1999, Ockenden et al. (2001) found that average values of PCB-28, PCB-138 and PCB-153 were similar, while concentrations of all other congeners were two to four times higher at Alert. This reflects higher inputs of PCBs to Arctic regions, and indicates that only lower chlorinated congeners are relatively more dominant in Antarctica. Concentrations of air PCBs have also been determined in some sub-Antarctic islands (Kallenborn et al. 1998; Montone et al. 2001a, 2003). Concentrations of individual PCB congeners ranged from not detected to a maximum value of 33.2 pg m^{-3} for PCB-52. The lower chlorinated congeners were predominant, and a rough comparison between literature data (Fig. 36) shows that average values at Halley Station were lower than those measured in sub-Antarctic islands.

As previously suggested for ^{222}Rn (Pereira et al. 1988), the relatively higher PCB concentrations on the Antarctic Peninsula and sub-Antarctic islands are probably due to long-range transport episodes from South America under the influence of cyclonic circulation. Kallenborn et al. (1998), for instance, explained peak concentrations of POPs in air samples from the northern Weddell Sea by relating them to the weather conditions which would have brought in contaminated air masses from South America.

In general, available data indicate that main source areas of POPs to Antarctica are located in the Southern Hemisphere; there is also evidence that

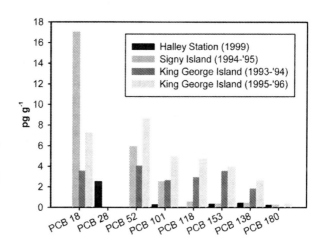

Fig. 36. Average atmospheric concentrations of PCB congeners (pg m^{-3}) at Antarctic (Halley) and sub-Antarctic (Signy Island, King George Island) stations (see text for sources of data)

this transport is quite rapid and somewhat reflects time-varying properties of the sources. Bacci et al. (1986) found that newly introduced insecticides dominated the HCH signal in lichens from the Antarctic Peninsula. Models which combine the transport of semi-volatile organic chemicals in air and water, and consider continuous exchange between the two compartments (due to deposition and re-volatilisation from water) show that the overall transport of POPs to remote regions is accelerated with respect to that in models treating air and water separately (Beyer and Matthies 2001). By sampling air over water near PCB monitoring sites at Moody Brook (Falkland Islands) and Halley Station, Ockenden et al. (2001) found that "over water" air concentrations at both sites (especially those of PCB-28) were greater than corresponding "over land" concentrations. Dachs et al. (2002) combined data from field measurements of air PCBs, PCDDs (polychlorinated dibenzodioxins) and dibenzofurans (PCDFs) with remote sensing estimations of oceanic temperature, wind speed and chlorophyll to model interactions between air–water exchange, phytoplankton uptake, and export of organic matter and POPs from the ocean surface layer. The models suggest that the deposition of POPs is enhanced at mid-high latitudes and is driven by phytoplankton uptake and the sinking of marine particulate matter. This hypothesis does not contradict the cold condensation effect, but suggests that the role of biogeochemical factors may be as important as temperature in controlling the global dynamics and ultimate sinks of POPs.

4.4 Antarctic Scientific Stations as Sources of Atmospheric Contaminants

One of the main values of Antarctica for science is the near-pristine environment, with snow, ice, soil, sediments, terrestrial and marine organisms providing an ideal archive of data on past and current trends in global atmospheric and marine contamination. However, tourism, scientific research and supporting logistic activities inevitably affect the Antarctic environment. Over the first half of the 20th century, the impact was limited because most expeditions had few participants and technological devices. However, beginning with the International Geophysical Year in 1958–1959 (with the involvement of 12 countries and over 5,000 persons occupying 55 stations on the continent and in the Southern Ocean), there was a significant increase in the geographical extent of research activities, and in the number of wintering stations, personnel and supporting infrastructures (Beltramino 1993). Concern about the localised detrimental impact of human activity in Antarctica began to be expressed in the 1970s (e.g. Parker 1978). At several stations, refuse was left to accumulate or dumped into the sea, and disused vehicles and bases were abandoned and left to deteriorate. Cameron (1972), for example, described many instances of poor waste management in the past at McMurdo Station (the largest in Antarctica, with about 1,200 persons in summer) and at remote field camps in the Dry Valleys (southern Victoria Land). The open burning of wastes contaminated the environment near scientific stations with soot containing PCBs (Risebrough et al. 1976).

Environmental awareness among the Antarctic Treaty states has greatly increased during the last decades, and significant improvements in pollution control and prevention have been made (or planned) under the requirements of the Protocol on Environmental Protection, which came into force in January 1998. With the exception of ship, aircraft and vehicle transport (to, from or within Antarctica), the release of persistent atmospheric contaminants is localised in scientific stations and is mainly due to the use of fuel (for the production of electricity, heating, water production, and equipment operation), waste disposal (including incineration), the use of chemical products in machinery, small engineering processes and scientific research. The risk of environmental impact is especially high in regions such as the Fildes Peninsula (King George Island), with a great concentration of multinational activities (10 scientific stations, various field huts and refuges, and some decades of overwintering persons; Harris 1991). In general, the most common persistent contaminants around stations are trace metals and POPs, which can be typically detected within a few hundred metres. Even at stations which have been occupied for long periods and were established when little attention was paid to environmental protection, no effects can be detected at distances greater than a kilometre (UNEP 2002a). Mazzera et al. (2001a) measured concentrations of PM_{10} (aerosol particles with aerodynamic diameter<10 µm) during the 1995–1996 and 1996–1997 austral summers at Hut Point (less than 1 km from the centre of McMurdo Station), and found that major contributors were soil (57 %), sea salt (15 %), fossil fuel combustion (14 %), and sources of secondary sulphate (10 %). Human activity at McMurdo Station consumes about 2×10^6 gallons of diesel fuel during summer, and it was estimated that the power-generating station for electricity and water production contributed to 69 % of combustion sources, while the remaining amount was due to heating. Ambient elemental carbon concentrations (129 ng m^{-3}; Mazzera et al. 2001b) were two orders of magnitude higher than background concentrations measured in Antarctica during austral summer (Bodhaine 1996; Wolff and Cachier 1998). The pollution plume at McMurdo contained concentrations of Pb (0.85 ng m^{-3}) and Zn (1.52 ng m^{-3}; Mazzera et al. 2001b) which were 17 and 46 times higher respectively than values measured during summer at the South Pole (0.051 and 0.033 ng m^{-3} respectively; Maenhaut et al. 1979). It seems likely that the deposition of atmospheric pollutants in McMurdo Sound can affect ecosystems at distances greater than that (1 km) indicated in the UNEP Report. Furthermore, metals and other atmospheric contaminants are released and widely dispersed from aircraft. Boutron and Wolff (1989), for instance, estimated that about 20 % (1,800 kg year^{-1}) of the total deposition of

Pb over continental snow was due to emissions from within Antarctica, and that a large proportion of the metal was from aviation gasoline.

The Environmental Protocol obliges signatories to the Antarctic Treaty to monitor the existing environmental impact, and the Antarctic Environment Officers Network (AEON) of COMNAP (Council of Managers of National Antarctic Programs) in July 2001 gave an account of environmental monitoring activities around Antarctic scientific stations. The release of atmospheric contaminants from Antarctic stations is trivial in a global context. However, monitoring is necessary to improve environmental management, to meet the legal requirements of the Protocol and to protect the scientific value of Antarctica. Monitoring data provide a means of defining the "fingerprint" of local releases, and to adopt suitable sampling strategies for studies concerning the assessment of global transport and deposition of persistent contaminants. Moreover, available data show that at present local inputs are the only ones with the potential to accumulate in organisms to levels which might induce biological responses (Venkatesan and Kennicutt 1996).

At McMurdo Station, besides soot, PCBs and trace metals, there is also evidence of local emission of PCDDs and PCDFs (Lugar et al. 1996). Trace levels of only a few PCDD/PCDF congeners were detected sporadically in air samples collected in 1992–1994 at a site about 500 m downwind of the station. The highest and most varied values were measured at a "downtown" location (total PCDDs varied in the range 0.12–1.8 pg m^{-3} and PCDFs in the range <0.02–2.77 pg m^{-3}), indicating that, in addition to the main solid-waste incinerator, there are various combustion sources which release dibenzo-p-dioxins and dibenzofurans at McMurdo. However, Antarctic air at remote site resulted free of PCDD/PCDF compounds, considering the detection limit in the sub-pg m^{-3} range (Lugar et al. 1996). Other atmospheric byproducts of combustion (also evaporating from hydrocarbon fuel spillages) are Polycyclic Aromatic Hydrocarbons (PAHs). For three summer seasons (1993–1995), Caricchia et al. (1995) measured particulate PAHs at a series of sampling stations set up within 200 m of the Italian "Baia Terra Nova" research station. Overall PAH concentrations ranged from 15–700 pg m^{-3} and the 11 analysed compounds showed very low concentrations (hundreds to thousands of times lower than particulate PAHs in urban areas). The plant for generating electricity was considered the most important source of PAHs at Baia Terra Nova Station.

4.5 Summary

Atmospheric transport is the main mechanism for bringing contaminants into Antarctica, while oceanic transport is probably less important and more likely to integrate over a longer timescale. Nevertheless, by virtue of its remoteness, patterns of atmospheric circulation in the Southern Hemisphere, and low-level and highly localised autochthonous anthropogenic sources of persistent toxic substances, Antarctica has the cleanest atmosphere in the world. Pesticides have neither been produced nor applied in Antarctica, and many other persistent atmospheric contaminants have only been used or released in very small quantities. Concentrations of many trace gases, such as tropospheric O_3, or of contaminants associated with atmospheric aerosols, are unusually low compared to other regions of the Earth. The Antarctic atmosphere therefore offers the opportunity to monitor the global occurrence and fate of persistent contaminants against a nearly pristine background. However, the Antarctic atmosphere has several features, such as high stability and low absolute humidity and precipitation, which make it particularly exposed to the impact of long-range transported contaminants from other continents. The recurring formation of the "ozone hole" during the austral spring is an example of this "fragility".

Sources of anthropogenic contaminants reaching Antarctica are mainly located in the Southern Hemisphere. The variability of sources and patterns of airflow and storm tracks may determine spatio-temporal variations in transport processes; however, most contaminants are probably carried to Antarctica in the upper atmosphere. In the lower troposphere, air masses reaching Antarctica from the outside must pass through the zone of cyclonic storms which surround the continent, and atmospheric precipitation removes a large proportion of airborne particles and reactive gases, depositing them in the Southern Ocean. The near-surface flow (radially outwards from the continent) of katabatic winds is another factor limiting low troposphere transport of contaminants to the Antarctic interior. However, the katabatic wind regime exerts its influence far beyond the shallow boundary layer. Near-surface airflows from the continent to the Southern Ocean are supplied by a mass of air extending through the entire troposphere to the boundary layer. This flow couples the troposphere with the stratosphere (i.e. with large mass exchanges between high and middle/low latitudes), and probably has major consequences for the supply of CFCs and low stratospheric temperatures in winter (i.e. the prerequisites for springtime depletion of stratospheric ozone over Antarctica).

The literature survey performed in this chapter clearly shows that spatial coverage for most atmospheric contaminants in Antarctica and the Southern Hemisphere is patchy and relevant data series are sparse. Although the occurrence of many atmospheric contaminants at very low concentrations challenges sample handling and analytical techniques, there is undoubtedly the need for more consistent measurement strategies if geographically extensive long-term patterns and changes related to human activity are to be identified. The long-range atmospheric transport of persistent contaminants in Antarctica is corroborated by the similarity of patterns to those observed in South America and Oceania or by chemical fingerprints. Furthermore, the geographic pattern of emission of many persistent atmospheric contaminants is changing on a global scale, with a shift from North America and Europe to Asia and many countries in the Southern Hemisphere. Owing to the growth of population, agricultural and industrial activities, the anthropogenic perturbation of the tropical troposphere in this hemisphere is expected to increase greatly over the next decades, with probable implications for the Southern Ocean and Antarctic environment.

This chapter highlights the effects of biomass burning and other human activities in the Southern Hemisphere, as well as that of volcanic activity on the climate and long-range transport of persistent contaminants to Antarctica. It discusses the tropospheric chemistry of Hg and recent findings indicating that polar regions may serve as cold traps for this metal. Notwithstanding an eventual decrease in anthropogenic emissions of Hg, natural sources of this metal such as volcanoes, and climate changes and their potential effects on atmospheric transport and deposition processes will probably enhance Hg deposition in Antarctic ecosystems during the next decades. Many long-range transport-persistent atmospheric contaminants such as artificial radionuclides, some pesticides and trace metals have shown decreasing deposition trends in Antarctica during the last two decades, therefore, there is an urgent need for further measurements to ascertain whether these declining trends are continuing.

In order to preserve the value of the continent as a data archive of past and currents trends in global atmospheric and marine contamination, it is necessary to minimise the environmental impact of human activity in Antarctica (scientific and supporting logistics, tourism, fisheries), and to assess the amount of trace metals and chemical compounds released by active volcanoes and other natural sources such as the Southern Ocean. The chapter was mostly devoted to persistent atmospheric contaminants and showed that they can be detected within a few hundred metres or a few kilometres around most scientific stations. At present, fossil fuel spills seem the most unpredictable and potentially most catastrophic contaminating events in the Antarctic region. Luckily, environmental awareness has greatly increased in Antarctica during the last decade, and significant improvements in pollution prevention and control have been made. The Protocol on Environmental Protection to the Antarctic Treaty includes provision for environmental impact assessment and monitoring of local activities. These procedures will likely help reduce the already very low release of some persistent pollutants in Antarctica and prevent environmental disasters. Although there is some scope for further reduction of pollutant emissions in Antarctica, such as the use of unleaded fuels or avoidance of Cl-containing species in incinerators, atmospheric contamination is inevitable as long as combustion takes place in scientific stations, ships, aircraft and vehicles.

5 Persistent Contaminants in Snow, Terrestrial Ecosystems and Inland Waters

5.1 Introduction

Antarctica is often considered a pristine environment and it may therefore come as a surprise that in recent years its snow and ice, the cleanest in the world, have become important environmental media for studying the nature and extent of atmospheric contamination in the Southern Hemisphere. Murozumi et al. (1969) provided the impetus for analysing snow and ice-core profiles with the aim of determining to what extent elements or compounds released by human activities become global contaminants. This work provided, for the first time, evidence that lead concentrations in Greenland snow have increased over 100-fold since ancient times. During the 1970s and 1980s, several researchers tried to assess profiles of anthropogenic contaminants in snow and ice from polar and mountainous regions but, due to sample contamination and the insufficient sensitivity of analytical techniques, many results were unreliable. Most of these problems were solved during the last decade, and this chapter will provide a survey of the most recent data on contaminant concentrations in Antarctic snow and in Southern Hemisphere glaciers, with the aim of depicting main source regions and transport pathways. Mechanisms involved in the atmospheric deposition of gases and particles, processes affecting the chemistry of snow and firn, and gaseous exchanges of the snow with the overlying atmosphere will be discussed. Unfortunately, the atmosphere/snow transfer of contaminants, and post-depositional processes affecting their incorporation into firn and ice are complex and largely unknown, thereby making it impossible to derive quantitative estimates of recent and past atmospheric concentrations of chemicals from snow and ice core data.

Less than 2% of the Antarctic continental area is permanently or seasonally free of ice and snow and, like deserts in warmer regions, such areas are mostly characterised by dry kettles, ventifacts, and surface and subsurface salt encrustations. However, owing to the very limited extent of chemical weathering processes, the environmental biogeochemistry of terrestrial and freshwater ecosystems is largely dominated by elements and compounds in snow and atmospheric dry deposition. Most of the ground free of ice and snow is barren but, at sites and microsites where liquid water is available for a few days or weeks in summer, cryptogamic organisms such as algae, lichens and mosses may develop. Lichens and many species of moss are perennial, lack a root apparatus and are largely dependent on atmospheric deposition for their metabolism. These organisms often constitute most of the biomass in terrestrial Antarctic ecosystems, and nutrients stored in their thalli may be a large proportion of the total budget. Cryptogams not only play an important role in the biogeochemical cycle of C and other essential elements but, due to their cation exchange capacity and slow growth rate, they also accumulate persistent atmospheric pollutants to levels well above those in atmospheric deposition. They have therefore been used worldwide as biomonitors of airborne metals, pesticides and radionuclides. Some species have wide geographic ranges, enabling the establishment of large-scale biomonitoring networks in remote regions. As the acquisition and maintenance of automatic instruments for sampling and analysing atmospheric contaminants is very expensive, and as some SCAR countries probably lack funding and/or technical skills to establish and operate sophisticated monitoring networks around scientific stations, this chapter highlights the opportunities for biomonitoring with lichens and/or mosses. These approaches may be useful adjuncts and/or substitutes for automatic devices to assess the deposition pattern of persistent contaminants around scientific stations, or on regional–continental scales (Bargagli 2001).

As discussed in Chapter 2, many lakes and ponds in continental Antarctica are located in poorly developed, rocky catchments, which usually lack outlets. These water bodies are the main sinks for water, solutes and contaminants deposited in the surrounding environment. Although scarcely investigated, the possible role of sediments and algal mats in Antarctic lakes in evaluating environmental contamination will be discussed.

5.2 Atmospheric Contaminant Deposition and Their Incorporation into Ice

Snow and ice are by far the most important components of the Antarctic environment. They affect the hydrologic cycle and energy balances and constitute an immense natural receptacle for atmospheric contaminants. There is little or no melting in the polar ice sheets and snow layers (including water molecules, particles and gases) build up year by year. The physico-chemical analysis of cores drilled in snow and ice at suitable locations yields data profiles which can be used to infer past and recent changes in atmospheric composition. In previous chapters we discussed the role of ice core studies in reconstructing variations in forcing factors and climate change, past volcanism, and atmospheric circulation through the input of marine ions, marine biological methanesulphonic acid (MSA), soil and cosmogenic dust. This chapter considers the deposition of organic contaminants and of particles bearing trace metals and radionuclides. However, depositional and post-depositional processes may affect the distribution of chemicals in snow layers, making it difficult to interpret the results of ice core profile analysis. Thus, before surveying literature data on past and present concentrations of persistent contaminants in the Antarctic environment, it seems opportune to discuss the processes governing their deposition, inclusion and preservation in snow, firn and, ultimately, ice.

Although post-depositional changes are usually less important for aerosol species (Wolff and Bales 1996), organochlorine compounds are not exclusively deposited in the aerosol phase, and important processes affecting the lower troposphere chemistry, such as O_3 concentrations and the rapid oxidation of gas-phase elemental mercury, can only be explained taking into account air–snow interactions. There is evidence, for instance, that due to physical and photochemical processes, snow can be a strong source of several compounds, such as formaldehyde, acetone and hydrogen peroxide, which affect local atmospheric photochemistry (e.g. Dominé and Shepson 2002; Houdier et al. 2002; Jacobi et al. 2002; Hutterli et al. 2003).

5.2.1 Dry, Wet and Occult Deposition in Polar Regions

Persistent atmospheric contaminants are deposited on Antarctic ice sheets through dry, wet (snow) and occult (fog) deposition. These processes are widely described in several books (e.g. Pruppacher and Klett 1980; Pruppacher et al. 1983; Sturges 1991; Schwartz and Slinn 1992; Kaimal and Finningan 1994; Delmas 1995; Wolff and Bales 1996; Baker 1997), and this chapter only provides a brief update of recent findings on main deposition mechanisms in polar regions (Fig. 37).

Wet deposition is often assumed to dominate in the removal of atmospheric contaminants. However, in cold regions with a low precipitation such as continental Antarctica, the continuous transport of particles and gases to the ice-sheet surface (dry deposition) constitutes the main deposition process for a large part of the year. In general, dry deposition is assumed to occur in three steps: (1) aerodynamic transport in the layer (few mm or less) of relatively calm air above the snow surface; (2) transport of particles and gases across the viscous boundary layer to the surface through sedimentation, interception or inertial impaction (gases may adsorb on snow crystals or on a layer of liquid water over crystals); (3) interactions with the snow, which determine whether contaminants remain there or return to the main-

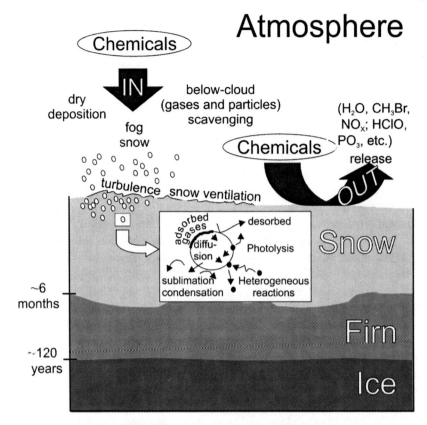

Fig. 37. Physico-chemical processes involved in the scavenging of atmospheric contaminants and air–snow exchanges

stream airflow. As larger particles undergo appreciable sedimentation and are subject to inertial influences, and very small particles have high Brownian diffusivities, those with a diameter of approximately 0.1 μm usually show the lowest deposition velocities (Davidson et al. 1996). However, dry deposition processes are much more complex and are affected by environmental and meteorological parameters. Wind pumping may significantly affect concentrations of chemical species in snow, and can facilitate rapid equilibration with the atmosphere for reversibly deposited species (Waddington et al. 1996). The effects of snow ventilation may be enhanced in environments with irregularities in the snow surface such as sastrugi, which create pressure gradients. The blowing and drifting of snow are ubiquitous in Antarctica, and determine dramatic physical and chemical transformations of snow covers (Pomeroy and Jones 1996). The re-suspension and transport of previously deposited snow and chemicals are associated with processes such as sublimation, which determines an increase in concentrations of conservative chemical species, and others such as volatilisation, which can reduce concentrations of non-conservative chemicals. Temperature profiles over the Greenland ice sheet reveal that the air near the snow surface is occasionally isolated from the free troposphere (Dibb 1996). During these events, dry deposition is expected to be negligible, and it has been hypothesised (Davidson et al. 1996) that the sporadic nature of tropospheric air advection to the snow surface might obscure ice core records.

Snow precipitation is an important cleansing mechanism for almost all airborne contaminants. A number of processes (in and below clouds), including nucleation, impaction, electrical effects, Brownian motion, dissolution and adsorption, are involved in the scavenging of aerosols and trace gases (e.g. Borys et al. 1993; Dixon et al. 1995). The individual contributions of the different processes to overall deposition depend on the nature of contaminants, ambient conditions and the precipitating system. Clouds in the Antarctic atmosphere contain ice crystals, and their nucleation has long been recognised to play a key role in precipitation and in the deposition of atmospheric contaminants. However, progress in elucidating the relative role of different types of ice nucleation is hindered by very complex issues. Nucleation can start from a variety of different nucleating chemical species and may proceed via different pathways; the activity of ice nuclei can be modified by various substrate coatings (Vali 1992; Koop et al. 1999). In the cold Antarctic atmosphere, soils and sea-salt particles, MSA and even many species of aerosols of anthropogenic origin can be effective ice initiators. As they sweep through the cloud, ice nuclei grow by water vapour diffusion at the expense of cloud droplets. Through this accretion process, known as "riming", crystals can achieve a diameter>100 µm (Hobbs and Deepak 1981). Ice crystals then aggregate to form snowflakes, which may be formed by a few or hundreds of ice crystals. As most of the aerosol mass within clouds is held in droplets, rimed snow accumulates much higher concentrations of atmospheric contaminants than unrimed snow (Franz and Eisenreich 1998). Studies by Borys et al. (1993) in Greenland yielded little evidence of riming, even in mid-summer. Riming is probably even less important in Antarctica, because atmospheric concentrations of water vapour are typically very low and liquid droplets in clouds are too small to be scavenged by ice crystals as rim. Besides the scavenging of atmospheric contaminants during ice nucleation and through in-cloud processes, additional particles and trace gases are intercepted by snowflakes falling to the ground surface (below-cloud scavenging). Like in ice crystals and cloud droplets, a viscous water film may cover falling snowflakes, allowing the dissolution of gases and ad/absorption of particles. Scavenged aerosols freeze upon contact with falling snowflakes, and this can determine a loss of contaminants, whereas the presence of organic surface films may enhance their accumulation. Below-cloud scavenging of particles is related to the amount, surface area and shape of crystals, which in turn depend on air temperature and motion, humidity and other meteorological parameters

(Voloshchuk et al. 1973; Podzimek 1999). As a rule, falling snowflakes, stellar plates and dendritic crystals act as filters for atmospheric particles – as they are porous, they allow air to pass, while particles are intercepted. In contrast, the lower porosity and reduced surface area of needles and columns make them less effective as particle scavengers. In general, below-cloud scavenging is less efficient than in-cloud scavenging because, unlike cloud particles, the particles below may be dry and may not experience growth by uptake of water (Davidson et al. 1996). However, by studying below-cloud scavenging of aerosol-bound radionuclides, Sparmacher et al. (1993) found that snow-out coefficients are about five times larger than washout coefficients (i.e. snow is a much more efficient scavenger than rain).

Any wet deposition process other than that due to the precipitation of snow is defined as occult deposition. This deposition in Antarctica probably occurs through fog droplets and during the growth of ice on the sheet surface by repeated hoar frosts (usually in winter and under clear skies; King and Turner 1997). Fog in polar regions consists of liquid droplets of water (super-cooled water, at temperatures well below freezing) or of ice crystals. Concentrations of chemicals in fog often exceed those in snow (e.g. Kalina and Puxbaun 1994), and there is evidence (Dibb 1996) that supercooled water droplets are greatly enriched in all soluble ionic species, with respect to snow falling at Summit (Greenland). Fog plays a prominent role in the deposition of contaminants (including trace metals) associated with sub-micron aerosol particles, through the same processes occurring in clouds and, above all, through direct impaction on snow, due to near-surface eddies.

Most of our knowledge on the transfer of chemicals and on post-deposition processes affecting the chemistry of surface snow derives from experiments performed in Greenland during the last decade, such as those within the framework of PASC (Polar Atmospheric and Snow Chemistry; http://www.igac.unh.edu/activities/pasc.php). Although seasonal and year-to-year variations in meteorological parameters make it difficult to estimate average air–snow fluxes of different chemical species, it has been estimated (e.g. Davidson et al. 1996) that in Greenland wet deposition during summer is the dominant process for all species of chemicals. For many chemical species, dry deposition and fog each contribute approximately 15–20 % to total deposition. These data indicate that all three deposition processes contribute significantly to the annual flux of chemicals to snow, and they should therefore be considered when using ice core chemistry to derive information on past atmospheric concentrations.

So far, most studies on ice core chemistry assume that changes in the chemical composition of different ice strata reflect changes in the composition of the atmosphere (or changes in snow deposition rates), and give only a qualitative picture of these changes. The chemical composition of glacial snow and ice is a proxy for the chemical composition of the overlying atmosphere and, without a better understanding of the processes involved in the transfer of contaminants to snow and firn (unconsolidated snow), there will probably be scarce improvement in the interpretation of ice core data.

5.2.2 Air–Snow Interactions and Post-Depositional Processes

Snow and firn form a porous medium, permeable to atmospheric gases for several metres. The permeability of snow generally increases with depth, with maximum values at depths of a few metres, below which the permeability and diffusion of gases decrease. Meteorological and environmental factors such as surface roughness can favour the ventilation of surface snow (within the top 1–2 m), and interstitial transport due to ventilation can occur at rates higher than those of diffusion (Albert et al. 2002). Investigations performed in recent years show that photochemical processes occurring within firn alter the pore air composition at a temporal scale of minutes to hours, and that the composition of surface and subsurface snow may change within days to months after deposition (Dibb and Jaffrezo 1997). The production of NO_x and HONO within firn (possibly as a result of nitrate ion photolysis; Honrath et al. 2000), and photochemical production of formaldehyde (HCHO), acetaldehyde (CH_3CHO) and acetone (($CH_3)_2CO$; Couch et al. 2000) clearly indicate that snow is chemically active.

After deposition, the snow undergoes progressive metamorphosis (snow-to-firn transition), a temperature gradient and water vapour flux is established, and the size and shape of crystals and, consequently, snow density change. The specific surface area decreases (from about 1,500 to about 400 $cm^2\,g^{-1}$ after 10 days; Cabanes et al. 2002), and this process contributes to the release of adsorbed trace gases. Besides the catalysis of photochemical reactions on crystal surfaces, many other processes occur during the ageing of snow, such as adsorption/desorption, condensation/sublimation, and diffusion of chemical species in the ice lattice (Dominé and Shepson 2002; Fig. 37).

As the release of reactive gases from snow modifies the composition of snow and that of the lower atmosphere, there is growing awareness of air–snow interactions. The importance of these interactions can be exemplified by reactions and fluxes of halogen species, which lead to the depletion of tropospheric O_3 and/or the rapid oxidation of elemental Hg to the gas phase and particulate phase, as discussed in the previous chapter. There is evidence (Swanson et al. 2002) that photochemistry associated with the snow surface environment in Greenland plays an important role in the oxidative capacity of the atmospheric boundary layer through the production of CH_3Br, CH_3I, C_2H_5I, alkenes and alkyl nitrates. A possible biological involvement was also suggested for the production of certain gases. Concentrations of CH_3Br in firn cores collected at the South Pole gradually decrease with depth, and this pattern is consistent with modelled changes in the global atmospheric mixing ratio of CH_3Br over the past century, estimated from increasing industrial emissions and biomass burning (Sturges et al. 2001). In contrast, the same authors found that the mixing ratio of CH_3Br and other trace gases in firn cores from Greenland and Canada increases with depth down to the firn close-off, suggesting a post-depositional change. The enrichment with depth is probably due to physical processes such as gravitational diffusion and temperature gradients. These results indicate that physical exchange processes and heterogeneous photochemical reactions in snow make it difficult to quantitatively interpret ice core data on reactive gas species and to evaluate changes in their concentrations from atmosphere to firn.

Most papers on interactions between atmospheric contaminants and snow focus on non-volatile components which do not partition between condensed and vapour phases after deposition. The study of snow scavenging of atmospheric semi-volatile contaminants such as POPs, which exist in the atmosphere in both gaseous and particulate forms, and of their interactions with surface snow is complex. The current understanding of interactions of organic vapours with snow and ice is probably still inadequate for a reliable assessment of fluxes and the fate of volatile chemicals, especially in Antarctica where POP concentrations are very low. Franz and Eisenreich (1998) collected snow, rain and air samples during winter at a suburban site in Minnesota to investigate atmospheric scavenging of PCBs and PAHs. Snow resulted one order of magnitude more efficient than rain in scavenging particulate matter from the atmosphere, and 80–88 % of S-PCBs and 93–96 % of S-PAHs in snow were associated with particulate matter. Gas scavenging was only important for low-molecular weight PCB congeners and PAHs. These results are only indicative, as the relative contributions of aerosol and gaseous scavenging are most likely time- and event-dependent.

The post-deposition fate of organic chemicals is largely unknown. However, as the chemical-sorbing capacity of snow is largely controlled by the available surface area, concentrations of S-POPs in snow pack probably decrease during snow metamorphosis, when some semi-volatile compounds volatilise from snow and diffuse into the atmosphere. This is corroborated by several field investigations (Hoff et al. 1995, and references therein), and suggests that caution is necessary when using POP concentrations in snow and firn to infer spatio-temporal deposition patterns.

Insoluble particles such as soil dust and volcanic ash hardly interact with gaseous species in the atmosphere. Trace metals can be emitted to the atmosphere as vapour by smelting activities, high-temperature combustion and volcanoes. However, all except Hg soon condense into particulate forms which persist until and after deposition. Almost all studies on aerosol or aerosol-associated chemicals in ice core records implicitly assume that the records provide unbiased data on the aerosol composition of air in which the snow formed. Dibb (1996) reviewed field experiments in Greenland, while investigating relationships between the composition of aerosols in the atmosphere and that of snow. The review shows that processes which remove aerosols and associated species from the atmosphere must determine fractionation between the composition of particles and snow. The larger, mainly crystalline particles are selectively removed by ice nucleation and dry deposition. As riming is thought to be a rare event over Antarctica, the abundant sub-micron ammonium sulphate aerosols, which carry many airborne trace metals, are probably under-represented in Antarctic snow. As fog droplets remove submicron particles and associated chemicals much more efficiently, spatio-temporal variations in meteorological conditions may affect the deposition of some trace metals, and their varying concentrations in ice cores do not necessarily reflect temporal variations in their atmospheric loading. The interpretation of ice core chemistry is further complicated by seasonal variations in snow accumulation over time, which determine significant changes in snow composition, even in the absence of any change in the aerosol concentration pattern over the study area. There is also evidence of significant spatial variations in snow chemistry, especially for aerosol-associated species (Dibb 1996). Thus, even for chemicals associated with insoluble particles, a better knowledge of air–snow exchange processes, and of bias variability in space and time, and how it can be identified in ice cores is needed to reliably estimate quantitative contaminant loading in the overlying atmosphere.

5.3 Snow and Ice Core Records of Airborne Trace Metals

In spite of the abovementioned difficulties, snow and ice cores represent one of the most valuable tools for reconstructing climate history and changes in atmospheric contamination. Other depositional environments such as peat bogs and lake sediments have been widely used for recording past changes in atmospheric aerosol chemistry (e.g. Shotyk et al. 1998; Chillrud et al. 1999; Weiss et al. 1999). These approaches may show limitations in temporal resolution due to physical mixing, low accumulation rates or bioturbation, and the interpretation of data can be complicated by the relative magnitude of other sources with respect to the atmosphere. The skeletons of annually banded corals have also been used to record surface ocean concentrations of metals, which largely reflect atmospheric sources. However, the dynamics of the upper ocean and the residence time of elements in the marine environment filter the atmospheric input signal and generally limit temporal resolution to a couple of years (Sherrell et al. 2000). Other long-term archives of atmospheric metal contamination such as tree rings or herbarium collections have been used; however, with respect to all these environmental matrices, the chemical composition of permanently frozen snow and ice is more directly related to that of the overlying atmosphere. Furthermore, snow and ice cores provide detailed records of atmospheric deposition (including individual precipitation events) over timescales of years to hundreds of millennia. For these reasons, in spite of difficulties in the quantitative evaluation of airborne contaminant concentrations, studies on snow and ice deposits have provided most of our knowledge on the history of climate (see Chap. 2) and atmospheric metal pollution. In addition to studies on snow and ice cores from Greenland and Antarctica to assess hemispheric or global-scale deposition of persistent atmospheric contaminants, during the last two decades research has also been performed on ice cores from the Alps (e.g. Wagenbach 1989) and high-altitude tropical glaciers. Studies in the latter regions developed after the pioneering work of Thompson and co-workers (1985), which showed the importance of data from ice cores collected near sources of moisture and atmospheric contaminants for reconstructing palaeoclimate and atmospheric contamination processes (e.g. Thompson 2000).

5.3.1 Lead as a Paradigm of Hemispheric-Scale Anthropogenic Impact

Concern for the possible effects of trace metal atmospheric pollution on global and/or hemispheric scales was raised when Murozumi et al. (1969) found evidence of extensive Pb contamination in ice cores from Greenland, due to anthropogenic sources in the Northern Hemisphere. This paper reported data from three polar locations, and meteorological and environmental differences among the sites may have affected time-series records; however, the finding remains substantially unchallenged and constitutes a landmark for research on atmosphere pollution. Although extremely low contaminant concentrations (usually pg g^{-1} or less) severely challenged environmental research on polar ice sheets for many years, during the last decade sample collection methods, contamination control and the sensitivity of analytical techniques have improved greatly. Table 2 reports literature data on Pb concentrations in snow deposited during the 1970s and 1980s in Antarctica (Görlach and Boutron 1992; Barbante et al. 1997; Planchon et al. 2001), Greenland (Boutron et al. 1991), the Andes (Correia et al. 2003) and the Alps (Rosman et al. 2000). Although different environmental conditions and large variations in Pb concentrations do not allow reliable comparisons between values measured in different regions, the data in Table 2 clearly show that, during the periods of maximum leaded-gasoline consumption, Pb concentrations in Antarctic snow were much lower than those in snow from all other remote areas of the world.

Historically, most anthropogenic sources of Pb and other airborne metals were located in the Northern Hemisphere, and evidence suggests that the impact of human activities began in ancient Greek and Roman times (e.g. Candelone et al. 1995). The analysis of ice cores and other environmental archives such as peat bogs, sediments or tree rings usually show that in more recent times, the atmospheric deposition of Pb in Europe peaked during the 18th century (from coal burning, ferrous and non-ferrous smelting, and open

Table 2. Range of Pb concentrations (pg g^{-1}) in snow deposited in the 1970s and 1980s in Antarctica, Greenland, the Andes and the Alps

Region	Location	Elevation (m)	Period	Pb range (pg g^{-1})	References
Antarctica	Coats Land	1,420	1980s	0.3–10.2	Planchon et al. (2001)
	Hercules Névé	2,960	1980/1991	1.1–14.3	Barbante et al. (1997)
	Adélie Land	2,000	1970/1998	2.1–11.0	Görlach and Boutron (1992)
Greenland	Summit	3,230	1967–1989	14–93	Boutron et al. (1991)
Andes	Nevado Illimani	6,350	1929–1999	82–1,080	Correia et al. (2003)
Alps	Mont Blanc	4,300	1981–1991	65–5,981	Rosman et al. (2000)

Region	Location
Antarctica	77° 34'S, 25° 22'W
	73° 06'S, 165° 28'E
	68° 00'S, 137° 46'E
Greenland	72° 35'N, 37° 38'W
Andes	16° 37'S, 67° 46'W
Alps	45° 50'N, 6° 48'E

waste incineration during the Industrial Revolution), and between the 1960s and 1980s, mainly due to the combustion of leaded gasoline by vehicles. During the last two decades, concentrations of Pb in Greenland snow have decreased in response to the remarkable reduction in automotive emissions (e.g. Candelone et al. 1995). These authors suggested that a 6.5-fold reduction in atmospheric Pb concentrations occurred between the 1970s and 1992, and concluded that current values are below those recorded at the beginning of the Industrial Revolution. However, high-resolution (subseasonal) studies on recent Greenland snow (Boyle et al. 1994; Cheam et al. 1998; Sherrell et al. 2000) show that, owing to high spatio-temporal variability in the deposition flux, which depends on the meteorological and environmental features of the sampling site, it is difficult to accurately quantify the magnitude of the decrease in Pb deposition. Sherrel et al. (2000), for instance, found that Pb and Cd concentrations in snow samples deposited in the period 1981–1990 at Summit (Greenland) show order-of-magnitude seasonal variability, with maxima in the spring of each year. The seasonal and inter-annual variability is so large as to complicate the assessment of decadal-scale trends and the effective reduction in Pb concentrations resulting from the phasing-out of leaded gasoline. During the 1981–1990 decade, a small decrease in Pb (<5 %) was estimated and no significant trend for Cd was found (Sherrel et al. 2000). In Greenland snow both metals were still dominated by anthropogenic sources, and Pb isotopic ratios ($^{206}Pb/^{207}Pb$ and $^{208}Pb/^{207}Pb$) indicated seasonally distinct source regions – from eastern Europe and the former Soviet Union during spring maxima, and a mixture of US and European sources for the seasonal Pb minima.

Although the interpretation of data from snow and ice cores collected in European glaciers is further complicated by periodic melting or percolation of meltwater, 1993–1996 records of atmospheric Pb deposition and Pb isotopes in snow at Jungfraujoch (Switzerland, about 3,500 m a.s.l.; Döring et al. 1997) indicated that the emission of Pb from traffic had decreased significantly, but was still detectable and in the same range as that from other anthropogenic sources such as waste incineration. Rosman et al. (2000) performed a detailed analysis of Pb concentrations and Pb isotopes in snow deposited at Mont Blanc and found large seasonal variations, especially in winter when a low-altitude inversion establishes in the area. $^{206}Pb/^{207}Pb$ ratios in snow from Mont Blanc and Greenland were considerably different, particularly in the period 1969–1980: while Mont Blanc samples were dominated by Australian Pb used in petrol in the Piedmont region of northwest Italy, those from Greenland were dominated by the highly radiogenic Mississippi valley-type Pb from the USA.

The low Pb concentrations in Antarctic snow make it difficult to carry out analyses and complicate the differentiation between natural and anthropogenic sources. Average annual precipitation on the Antarctic plateau is very low (usually 2–4 g cm^{-2} year^{-1}), with the advantage that a 1-m core in Antarctic ice covers many years; however, this makes it difficult to use stratigraphic methods, because a single severe storm could blow away an entire annual layer (Hammer 1982). As a result, trace metal concentrations in snow and firn samples usually show large intra- and inter-annual variations. In eight snow–ice sections of two well-dated cores from a small, coastal ice cap (Law Dome, Wilkes Land) with high snow-accumulation rates (64–116 g cm^{-2} year^{-1}), Pb and Cd concentrations in winter were two- to fourfold greater than in spring–summer (Hong et al. 1998). A very high metal concentration variability was also found in samples from two snow pits in Coats Land, covering the 1920–1990 period (Planchon et al. 2002b); the highest measured Pb, Cr, Mn and U concentrations, for instance, were approximately 100 times higher than the lowest ones. Thus, while Antarctic ice cores allow the assessment of average Pb concentrations during a climatic cycle, with low values during the Holocene (about 0.4 pg g^{-1}) and relatively high values in the cold terminal stage of the last ice age (about 14 pg g^{-1}; Boutron et al. 1987), metal deposition trends during past decades or centuries are difficult to assess.

5.3.2 Natural and Anthropogenic Inputs of Lead to Antarctic Snow

The main natural sources of trace metals in Antarctica are rock and soil dust, sea-salt spray, volcanic emissions and marine biogenic activity. Even in Antarctic snow from the 1970s (Table 2), concentrations of Pb, the most important anthropogenic metal pollutant during this period and the 19th century, were very low and in the same range or lower than those measured in Antarctic ice from the terminal stage of the last ice age (Boutron and Patterson 1987). Thus, natural sources have always made an important contribution to Pb deposition in Antarctica. Boutron and Patterson (1987) estimated Pb contributions in the Antarctic environment and concluded that, apart from local increases in Pb concentrations due to emissions from scientific stations, the globally significant concentration of Pb in Antarctic snow was about 2 pg g^{-1} (i.e. about five times higher than values measured through most of the Holocene). The corresponding value in Antarctic air was estimated to be about 7 pg Pb m^{-3}: 20 % from soil dust, volcanoes and sea salt, and the remainder from anthropogenic sources in the Southern Hemisphere. Studies of samples from snow pits in Coats Land (Wolff and Suttie 1994) and Victoria Land (Barbante et al. 1997) showed a quick rise in Pb concentrations from the 1950s up to peak values (about 8 pg g^{-1}) in the mid-1970s in Coats Land, and in the mid-1980s in Victoria Land, which were mainly attributed to the consumption of leaded gasoline in Southern Hemisphere countries. After reconstructing temporal trends in annual Pb consumption in these countries, Barbante et al. (1997) hypothesised that Pb concentrations in Coats Land snow (Atlantic sector) began to decrease about 10 years earlier than in Victoria Land snow (Pacific sector) because Brazil adopted alcohol fuels about 10 years before the

introduction of unleaded gasoline in Australia and other countries in Oceania. Figure 38 shows variations in Pb concentrations (period 1940-1991) in
snow samples from Adélie Land, Coats Land and Victoria Land, three sites
located at different altitudes (2,000, 1,420 and 700 m respectively), distances
from the sea (180, 150 and 40 km respectively), and characterised by different
mean snow-accumulation rates (8.0, 5.6 and 27 g H_2O cm^{-2} year^{-1} respectively;
Görlach and Boutron 1992; Barbante et al. 1997; Planchon et al. 2003).

As a rule, anthropogenic contributions to total concentrations of Pb in
Antarctic ice are estimated by calculating the ratio between concentrations in
ice and those of proxies such as Al for rock and soil dust, Na for sea-salt spray,
nssSO$_4^{2-}$ (non-sea salt sulphate) for volcanoes, or enrichment factors (EF
Pb=(Pb/Al)$_{snow}$/(Pb/Al)$_{mean\ Earth\ crust}$). Most studies on Antarctic ice cores
which adopt these rough estimates assume that contributions from rock and
soil dust are almost adequate to account for trace metal in ancient atmospheric depositions. Boutron et al. (1987), for example, attributed the remarkable increase (about 30-fold) in Pb concentrations in ice from the last cold climatic stage to the enhanced flux of rock and soil particles determined by
aridity, strong winds and lower sea levels. In contrast, Matsumoto and Hinkley (2001) sustain that rock and soil dust accounts for only a small percent of
Pb in most pre-industrial ice samples from Taylor Dome (West Antarctica),
and that the Pb deposition rate matches masses, proportions of metals, and Pb
isotopes in worldwide emissions from quiescently degassing volcanoes. There
is also evidence for the impact of atmospheric deposition of Pb from volcanic
eruptions in polar regions. Hong et al. (1996a) reported that the 1783-1784
Laki eruption (Iceland) determined a three-fold increase in Pb concentrations with respect to background values in Greenland ice. A recent study by
Vallelonga et al. (2003) on Law Dome (66.8° S, 112.4° E) ice for the period
1814-1819, which aimed to investigate the possible impact of the Tambora

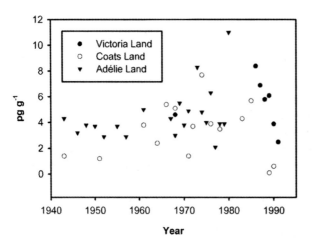

Fig. 38. Temporal varia-
tions (1940-1991) of
average annual concen-
trations (pg g^{-1}) of Pb in
snow samples from
Adélie Land (Görland
and Boutron 1992), Victo-
ria Land (Barbante et al.
1997), and Coats Land
(Planchon et al. 2003)

(8.5° S, 117.4° E, Indonesia) eruption in 1815, yielded rather complex and unexpected results. Only the S emission from the Tambora eruption was recorded, while increased Pb and Bi concentrations in the ice core were primarily due to emissions from Mount Erebus (Ross Island). As discussed in the previous chapter, Mount Erebus emissions have an unusually large metal-to-sulphur ratio, and Vallelonga et al. (2003) found that such emissions do not increase $nssSO_4^{2-}$ concentrations in Law Dome ice. Having found a rather subtle deposition of metals in the continent from active Antarctic volcanoes, the study highlighted the inadequacy of $nssSO_4^{2-}$ as a proxy for volcanic emissions and, in general, our scarce knowledge of the impact of metals from natural sources on Antarctic ice.

One of the most reliable ways to investigate the changing anthropogenic vs. natural origin of Pb is through the analysis of Pb isotopes (Rosman 2001). A 150-year record of Pb isotopes was recently reported by Planchon et al. (2003) in snow and firn samples from Coats Land, dated 1840–1990. Significant intra-annual variations in Pb concentrations (from 0.1–9.3 pg g^{-1}) and isotopic compositions ($^{206}Pb/^{207}Pb$ ratios of 1.096–1.208) were found. The latter variations are likely indicative of changing Pb inputs to Antarctic snow. In pre-1880 samples the average Pb concentration was 0.4 pg g^{-1} (0.1–0.9 pg g^{-1}), and the average $^{206}Pb/^{207}Pb$ ratio was 1.194. By combining Ba concentrations in samples with the mean Pb/Ba ratio in the Earth's crust, the soil and rock dust contribution was estimated to account for about 30% of measured Pb concentrations. The remaining 70% was attributed to volcanic emissions (particularly from active volcanoes in the Antarctic Peninsula), although a shift of isotopic signatures towards less radiogenic ratios could also indicate a small contribution from anthropogenic sources (Planchon et al. 2003). In the following 40 years, Pb concentrations increased (average 2.8 pg g^{-1}), with two maxima, during the early 1900s and the 1920s, showing different radiogenic signatures. These results were interpreted as an indication of early atmospheric contamination in Antarctica, from a mix of local (whaling, sealing and exploration) and remote (above all, non-ferrous metal production in the Southern Hemisphere) anthropogenic sources. The opening of the Panama Canal, economic recession and World War II probably determined a decrease in anthropogenic inputs of Pb to Antarctica; during the 1940s–1950s the average concentration in snow was 1.4 pg g^{-1}. In the following decades up to the early 1980s, there was a significant increase in Pb contamination (Fig. 38), mainly due to the use of leaded gasoline and non-ferrous metal production in South America, South Africa and Oceania. Increasing human activity in Antarctica, especially during the International Geophysical Year and after the establishment of several scientific stations in the mid-1950s, probably contributed to the increase in levels of Pb in the Antarctic atmosphere. However, the analysis of snow samples collected at increasing distances from the scientific stations (e.g. Boutron and Wolff 1989; Suttie and Wolff 1993) indicated that emissions from Antarctic stations probably account for only a limited percentage of metal fallout to the continent. Although Coats Land lies about 3,000 km from Mount Erebus, some radiogenic values measured by Planchon et al. (2003) in a snow sample dated 1950.8 were consistent with a larger proportion of Mount Erebus-type Pb.

In January 1841, during the Ross expedition, this volcano was described as emitting "flame and smoke in great profusion"; it was probably also in an eruptive phase in 1815, as Vallelonga et al. (2003) detected Pb and other metals from Mount Erebus in snow deposited at Law Dome in that year. However, the detection of Pb deposition in the 1950s in Coats Land, which is very far from Mount Erebus, indicates that the impact of the volcano on Antarctic atmosphere and snow chemistry could be greater than previously supposed by several scientists.

5.3.3 Copper, Cadmium and Zinc

Depending upon their bioavailability and toxicity in various environmental compartments, trace metals may determine adverse effects on the environment and human health. During the last 40 years, many books have been published on the behaviour, fate and effects of metals on the environment (e.g. Fergusson 1990; Bargagli 1998; Markert and Friese 2000; Adriano 2001). However, the reliable determination of trace element concentrations in some abiotic compartments of ecosystems, such as atmospheric deposition, freshwater, seawater, snow and ice, challenged analytical chemistry for many years. Powerful analytical techniques, such as laser excited atomic fluorescence spectrometry (LEAFS), differential pulse anodic stripping voltammetry (DPASV), and quadrupole and double-focusing magnetic sector inductively coupled plasma mass spectrometry (ICP-MS), were developed in the last two decades. These techniques may even allow the direct determination of several elements in snow and ice samples, with no need for pre-concentration prior to analysis (e.g. Bolshov et al. 1994; Scarponi et al. 1994; Barbante et al. 2001a; Planchon et al. 2001). The ultrasensitive (in most environmental matrices and for many elements at the sub-picogram per gram level) ICP-MS technique has multi-element capability, with low sample consumption and high daily sample throughput. At present, the main problem in studying the elemental composition of environmental matrices with very low element concentrations is the maintenance of sample integrity before analysis. A rigorous control is necessary during sampling, storage and pre-treatment of specimens in order to prevent contamination and loss of elements. However, suitable procedures have been developed for sampling snow and ice, and decontamination of samples usually consists in the elimination of the outside concentric layers and recovery of the presumably uncontaminated inner core. Through field and laboratory procedures carried out in strict accordance with ultra-clean room protocols, and by using ultra-pure reagents and very sensitive analytical techniques, ice cores from Greenland were used to reconstruct the production history of some metals in the Northern Hemisphere. It was found, for instance, that anthropogenic emissions of trace elements in Europe began as early as 7,000 years ago, and that some metals probably became early hemispheric-scale pollutants already in the Bronze Age (Candelone et al. 1995). Estimated cumulative Pb deposition in the Greenland ice cap (about 400 metric tonnes) at the height of the Greek and Roman civilisations is about one order of magnitude lower than Pb deposition after the post-Industrial Revolution (Candelone et al. 1995). However, in the past 2,500 years the deposition of Cu in Greenland ice (Hong et al. 1996b) has exceeded natural background fluxes, and cumulative ancient atmospheric deposition was probably one order of magnitude greater than that since the Industrial Revolution. This ancient large-scale atmospheric pollution by Cu was ascribed to uncontrolled and wasteful smelting procedures during Roman and medieval times, especially in Europe and China. The quantitative reconstruction of the history of metal production is just one possible application of metal studies in ice core layers. These natural archives of airborne trace metals may also provide a better understanding of spatio-temporal changes in metal emissions from natural sources such as active volcanoes, or of changes in atmospheric transport and deposition processes, in connection with climatic changes. The analysis of recent snow may provide an early indication of large-scale atmospheric contamination by "new pollutants", such the release of Pt group elements from cars equipped with catalytic converters (e.g. Barbante et al. 1999, 2001b), or may be used to assess the effects of measures for reducing atmospheric pollutants such as Pb, Hg and other elements.

Although Antarctica is the remotest continent on Earth, and although its ice and snow are very clean and undoubtedly constitute one of the most important environmental matrices to establish global background concentrations of trace metal deposition, it cannot be excluded that the Antarctic environment is affected by enhanced deposition of other metals besides Pb from anthropogenic sources in the Southern Hemisphere. A recent inventory of global and regional emissions of trace metals to the atmosphere (Pacyna and Pacyna 2001) indicates that, owing to the relatively recent increase in industrial production, Asian and South American countries were among the largest emitters of metals in the mid-1990s. As measures for the abatement of environmental pollution in most of these countries do not always follow industrial growth, their relative contribution to global emissions of trace elements has probably further increased in recent years. South America is the continent most involved in the transport and deposition of trace elements in the Antarctic Peninsula and continental Antarctica and, on a global scale, it is also one of the most important emitters of Cu, Cd, As, Zn and other metals from the smelting of non-ferrous metals (Fig. 39). For instance, Chile, which is quite close to the Antarctic Peninsula, is responsible for nearly 15% of the world production of Cu (Wolff et al. 1999) and is one of the largest emitter countries

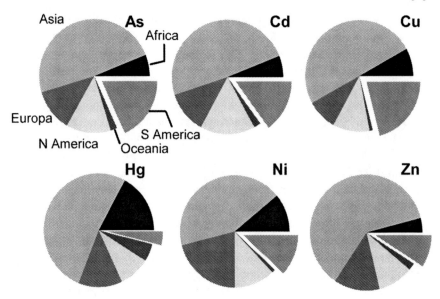

Fig. 39. Estimated emissions of trace elements to the atmosphere in different regions of the world. (Data from Pacyna and Pacyna 2001)

of Cu, As and Cd (Hong et al. 1996 c; Pacyna and Pacyna 2001). Although in these inventories emissions from Chile and other South American countries were estimated on the basis of emission factors and available statistical data on metal production, the estimates were recently corroborated by the results of studies on ice cores from the Andes. Figure 40 reports indicative values of average Cd, Cu and Zn concentrations in snow samples from Antarctica, Greenland, the Alps and the Andes. Values are only indicative, because they were calculated from literature data and refer to sampling sites with different meteorological and environmental conditions, and to snow samples deposited in very different periods: Antarctic data refer to snow samples from Adélie Land (Görlach and Boutron 1992) and Coats Land (Wolff et al. 1999; Planchon et al. 2002b) deposited in the 1920–1990 period; snow samples from central Greenland (about 72° 20'N, 38° 45'W, elevation 3,200–3,300 m; Savarino et al. 1994; Candelone et al. 1996) were collected in the period spring 1989–summer 1992; data from the Alps refer to snow deposited from summer 1960 to winter 1967–1968 on Mont Blanc (Dôme du Goûter, 4,304 m; van de Velde et al. 1998) and to fresh snow samples (February–March 1998) collected in the French Alps between 1,150 and 3,532 m a.s.l. (Veysseyre et al. 2001); finally, Andean values were calculated from those reported by Ferrari et al. (2001) for snow deposited in 1897 in the Sajama ice cap (Bolivia, at 6,452 m a.s.l.), and by Correia et al. (2003) for snow deposited in the period 1919–1999 at Nevado Illimani (eastern Bolivian Andes, at 6,350 m a.s.l.). In

Fig. 40. Average concentrations (pg g^{-1}) of Cd, Cu and Zn in snow and ice from Antarctica, Greenland, the Alps and the Andes (see text for sources of data)

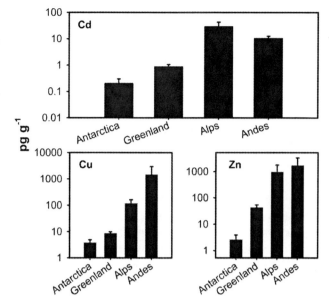

spite of large approximations inherent to the calculation of average metal concentrations in snow from the four continents, Fig. 40 shows that, during the last century, Cu concentrations in the South American atmosphere at altitudes exceeding 6,300 m a.s.l. were significantly higher than those in snow deposited in the 1960s and 1998 in the French Alps, at sampling sites located at much lower altitudes and quite close (often less than 50 km) to urban and industrial areas. On the basis of these considerations, average Cd and Zn concentrations in the Andes also seem to indicate widespread atmospheric pollution by metals in South America.

Antarctic snow has the lowest concentrations of trace metals, but Cd, Cu and Zn concentrations show different patterns with respect to those in samples from Greenland and the Alps. Atmospheric deposition of the three metals in the Northern Hemisphere is dominated by Zn, while in the Southern Hemisphere and especially in Antarctica the deposition of Cu is in the same range or higher than that of Zn. Moreover, while Zn concentrations in Greenland are about 15 times higher than those in Antarctica, this difference is smaller for Cd and Cu (about 4 and 2 times respectively). Thus, data summarised in Fig. 40 suggest that Antarctic snow could be contaminated by atmospheric deposition of Cu and Cd from Chile and other countries in South America. Wolff et al. (1999) estimated that these countries account for about 30 % of the global production of Cu, and for only 10 % of that of Zn, which is smelted in tropical countries. In Coats Land these authors found significantly higher concentrations of Cu in snow deposited in the 1970s and 1980s than in older snow samples; by estimating possible natural inputs of metals, they came to the conclusion that natural sources cannot explain measured concentrations of Cd, and especially not those of Cu. Görlach and Boutron (1992) stated that their data on Adélie Land snow indicated no temporal trends in element concentrations, but a re-examination of their data by Wolff et al. (1999) revealed that Cu concentrations in samples from 1970 onwards were about 2 times higher than those in samples of the previous three decades. Enhanced concentrations of Cu and other metals in snow samples deposited in Coats Land during the late 19th–early 20th centuries have also been reported by Planchon et al. (2002a). This contamination is likely due to anthropogenic emissions of Cu in the Southern Hemisphere, particularly from South America. From 1925 to 1986 the production of Cu in Chile and Peru increased six- to seven-fold; the average content of this metal in snow from Nevado Illimani (Correia et al. 2003) in the 1919–1999 period was three to four times higher than average values for the 1919–1925 period, and about 20 times higher than concentrations measured by Ferrari et al. (2001) in 1897 snow from the Sajama ice cap (about 200 km SW of Nevado Illimani).

Although there is evidence of Cd emissions in South America, no clear temporal trends in Cd concentrations have been observed in Antarctic snow and firn. While studying the elemental composition of surface snow in Queen Maud Land, Ikegawa et al. (1999) measured higher Cd concentrations in samples from the inland plateau than in those from coastal katabatic wind areas, and attributed this difference to the involvement of stratospheric precipitation. Boutron et al. (1993) analysed various sections of deep ice cores from Dome C and Vostok and found highly variable concentrations of Cd, with maximum values occurring during cold climate stages. The Cd content in surface snow from Adélie Land and the South Pole did not indicate recent increases in tropospheric (or stratospheric) inputs of Cd; values were comparable to those of the Holocene, although some samples showed an excess of Cd with respect to the estimated contribution from rock and soil dust, sea salt and volcanic emissions. Wolff et al. (1999) in Coats Land firn pointed out a weak but significant relationship between concentrations of Cd and Zn, but neither metal concentration increased significantly between 1923 and 1986. When analysing dated snow samples (1834–1990) from the same area, Planchon et al. (2002a) found no clear temporal trends in Cd concentrations, and enrichment factors (with respect to the Earth's crust) in recent decades resulted similar to those in older samples.

Available data seem to indicate that there are important natural sources of Cd in continental Antarctica which "mask" eventual inputs from remote anthropogenic sources. In addition to volcanoes and soil particles, marine aerosol can also play an important role in the transfer of Cd to Antarctica. Bubbles bursting through the sea surface can concentrate metals with respect to seawater. Although the enrichment factor of 10^4 for Cd in sea spray, adopted by Boutron et al. (1993), is probably unrealistic (Matsumoto and Hinkley 2001; Planchon et al. 2002a), it is well known that high concentrations of Cd occur in waters and organisms of the Southern Ocean (Westerlund and Öhman 1991b; Bargagli et al. 1996b). There is also evidence of decreasing concentrations of Cd in snow samples from the Filchner Ice Shelf with increasing distance from the ice edge, and of extremely high enrichment factors of the metal compared to the abundance of Cd in the Earth's crust (Heumann 1993). Southern Ocean bacteria can methylate Cd, and 15–30% of the total metal may occur as $MeCd^+$ (Pongratz and Heumann 1999; Heumann 2001). Biomethylation may play a significant role in the transfer of Cd to the continent. In contrast to Hg, which forms volatile permethylated Me_2Hg, metals such as Cd and Pb are transformed by bacteria into ionic methylated compounds ($MeCd^+$, Me_2Pb^{2+} and Me_3Pb^+). These compounds are more soluble in water and, because they have both hydrophobic (methyl group) and hydrophilic (positive charge) properties, they are efficiently enriched at the water/air interface and in sea-spray aerosols. However, quantitative estimates of Cd inputs to the continent from the marine environment are lacking.

Although studies from the 1980s and 1990s report no Zn inputs from anthropogenic sources in Antarctic snow, and although data summarised in Fig. 40 suggest a reduced impact of Zn compared to that of Cu and Cd, Planchon et al. (2002a) found increased enrichment factors for Zn and other elements in Coats Land snow samples from the second half of the 20th century.

5.3.4 Mercury and Other Trace Metals

Using the ICP-MS technique, Ikegawa et al. (1997) determined the concentrations of 36 elements in drifting snow samples collected from July to December 1991 at Asuka Station (71° 32'S, 24° 08'E). They found peak concentrations of most elements and $nssSO_4^{2-}$ in late September to early October, and hypothesised that volcanic eruptions on Mt. Pinatubo (June 1991) and Mt. Hudson (August 1991) may have been responsible for enhanced Pb, Cd, Cu, Zn, Se and $nssSO_4^2$ concentrations in the 1991 Antarctic spring. The analysis of enrichment factors suggested that Na, Mg, Ca, K and Sr were of marine origin, while Al, Fe, Mn, Rb, Cr, Ni, Ga, V and all rare earth elements were of crustal origin. A subsequent study (Ikegawa et al. 1999) on 37 element concentrations in surface snow samples collected between 1991 and 1993 during an over-snow traverse in East Queen Maud Land showed an increased fallout flux for elements such as Co, Ni and Cd at altitudes above 2,500–3,000 m a.s.l. It was suggested that the distribution of elements could reflect polar stratospheric precipitation or long-range tropospheric transport from the Southern Hemisphere. With the aim of providing an indicative picture of the elemental composition of recent Antarctic snow, Table 3 summarises data on concentrations of rare earth elements (Ikegawa et al. 1997) and of many other elements (Barbante et al. 1999; Planchon et al. 2002a). Average values measured by Veysseyre et al. (2001) in fresh snow collected at different altitudes in the French Alps are also

Table 3. Elemental composition (pg g^{-1}) of recent snow from Antarctica and the French Alps

Element	Antarctica 1991–1993 (Ikegawa et al. 1999)	Antarctica 1989–1990 (Barbante et al. 1999; Planchon et al. 2002a)	French Alps 1999 (Veysseyre et al. 2001)
Ag		0.14	0.88±0.28
Al		130,000	29,112±28,000
Au			0.76±0.04
B			498±116
Ba		1.2	232±215
Bi		0.052	6.0±6.5
Cd		0.14	26±18
Ce	1.6±0.8		
Co			4.4±2.4
Cr		1.6	
Cu		2.4	79±86
Dy	0.08±0.007		
Er	0.041±0.035		
Eu	0.021±0.002		
Fe			4,879±4,512
Gd	0.09±0.07		
La	0.86±0.45		
Li			229±10
Mn		4.5	413±442
Mo			19±7
Nd	0.69±0.034		
Pb		1.6	472±579
Pd		0.54±0.12[a]	2.6±0.9
Pr	0.23±0.021		
Pt		0.37±0.18[a]	0.50±0.04
Rh		0.04±0.01[a]	
Sb			34±27
Sm	0.11±0.008		
Sn			215±32
Th	0.68±0.063		
Ti			71±87
U		0.016	5.1±7.7
V		0.4	36±35
Yb	0.097±0.190		
Zn		0.8	340±247

[a] Data from Barbante et al. (1999)

reported to point out the different magnitude of element concentrations in remote regions and in ones impacted by human activities. These data show that only Pt, Pd and Ag concentrations in Antarctic and European snow are in the same range. Indications of atmospheric contamination by Ag and other elements such as Cr, Zn, Bi and U (in addition to the above-reported Pb and Cu contamination) were detected by Planchon et al. (2002a) in Coats Land snow deposited during the 1970s–1980s. Ag, Cr, Pb and U concentrations began to decrease at the end of the 1980s. It was suggested that natural sources, such as multiple eruptions in the Deception Island volcano, and anthropogenic activities in the Southern Hemisphere, such as increased U mining and milling operations, or mining and smelting of non-ferrous metals, contributed to the temporary increase in metal deposition. Although production of other elements such as Mn and V also increased in the same period, their enrichment factors did not significantly increase in Antarctic snow. These elements, like lithophilic Al, Ba, Fe and Ti, are rather abundant in airborne rock and soil dust and, as in the case of Cd, their natural inputs are probably so large as to mask anthropogenic ones.

As discussed in the previous chapter, the atmospheric transport of Hg and its rapid oxidation and deposition in polar regions after spring sunrise is a critical contamination issue. Nevertheless, there are only few investigations on the dynamics of Hg in polar snow, and it is still unknown whether polar regions are important sinks in the global cycle of Hg. Vandal et al. (1993) measured Hg concentrations in 14 sections of an ice core from Dome C (77° 39'S, 124° 10'E, 3,240 m a.s.l.), covering the past 40,000 years. They found that the Hg content was less than 1 pg g^{-1} 34,000 years B.P.; it increased to 2.1 pg g^{-1} and remained elevated during the last glacial maximum (from 28,000 to 18,000 years B.P.), decreased during the transition from the last ice age to the Holocene (17,000–13,000 years B.P.), and reached concentrations<0.5 pg g^{-1} during the Holocene. The emission of gaseous Hg from productive ocean regions was considered to be the main source of metal to Antarctic snow, and the threefold increase in Hg deposition during the last glacial maximum was mainly attributed to climatic conditions and enhanced primary productivity in the Southern Ocean. There is evidence (Petit et al. 1981) that aridity, higher winds and stronger poleward atmospheric transport enhanced the accumulation of aerosols in Antarctic ice. An increased input of Fe and other essential lithophilic elements in the Southern Ocean probably enhanced marine productivity in the Southern Ocean; although mechanisms are poorly known, there is evidence that marine organisms can promote the reduction of reactive Hg species to volatile Hg° (Mason et al. 1994). Research by Heumann (2001) shows that pure cultures of psychrotrophic bacteria (i.e. having a minimum growth temperature in the range 0–5 °C) from the Southern Ocean can methylate inorganic Hg^{2+}, further contributing to the transfer of this metal from the ocean to the atmosphere. Considering the rapid equilibrium between surface oceans and the atmosphere, and the scarce sedimentation of

Hg in oceans, it seems probable that Antarctica and the other continents are the main sink for atmospheric Hg released by oceans. In samples of surface snow collected at the South Pole and other sites at different distances from the sea, Vandal et al. (1993) detected concentrations in the 0.13–0.50 pg g^{-1} range, with lower values occurring in samples from inland sites. Sheppard et al. (1991) reported values in the 0.1–1.5 pg g^{-1} range and no temporal trend in firn samples from Southern Victoria Land.

5.3.5 Persistent Organic Contaminants

The first two papers reporting the presence of DDT compounds in Antarctica were published in 1966. George and Frear (1966) failed to detect DDTs in snow and water samples collected around McMurdo Station, but they found the chemicals in Antarctic organisms at higher levels of the marine food web, such as marine mammals and seabirds. In the same year, Staden et al. (1966) reported the presence of DDT residues in Adélie penguins and a crabeater seal. The first record of DDT in Antarctic snow was published by Peterle (1969), although a subsequent review of absolute amounts and ratios of DDT compounds revealed that, due to the probable contamination of samples, the reported values were too high (Peel 1975; Risebrough 1977). However, the analysis of snow samples collected in austral summer 1975 from a pit on a permanent snowfield at Doumer Island (64° 51'S, 63° 35'W) confirmed the presence of DDT residues and PCBs (Risebrough et al. 1976). Average concentrations in surface snow of pp'-DDT (about 0.5 pg g^{-1}), pp'-DDE (about 0.1 pg g^{-1}) and total PCBs (about 0.15 pg g^{-1}) increased in samples collected at 2–4 m depth (up to 4, 0.27 and 1.2 pg g^{-1} respectively), and then decreased (at 5.5–6 m depth, the measured values were 2.1, 0.21 and 0.28 pg g^{-1} respectively). The distribution pattern of pp'-DDT and pp'-DDE was similar, and the quantitative ratio between the two compounds was approximately 10 along the profile. Although the samples were not dated, these results seem to indicate that enhanced deposition of DDTs and PCBs probably occurred in the decade before 1975. In the Canadian Arctic region (Agassiz Ice Cap, Ellesmere Island; Gregor et al. 1995), for instance, small amounts of PCBs were first evident in ice from 1957 to 1963, and deposition increased thereafter.

Besides the analysis of surface snow, Risebrough et al. (1976) also determined DDT and PCB concentrations in penguin eggs from the Antarctic Peninsula, and in the blubber of a leopard seal killed by a killer whale; they came to the conclusion that the atmosphere is the main pathway for DDT and PCB transport to Antarctica. However, owing to the presence of the small Chilean Station of Yelcho on Doumer Island, and given the frequent burning in the past of waste materials at the nearby and larger Palmer Station (on Anvers Island, 64° 46'S, 64° 03'W), it was supposed that a portion of PCB residues also derived from local sources.

Tanabe et al. (1983a) analysed PCBs and chlorinated hydrocarbons such as DDTs and HCHs in snow samples collected close to Syowa Station. In contrast to Risebrough and co-workers, they found no significant variations in total DDTs, HCHs or PCBs between samples of surface and deep snow. This study suggests that the deposition rate of pesticides (never used in Antarctica) and industrial compounds such as PCBs (these have been used in the region) has remained the same since the 1960s.

More recently, Fuoco and Ceccarini (2001) have determined concentrations of selected PCB congeners in samples of recent snow and firn, collected during the austral summer 1993–1994 in seven snowfields throughout northern Victoria Land. The pattern of PCB isomers in snow (Fig. 41) showed a predominance of lower chlorinated congeners and, although in this study PCB-18 was not considered, the pattern in Fig. 41 roughly reflects that reported for atmospheric PCBs at Signy Island and King George Island (Fig. 36). The numerous studies performed on PCB congener patterns in Arctic air and snow show that their distribution is very similar, with lighter tri-chlorinated PCB homologues largely dominating the atmosphere and snow of northern sites (Macdonald et al. 2000). The sampling sites of surface snow throughout northern Victoria Land were located at different altitudes (from sea level to 3,000 m a.s.l.) and at varying distances from the sea, but total PCB concentrations measured by Fuoco and Ceccarini (2001) showed no significant spatial variations (range 0.28–0.73 pg g^{-1}; mean=0.52 pg g^{-1}). Samples from a 2.5-m-deep pit at the Hercules Névé collected in summer 1993–1994 and 1994–1995 showed higher total PCB concentrations (about 1 pg g^{-1}) in the deepest samples (presumably deposited in 1987) than in surface snow (about 0.65 pg g^{-1}). This result seems to corroborate previous findings by Risebrough et al. (1976), and agrees with the general decreasing trend in POP concentrations in the atmosphere of Antarctica and the sub-Antarctic islands during the 1980s and 1990s (Fig. 35).

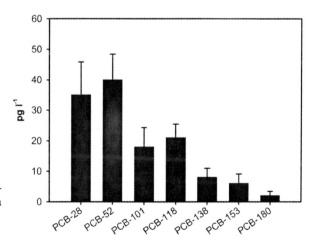

Fig. 41. Mean concentrations (pg l^{-1}) of PCB isomers in samples of snow and firn from seven snowfields in northern Victoria Land. (Data from Fuoco and Ceccarini 2001)

Estimates of spatio-temporal variations in the rate of POP atmospheric deposition based on snow and firn data may contain large uncertainties due to meteorological conditions – which make it very difficult to select sampling sites representative of local atmospheric deposition, especially in terrains with blowing and drifting snow – and to the possible post-depositional release of some contaminants. As deeper snow retains its burden of more volatile POPs such as HCH better than shallow snow, sampling sites in gullies or hummocks can produce significantly different results. Moreover, under the same conditions of atmospheric contamination, the scavenging of atmospheric POPs may be enhanced by lower temperatures. For instance, higher average concentrations of PCBs and DDTs have been measured in Arctic snowfall than in Lake Superior (Macdonald et al. 2000). Temperature variations may also affect the post-depositional loss of more volatile compounds, and this process may be accentuated under warmer climatic conditions.

Despite the difficulty in making reliable comparisons between POP concentrations in snow from regions with different climatic and environmental conditions, Table 4 reports the results of research performed on snow from Antarctica and the Arctic (Macdonald et al. 2000; Melnikov et al. 2003), Alps and Pyrenees (Carrera et al. 2001). Comparisons among the data in Table 4 are complicated not only by the different sampling years and very different climatic and environmental conditions, but also by the different approaches adopted for calculating the total amount of PCBs and HCHs (e.g. sum of dif-

Table 4. Ranges of POP concentrations (pg g^{-1}) in snow from Antarctica, the Arctic, Alps and Pyrenees

Region		Period	ΣDDT	ΣHCH	ΣPCB	Reference
Antarctica	Doumer Island	1975	0.5–4.3	–	0.03–1.2	Risebrough et al. (1976)
	Syowa Station	1982	0.16–1.0[a]	1.5–4.9[a]	0.16–1.0[a]	Tanabe et al. (1983a)
	Northern Victoria Land	1994	–	–	0.28–0.73	Fuoco and Ceccarini (2001)
Arctic	Canada	1991–1992	<0.01–0.42	0.02–3.7	0.9–13	Macdonald et al. (2000)
	Ob-Yenisey River watershed	1992–1993	0.5–0.7	1.2–3.4	0.4–0.6	Melnikov et al. (2003)
Europe	Alps and Pyrenees	1997–1998	n.d.–330	0.49–1.1	0.22–2.2	Carrera et al. (2001)

[a] Concentrations expressed as pg dm^{-3}; n.d., not detectable

ferent number of congeners; quantified as Aroclor mixtures; sum of α-HCH and γ-HCH). However, in the first half of the 1990s, average PCB concentrations in Victoria Land snow (about 0.5 pg g^{-1}) corresponded to the mean value measured by Melnikov et al. (2003) in snow from the Ob-Yenisey River watershed, and were about 4 times lower than the average 4.1 pg g^{-1} reported by Macdonald et al. (2000) for Canadian Arctic snow.

5.4 Monitoring of Persistent Contaminants Around Scientific Stations Through Snow

In 1991, with the adoption of the Madrid Protocol to the Antarctic Treaty, promoting international scientific cooperation and environmental protection, Antarctica finally emerged from the geopolitics of the Cold War period, which had focused on territorial claims and the possible exploitation of natural resources. These changes were accompanied by a growing number of nations acceding to the Antarctic Treaty and, consequently, by a growing number of new scientific stations. In the 1980s only 23 nations had acceded to the Antarctic Treaty but, between 1981 and 1991, they became 42 and their number has further increased during the last decade. Currently, more than 30 stations (south of 60° S) operate during the Antarctic winter, and a much larger number operate during the austral summer. Data on total presence in terms of person-days in Antarctica are lacking or vary from year to year, but at least 4,000–5,000 science and support personnel work each year in Antarctica. Thus, although the Protocol on Environmental Protection to the Antarctic Treaty has the objective of "comprehensive protection of the Antarctic environment and dependent and associated ecosystems", and recognises "its value as an area for the conduct of scientific research, in particular research essential to understanding the global environment", scientific activities and supporting logistics accidentally or incidentally determine significant environmental impact on a local scale.

Snow in Antarctica has a determining influence on the deposition of contaminants, and constitutes the main environmental matrix to monitor changes in the composition of the atmosphere. Scientists studying the chemistry of snow and ice to understand global or hemispheric contamination processes are deeply concerned by the possible impact of chemicals released within Antarctica itself. To evaluate whether metals released by human activities in Antarctica can compromise studies on atmospheric contamination by metals in the Southern Hemisphere, Boutron and Wolff (1989) made a comprehensive inventory of Cd, Cu, Pb, Zn and S emissions from the burning of fossil fuels and waste south of 60° S. Assuming a uniform deposition of elements, a mean snow-accumulation rate of 156 kg m^{-2} year^{-1}, and using available data on element concentrations in Antarctic snow, they estimated the fallout fluxes of contaminants. Comparisons between estimates of emissions and deposition fluxes in Antarctica showed that the impact of gasoline (particularly aviation gasoline) accounted for up to 19% of Pb fallout. According to Boutron and Wolff (1989), anthropogenic emissions of Pb within Antarctica in the second half of the last century were higher than natural fallout fluxes throughout the Holocene. They concluded that, of the studied elements, only Pb concentrations in snow over vast areas were affected by aircraft and other manmade emissions. This impact has been significantly reduced in recent years by the progressive phasing-out of leaded gasoline for aircraft and vehicles.

In order to evaluate the impact of combustion processes in Antarctica on different spatial scales, Suttie and Wolff (1993) determined Cd, Cu, Pb and Zn concentrations in samples of surface snow collected at increasing distances from an isolated (leaded petrol) running generator, and from Halley 4 and Halley 5 research stations (Brunt Ice Shelf). It was estimated that about 40% of the total emitted Pb from the isolated generator was deposited on the surrounding snow surface, while the remainder was retained in the atmosphere. As for the four considered metals, the influence of the generator could not be detected in snow beyond 40 m downwind. The interpretation of data from snow samples collected on two traverses departing from the two stations was more difficult; however, the influence of the stations could not be detected beyond 10 km, with possibly longer tailing effects on the downwind side. While Pb, Cu and Zn concentrations fell significantly with increasing distance from Halley 4, those of Cd showed little variation along the whole traverse (300 km inland). Again, these results seem to corroborate the previous hypothesis that natural inputs of Cd in Antarctica from volcanoes and the marine environment largely prevail and probably mask any anthropogenic inputs.

In general, almost all studies on the presence of anthropogenic contaminants around Antarctic stations reveal a small and highly localised impact. Warren and Clarke (1990) measured soot in air and surface snow at Amundsen-Scott Station (South Pole) to evaluate the impact from the burning of diesel fuel. Soot became undetectable at a 2-km distance, except in the downwind direction. Stenberg et al. (1998) measured concentrations of major ions and light-absorbing materials in snow samples collected near the Swedish Station Wasa and the Fennish Station Aboa (73° 03′S, 13° 25′W), at the beginning (December) and end (February) of two summer expeditions. After the periods of occupation, the concentrations of ions and light-absorbing materials in surface snow increased, probably as a consequence of natural and anthropogenic inputs. However, no significant increase in concentrations was observed 1.5 km upwind or 3 km downwind of the stations.

Besides inorganic particulates and ions, human activities in Antarctica release a number of persistent organic atmospheric contaminants. To ascertain whether Palmer Station (Anvers Island) was a significant source of PCBs, in February 1975 Risebrough et al. (1976) collected four samples of surface snow at distances of 0.5–1.5 km from the station. They found high particulate contents, principally soot, and concentrations of total PCBs ranging from 4 to 10 pg g^{-1} (i.e. 8–20 times higher than those measured in snow from the same region at Doumer Island; Table 4). PAHs are probably among the commonest contaminants near Antarctic research stations, because they are released as by-products of combustion and through hydrocarbon fuel spillage (UNEP 2002a). Although data on Antarctic snow are lacking, these compounds are reliable tracers of combustion, and their concentrations can be usefully com-pared with those of other tracers (e.g. soot or ammonium ion) to evaluate the impact of local or remote combustion sources (Masclet et al. 2000). Many data are available on PAHs in Arctic snow and ice (for a review, see Macdonald et al. 2000). Time series of typical anthropogenic compounds such as fluorene or phenanthrene in the Agassiz Ice Cap reveal peak concentrations in the 1950s and 1960s, which then decrease to relatively constant, but still elevated levels in the1980s and 1990s (mean flux=11±6 µg m^{-2} year^{-1}; Peters et al. 1995). As the global use of fossil fuels has increased more than twofold during the last two decades, the relatively constant concentrations of PAHs in Arctic snow are rather surprising. As suggested by Macdonald et al. (2000), concentrations measured in Arctic snow are probably not simply related to total combustion but also reflect changes in fossil fuel use, such as the progressive substitution of coal with liquid fuels.

5.5 Contaminants in Antarctic Soils

Despite local and microscale variations, most Antarctic soils are dry and coarse in texture, with low contents of silt, clay and organic matter. Soil hori-zons are poorly developed and the boundary of the active layer (the zone from 1 to several decimetres subject to diurnal and annual freeze–thaw cycles) is sometimes difficult to recognise, because interstitial moisture may be insuffi-cient for cementation (dry permafrost). Considerable amounts of water-solu-ble salts accumulate linearly with time on the soil surface, under surface stones or in discrete horizons dispersed throughout the soil profile. Atmos-pheric deposition is the main source of ions, and salt composition largely reflects that of seawater, especially in coastal ice-free areas (e.g. Claridge and Campbell 1977; Bockheim 1997; Bargagli et al. 2001).

As a rule, the mobility of trace metals and other contaminants in soils is reduced by the presence of organic materials and clay. Although Antarctic soils contain negligible amounts of organic matter and clay, soluble and insol-uble chemicals move very slowly due to the low moisture content (often <1 %). For instance, by analysing Pb, Zn and Cu concentrations in soil profiles from sites at Marble Point and Scott Base (southern Victoria Land) polluted by human activities (crushed batteries, scattered rubbish, and buildings) about 30 years ago, it was found that in this period Pb moved about 50 cm vertically from the source (Claridge et al. 1995). Lead and Zn were mainly associated with carbonates and oxides, and ionic forms of the metals had accumulated at the interface between the loose soil and the underlying ice-cemented material.

The construction of Scott Base began in 1957; since then the base has accommodated between 10 and 100 people each winter and summer respectively. A detailed study on water and 0.01 M HNO_3 leachates from soil samples collected around the station showed appreciable contamination by Ag, Cd, Cu, Pb and Zn, especially at sites where materials such as oils and fuels, chemicals or refuse have been dumped or stored, or in areas more affected by emissions from the incinerator, generator and vehicles. Surface water flows, water percolation through soils, wind and earthmoving operations redistributed contaminants around the station. At times of high water flow, such as during snowmelt in early summer, or in disturbed soil with melting permafrost and ice, some metals such as Zn had moved through the soil profiles and into drainage areas; Ag had probably moved to the shoreline and into the marine environment (Sheppard et al. 2000).

Very high average concentrations of Ag, As, Cd, Cr, Cu, Pb and Zn (13, 62, 22, 140, 2,100, 1,600 and 3,600 $\mu g \ g^{-1}$ dry wt. respectively) were measured by Kennicutt et al. (1995) in a couple of abandoned dump sites at Palmer and Old Palmer stations (Anvers Island, Antarctic Peninsula). Nevertheless, compared to sediments from remote areas, subtidal sediments collected near the two stations showed little evidence of metal contamination.

McMurdo Station (USA), on the southern tip of Ross Island, was constructed in 1955–1956 and since then, it has been the largest, permanent year-round station in Antarctica, with a population ranging from 250 to 1,200 people. Prior to major changes in station management in the 1980s, wastes were dumped and open-burned (NSF 1991). These practices have since ceased, and efforts have been made to assess the extent of environmental contamination and to remediate contaminated sites. Although less numerous than those in coastal marine environments, several studies have determined levels of trace metals in soils (e.g. AECOM 1992; ANL 1992; COMNAP-AEON 2001). In order to determine background concentrations of trace metals in the McMurdo area, Crockett (1998) analysed soil from control areas near the station, using two different extraction procedures: heated nitric/perchloric acid, followed by concentrated hydrofluoric acid, and hot 1:1 nitric acid with additions of 30 % hydrogen peroxide. The results obtained with the latter USEPA method are summarised in Table 5, and compared with reference values in surface soils from granitic rocks in the McMurdo Dry Valleys (Webster et al. 2003) and from granitic and volcanic rocks in northern Victoria Land (Bargagli et al. 1995, 1999). Although comparable chemical extraction procedures were used in these studies, different grain-size fractions of soil samples were selected for analytical determinations. As element concentrations in soils increase with

Table 5. Typical concentrations of elements (mean±SD; μg g^{-1} dry wt.; Al and Fe, %) in surface soils from Victoria Land (hot HNO$_3$ extraction)

Site	Soil	Ag	Al	As	Ba	Cd	Cr	Cu	Fe	Hg	Mn	Ni	Pb	Se	Zn
McMurdo Station (77° 50'S, 166° 40'E)[a]	Sandy grey basalt	<0.20	2.5±0.3	1.1±0.3	109±19	<0.62	16±4		3.2±0.5	<0.05	-	57±19	2.2±0.4	<0.35	47±8
	Coarse red scoria	<0.20	1.9±0.4	1.9±0.4	145±41	<0.72	5.0±1.0		2.2±0.4	<0.05	-	49±19	1.3±0.3	<0.34	29±5
Northern Victoria Land (from 73–74° S, 164–168° E)[b]	Volcanic (<250 μm)	-	4.2±2.3	4.2±2.3	310±187	0.13±0.05	38±19	21±8	3.1±1.2	0.04±0.02	903±487	12±9	8.8±3.5	-	84±27
	Granitic (<250 μm)	-	3.8±1.8	3.8±1.8	549±222	0.22±0.08	24±14	16±9	2.6±1.5	0.05±0.03	429±222	10±6	11±6	-	66±25
Ex–Vanda Station (77° 32'S, 161° 32'E)[c]	Granitic (<250 μm)	-		-	-	0.02±0.01		26±5	0.9±0.3	-	110±28	12±2	2.3±0.3	-	19±6

[a] Data extracted from Crockett (1988)
[b] Data extracted from Bargagli et al. (1995, 1999a)
[c] Data extracted from Webster et al. (2003)

decreasing grain size, the different characteristics of samples (sandy grey basalts and red coarse scoria) analysed by Crockett (1998) probably contributed to the different element concentrations in volcanic soils from McMurdo (Table 5). Comparisons between the same grain-size fraction (<250 μm) of volcanic and granitic soils from northern Victoria Land show that, with the exception of Ba, Cd and Pb, volcanic materials contain similar or higher contents of elements, with a pattern reflecting that usually reported for total concentrations of elements in soil-forming basaltic and granitic rocks (Adriano 2001).

New Zealand's mainland station on the shore of Lake Vanda (McMurdo Dry Valleys) was used for summer field activities from 1968–1993. It was removed in the summer of 1993–1994, when rising lake levels threatened to flood the site (Waterhouse 1997). Some contaminated soils were removed, but in 1995 lake waters encroached the zone disturbed by the removal of the station and soil. Although the lake level then stabilised (Hawes 2001), it cannot be excluded that further flooding of contaminated areas will occur in the future. To assess the potential for adverse environmental effects in Lake Vanda, Webster et al. (2003) studied residual metal contamination at the site. The soils derived from granitoid bedrocks had low moisture contents (0.5–3.9%) which increased with depth up to the ice-cemented layer (25–65 cm depth). With respect to metal concentrations in samples from control sites (Table 5), those in soils at all depths from a gully which had received domestic wash waters and from a site near the former wet-chemistry laboratory had higher Ag, Cd, Pb and Zn concentrations (from 2 to 20 times the average background value). Despite the fact that Cu, Ni and Co concentrations in soil profiles were not elevated with respect to background values, their contents, together with those of phosphate, Pb and Zn, increased significantly in suprapermafrost fluids. The results of this study indicate that some contaminants could be selectively leached from flooded soils if the lake level were to rise. However, Webster and co-workers concluded that, owing to the small size of the contaminated area and the low level of contamination, the adverse effects on Lake Vanda would be minimal, such as the phosphate-enhanced growth of cyanobacteria.

Most of Antarctica is covered by ice, and the use of soil samples to estimate the deposition of artificial radionuclides on the continent has a limited significance. However, several radio-ecological surveys have been performed in Antarctic ice-free areas (e.g. Triulzi et al. 1990; Roos et al. 1994; Tubertini et al. 1995; Jia et al. 1999), and a mean concentration of 2.2 Bq kg^{-1} was calculated for ^{137}Cs in soils at the 60–70° S latitude band (Godoy et al. 1998). Based on this content, the ^{137}Cs inventory for the 60–70° S latitude band was estimated (decay-corrected for 1992) to be 143 Bq m^{-2}.

Fuel oil spills are one of the most widespread sources of environmental contamination near Antarctic stations and in remote zones for refuelling of aircraft and vehicles. A number of remediation strategies have been considered, and bioremediaton through autochthonous bacterial flora from Antarctic soils has often been proposed as the only viable option which can be implemented on a large scale (e.g. Snape et al. 2001; Ferguson et al. 2003; Ruberto et al. 2003). In soils where oil spills have occurred, toxic PAHs may accumulate in very high concentrations. In a sample collected near Old Palmer Station, Kennicutt et al. (1992) measured a concentration of 345,000 ng g^{-1} dry wt. In contaminated soil samples from an area near the Scott base for the storage of drums of engine oil, PAH concentrations ranged from 41 to 8.105 ng g^{-1} dry wt. (Aislabie et al. 1999). As reported in other surveys performed in Antarctic soils, sample concentrations of naphthalene and other PAHs indicative of spilled fuel always exceed those of pyrogenic PAHs (from high-temperature combustion); nevertheless, 11 of the 16 PAHs on the USEPA Priority Pollutant List were detected. Hydrocarbons had migrated to lower depths, and the highest concentrations occurred in 2–10 cm deep subsurface soils. Mazzera et al. (1999) surveyed PAH concentrations in surface soils at McMurdo Station during peak summer activity. The maximum values were measured for naphthalene, acenaphthalene, acenaphthylene and fluoranthene (27,000±2,600, 17,800±1,270, 15,700±6,300 and 13,300±430 ng g^{-1} dry wt. respectively), especially in soils from unpaved roadbeds near the gasoline pumps for seawater intake. This polluted site was a potential source of the high levels of PAHs found in sediments in McMurdo Sound by Kennicutt et al. (1995). However, available data on Antarctic soils show that PAH pollution is highly localised, and concentrations in samples from control areas are usually below detection limits. In the environment around Davis Station, Green and Nichols (1995) found very low concentrations, and the content of individual PAHs around a fuel depot only exceeded 1 ng g^{-1} dry wt.

When deposited on soils, PAHs may have a number of possible fates such as volatilisation, photooxidation, leaching or microbial degradation; it therefore seems likely that the risk of possible adverse effects on functional properties of Antarctic soils are negligible, except at sites directly affected by spillage of fuels.

In contrast to PAHs, a large percentage of PCBs deposited on Antarctic soils are produced in other continents of the Southern Hemisphere. Thus, PCBs do not usually accumulate at localised sites near scientific stations, but show a rather homogenous distribution over large areas. Fuoco et al. (1996) measured an average concentration of 60±38 pg g^{-1} dry wt. in soil samples collected near Baia Terra Nova Station and throughout northern Victoria Land. As previously discussed for trace metals in soil, the variability of PCB data was probably due to the different grain size of samples. Trace metals and PCBs are generally adsorbed on the surface of soil particles, and 1 g of clay or silt develops a surface area much larger than that of 1 g of coarse sand; for reliable comparisons among concentrations of contaminants measured in different soils, analysed particles should have the same grain size, or contaminant concentrations must be expressed in relation to the unit of surface area rather than the unit of mass (Bargagli et al. 1988). Fuoco et al. (1996) analysed the particle size of soil samples and calculated the specific surface area (expressed in square metre of surface per cubic centimetre of dry soil). PCB concentrations normalised to the specific surface area showed very low variation coefficients, indicating a quite homogenous deposition of PCBs throughout northern Victoria Land.

5.6 Cryptogamic Organisms as Biomonitors of Atmospheric Contaminants

Although lichens and mosses are completely unrelated groups of cryptogamic organisms, they have a number of features in common. They both lack root systems, have a high cation-exchange capacity (CEC), and depend largely on atmospheric deposition for their nutrient supply. Mosses and lichens therefore have elemental compositions which reflect in an integrated way gaseous, dissolved and particulate elements in atmospheric deposition (Bargagli 1998). Owing to their ability to tolerate long periods of desiccation and extreme conditions of light and temperature, lichens and mosses are the principal component of terrestrial flora in many ecosystems of continental Antarctica. Although scarcely considered by many Antarctic environmental managers, these organisms can play a very important role as biomonitors of persistent contaminant deposition around scientific stations and in field camps.

5.6.1 Accumulation of Persistent Contaminants in Antarctic Lichens

In almost all climatic regions, lichens are among the first colonisers of exposed rocks and other substrata, except snow and ice. The ecological success of these symbiotic associations of fungi and green or blue-green algae in extremely cold, dry and nutrient-poor habitats of Antarctic is the result of their ability to tolerate long periods of desiccation and to uptake available water and essential elements over the entire thallus surface, by rapid passive processes. Cations are passively bound to anionic sites on the cell wall or outer surface of the plasma membrane, and may also enter and accumulate in mycobiont and photobiont cells through slower and more selective mechanisms (Bargagli and Mikhailova 2002). Besides gaseous and soluble elements and compounds, lichens can also trap airborne particles in the loose hyphal weft of the medulla (Garty et al. 1979). Lichens are perennial and, because of their slow growth rate, persistent atmospheric pollutants accumulate in the thalli to levels well above those of atmospheric deposition. These symbiotic organisms are used worldwide as biomonitors of atmospheric deposition of trace metals, pesticides and radionuclides (e.g. Smith and Clark 1986; Bargagli 1989, 1998; Muir et al. 1993; Nimis et al. 2002). Some species of macrolichens have wide geographic ranges and enable the establishment of large-scale bio-monitoring networks, especially in remote regions. Many lichen surveys have been performed in the Arctic (e.g. Nash and Gries 1995; Chiarenzelli et al. 1997; Kelly and Gobas 2001); these studies generally indicate that, unlike arti-ficial radionuclides and POPs, it is difficult to assess trace metal deposition from remote anthropogenic sources with lichens, due to the contribution of natural sources. Considering the remoteness of Antarctica, the most wide-spread species of macrolichens such as *Umbilicaria decussata* and *Usnea antarctica* can probably be used to detect atmospheric deposition of radionu-clides, and of several POPs from other continents in the Southern Hemi-sphere. An important application is in monitoring the deposition of myco-phytotoxic compounds, trace metals, PCBs and PAHs around scientific stations.

The elemental composition of Antarctic lichens shows that the marine environment is the main source of major and trace elements, both directly, through the deposition of aerosols and the melting of snow, and indirectly, through the solubilisation of salt encrustations on rock and soil surfaces (Bargagli et al. 2003). In contrast to other metals, Hg mostly occurs in the atmosphere in gaseous forms and, from several large-scale biomonitoring surveys performed in Italian regions with anthropogenic and/or natural sources of Hg (active volcanoes, geothermal fields and mineral deposits), lichens appear to be the most reliable biomonitors of atmospheric Hg (Bar-gagli 1990; Bargagli and Barghigiani 1991). Thus, owing to the inconsistency of the little available data on Hg concentrations in Antarctica in the early 1990s, a survey was performed on the distribution of Hg in surface soils and macrolichens of northern Victoria Land (Bargagli et al. 1993). Results were rather surprising, because Hg concentrations in surface soils were among the lowest ever recorded in terrestrial ecosystems (mean 0.034 ± 0.023 μg g^{-1} dry wt.), while average concentrations in lichens (0.386 ± 0.190 and 0.344 ± 0.204 μg g^{-1} dry wt. in *U. antarctica* and *U. decussata* respectively) were higher than in samples of related species from other remote areas, and corresponded to average values measured in several urban and industrial areas of the Northern Hemisphere. As discussed in the previous chapter, the recent finding of photochemically driven oxidation of boundary-layer Hg° after polar sunrise, which determines a rapid deposition of oxidised Hg species in snow during the polar spring, probably contributes to the accu-mulation of the metal in lichen thalli. Obviously, as suggested 10 years ago, ecophysiological and environmental factors, such the very slow growth rate of Antarctic lichens (concentrations are expressed in μg per g of lichen, and are therefore affected by the growth rate of each sample specimen), and active volcanoes can also contribute to concentrations of Hg and other trace metals in Victoria Land lichens.

Predicted changes in the Antarctic climate, atmospheric deposition and environmental biogeochemistry will probably at first affect the growth rate and concentrations of major and trace elements in lichen thalli. Considering that temporal variations in the elemental composition of lichens can be used as an early warning system to detect climatic and environmental changes in Antarctic terrestrial ecosystems, concentrations of major and trace elements were determined in samples of *U. decussata* collected at the same sites, throughout northern Victoria Land, in January 1989 and 1999 (Bargagli et al. 1998b, 2000). Statistically significant variations were not detected, and overall average values from the two surveys are summarised in Table 6 and compared with those measured in lichens of the same genus from reference areas in Europe (Seaward et al. 1981) and the North West Territories (Canada, Chiarenzelli et al. 1997). A study on other lichen species from northern Quebec (Canada; Crête et al. 1992) reported average concentrations of 0.17, 0.09 and 4.1 μg g^{-1} dry wt. for Cd, Hg and Pb respectively, and 378 Bq kg^{-1} dry wt. for ^{137}Cs. Although it is difficult to compare between element concentrations measured in different samples (even between those belonging to the same species and collected in the same region; Bargagli 1995), due to differences in analytical procedures, lichen growth rates and amounts of rock and soil particles adsorbed to samples, the main difference between baseline concentrations of trace metals in lichens from both hemispheres is a much lower Pb content in Antarctic samples. Concentrations of essential (e.g. Cu, Zn, Mn) and non-essential (e.g. Cd and Hg) elements in Antarctic lichens are comparable to those in samples from the Northern Hemisphere. These comparisons indicate that, in contrast to Antarctic snow, lichens probably cannot give reliable indications on metal deposition in Antarctica from anthropogenic sources in the Southern Hemisphere. However, several small-scale surveys performed around Antarctic scientific stations show that these organisms are reliable biomonitors of metal deposition from local sources. Samples belonging to the genus *Usnea*, for instance, were used to evaluate the impact of the construction of a crushed-rock airstrip at Rothera Point (BAS 1989), and to assess the impact of human activities at King George and Livingston islands in the areas occupied by Polish, Brazilian, Russian, Chilean, Spanish, Bulgarian and Korean stations (Poblet et al. 1997; Olech et al. 1998; Hong et al. 1999; Yurukova and Ganeva 1999).

Owing to their slow growth rates and long life cycles (probably of up to some 100 years), lichens behave as long-term integrators of persistent atmospheric pollutants; the analysis of 50- to 60-year-old specimens may yield information about total radioactive, DDT and other POP deposition since the beginning of nuclear tests, and about the large-scale production and use of pesticides. As a rule, concentrations of local and long-range transported contaminants in lichen thalli are much higher than those in the atmosphere or snow, and this makes analytical determinations easier and cheaper. The average content of HCB, HCHs, DDTs and PCBs in lichens and mosses collected in 1985 and 1988

Table 6. Baseline concentrations (mean±SD; µg g^{-1} dry wt.) of major and trace elements in *Umbilicaria decussata* from Northern Victoria Land compared with literature data for lichens of the same genus collected in reference areas of the Northern Hemisphere

Region	Species	Al	Ca	Cd	Cr	Cu	Fe	Hg	K	Mg	Mn	Na	P	Pb	Zn
Antarctica, Victoria Land[a]	*U. decussata*	727±515	517±324	0.18±0.10	1.6±0.8	5.3±3.8	812±536	0.39±0.27	2,170±829	508±415	19±7	201±118	865±459	0.65±0.41	20±6
SW Poland[b]	*U. cylindrica*	-	-	-	4.4±0.8	7.6±1.1	1,372±438	-	-	-	19±6	-	-	34±12	39±5
SE Ireland	*U. cylindrica*	-	-	-	3.6	7.7	799	-	-	-	31	-	-	41.6	63
NW Canada[c]	*U. polyphylla*	-	-	0.19±0.05	3.0±0.9	5.2±3.1	-	-	-	-	-	-	-	10.4±5.6	22.2±3.4

[a] Data extracted from Bargagli et al. (2000)
[b] Data extracted from Seaward et al. (1991)
[c] Data extracted from Chiarenzelli et al. (1997)

in the Antarctic Peninsula and northern Victoria Land ranged from 0.2–9.9 ng g^{-1} dry wt. (Bacci et al. 1986; Focardi et al. 1991; i.e. three orders of magnitude higher than those in Antarctic snow; Table 4). The mean content of ^{137}Cs in samples of *U. antarctica* from King George Island (22.5±11.7 Bq kg^{-1} dry wt.) was about seven times higher than that in surface soils (3.3±2.1 Bq kg^{-1} dry wt.) from the same sites (Godoy et al. 1998). As lichens are temporal integrators of persistent atmospheric contaminants, and unlike expensive automatic monitoring networks, biomonitoring does not require repeated sampling to achieve significant information on atmospheric contamination. In areas where indigenous cryptogams are lacking or available species of lichens and mosses are unsuitable for biomonitoring (e.g. crustose lichens or moss with very short turfs), short-term "active" monitoring can be performed by transplants (i.e. the moss or lichen bag technique; Bargagli 1998).

5.6.2 The Potential Role of Bryophytes as Biomonitors

Antarctic bryophytic vegetation is often dominated by small acrocarpus mosses belonging to the genus *Bryum*, which form flat, short turfs. These mosses have an ephemeral life strategy and produce vegetative propagules capable of rapid development on unstable and wet substrata; under dry conditions the moss turfs form plate crusts, which are readily detached and redistributed by winds. Owing to their short lifetime and scarce biomass interspersed with large amounts of soil particles, these mosses cannot be used to investigate the relative composition of dry, wet and occult atmospheric deposition. However, in sheltered and more stable substrata, which receive enough water from melting snow during summer, several moss species tend to have a persistent life strategy and may survive for many years. These mosses can form large stands (up to some square metres), and cushions may be up to several centimetres thick, sometimes showing a thin layer of peat. Some cosmopolitan or circumpolar species were first investigated in Sweden (Rühling and Tyler 1968), and there is evidence that these organisms can be very useful for large-scale biomonitoring surveys.

As a rule, surveys for trace metal deposition are based on the analysis of total element concentrations in unwashed mosses (cryptogam samples are not washed so as not to remove soluble intercellular elements; Brown 1984). The high surface-to-mass ratio of mosses provides a highly effective trap for airborne particles, and this property is very useful in studying long-range atmospheric transport and deposition of pollen and spores in Antarctica (e.g. Linskens et al. 1993). However, the same property complicates the interpretation of data from trace element biomonitoring surveys. In dry and barren Antarctic terrestrial ecosystems, "raw" concentrations of elements in moss samples often reflect the biogeochemical nature of soils and rocks rather than atmospheric input of elements. As a rule, levels of Al, Fe, Cr, Fe and other lithophilic elements are higher in Antarctic mosses than in samples of related species from forest ecosystems in the Northern Hemisphere. By comparing regional background values in the literature, Bargagli et al. (1995) found that concentrations of many elements increase from Amazonian or temperate rainforests to agricultural or barren and desert areas, independently of the adopted moss species or analytical procedure. The more evident exceptions were Cd and Hg (elements with a low crustal abundance) and sometimes Pb, which arises mostly from anthropogenic sources. For a more reliable comparison between the elemental composition of mosses from sites with different climatic and environmental conditions, and to establish reliable background concentrations, it was necessary to minimise the effects of soil contamination. This was done (e.g. Bargagli et al. 1995) by assuming that the lowest concentration of Al measured in Antarctic mosses (about 250 μg g^{-1} dry wt.) was the non-particulate fraction of the element. By ascribing the excess of Al above this value to entrapped soil particles, the amount of soil contaminating each sample was then evaluated (taking soil samples at the same site, and digesting and analysing them following the same procedure used for mosses). More realistic background concentrations of trace metals in Antarctic mosses were subsequently estimated by subtracting the amount attributed to entrapped soil particles from the total measured concentration. Table 7 reports total concentrations of trace elements in the Antarctic moss *Bryum pseudotriquetrum* from northern Victoria Land, together with estimated background concentrations (assuming 250 μg g^{-1} dry wt. of Al as a standard for the non-particulate fraction). Baseline total concentrations of trace elements in samples of *Hylocomium splendens* from Alaska (Wiersma et al. 1986) and Greenland (Pilegaard 1987) are reported for comparison. The amount of soil particles in mosses from Alaska were probably slightly higher than in Antarctic samples, while the average Fe concentration in Greenland samples seems to indicate a low element contribution from soils and rocks. If we bear in mind the different input of lithophilic elements, the data summarised in Table 7 show that concentrations of most trace elements in Antarctic mosses are in the same range as those measured in the Arctic. As for Antarctic lichens, the main difference is a significantly lower Pb content (the most widespread metal pollutant in the Northern Hemisphere). Concentrations of Hg and Cd in *B. pseudotriquetrum* are in the same range or slightly higher than those in mosses from control areas in Europe (Bargagli et al. 1998 c). Through a detailed study of the accumulation of these metals in Antarctic mosses growing along nutrient and moisture gradients in coastal ice-free areas, it was found that the marine environment is the main source of Cd. Concentrations of this metal and those of P increased in samples collected near a beach with a penguin rookery. Besides direct deposition through marine aerosol, it was found that seabird excrements increased the environmental bioavailability of Cd and P. On the contrary, Pb and Hg contributions from guano were negligible. The moss *Campyloplus pyriformis* growing on the warm fumarolic

Table 7. Total concentrations and estimated background concentrations ($\mu g\,g^{-1}$ dry wt.) of trace elements in Antarctic mosses compared with data for mosses collected in other remote areas

Region	Species	Al	Ba	Cd	Cr	Cu	Fe	Hg	Mn	Ni	Pb	Zn
Northern Victoria Land[a]	Bryum pseudotriquetrum	5,100	138	0.10	8.2	12	3,980	0.12	268	5.5	1.9	68
	Estimated background	250	47	0.09	2.3	7.1	205	0.12	134	2.5	1.1	35
Alaska[b]	Hylocomium splendens	4,300	200		20	15	4,900		530	9.0	6.8	65
Greenland[c]	Hylocomium splendens	–	–	0.15	3.8	–	1,450	0.09	–	–	6.8	47

[a] Data extracted from Bargagli et al. (1995)
[b] Data extracted from Wiersma et al. (1986)
[c] Data extracted from Pilegaard (1987)

ground of the Mount Melbourne crater had a rather high Hg concentration (1.52 μg g^{-1} dry wt.), thus sustaining the hypothesis that active volcanoes and fumaroles contribute to the enhanced accumulation of Hg in Victoria Land cryptogamic organisms.

As in the case of lichens, the average concentration of ^{137}C in mosses from King George Island (23±14 Bq kg^{-1}) was much higher than that in surface soils (Godoy et al. 1998). During summer, sublimating snow and high evaporation determine deposition of soluble salts on moss and lichen surfaces. Electrical conductivity measurements (Bargagli et al. 1999) in moss-supporting soils (protorankers) show that during summer salts tend to migrate and concentrate in mosses, which have a much higher CEC and evaporation surface than soil. In addition to major and trace elements, radionuclides may also be transferred from soil to mosses, as there is evidence of their mobilisation from radioactive particles deposited on soils (Salbu et al. 1998).

5.7 Anthropogenic Impact on Lakes and Streams

Continental Antarctica does not have stream–river drainage systems, and ephemeral streams and ponds only exist in coastal regions in summer. Most of the lakes have no outlets, and water gained by surface stream inflow or groundwater is mainly lost through evaporation or sublimation of surface ice (in perennial ice-covered lakes). Antarctic lakes are thus the main sinks for water and solutes from the surrounding environment. Depending on their origin, the presence of a perennial ice cover, exposed rocks and soils in the watershed, seabirds and distance from the sea, the water may show very different characteristics – from almost distilled to salt-rich brine which does not freeze in winter (e.g. Don Juan Pond in the upper Wright Valley). Despite 40 years of study, there is still disagreement over the origin and evolution of major and trace elements in Antarctic lakes. Marine aerosols, snow precipitation, and leaching of soils, rocks and widespread soluble salt encrustations are among the most likely sources of solutes. Solar radiation, wind, snow precipitation, and frequent freeze–thaw cycles determine large daily, seasonal and inter-annual variability in the amount and composition of Antarctic inland waters. Despite this variability, waters, sediments and the biota are natural integrators of soluble elements and compounds deposited in the watershed, and these matrices may play a very important role as indicators of the impact of local human activities and/or of long-term changes in climatic conditions and biogeochemical processes.

Although there has been some effort in recent years to undertake impact assessment studies in the vicinity of Antarctic stations (e.g. COMNAP-AEON 2001), very few monitoring surveys have been performed on freshwater ecosystems.

The previously reported data on trace metals and nutrient concentrations in soils from the shore of Lake Vanda, formerly occupied by a field research station and now at risk of water inundation, is one of the few examples of environmental impact assessment and cooperative efforts between scientists and environmental managers to minimise the possible effects of pollutants released during 26 years of human presence. Lake Vanda is a closed basin (about 5.6 km long and 1.5 km wide) receiving water from the Onyx River; it is permanently ice-covered and the 68-m water column has been stratified for some 1,200 years (Wilson 1964). The water above about 45 m is cool, well mixed and rich in dissolved oxygen – one of the world's clearest and most oligotrophic. Below 45 m, chlorinity and temperature increase abruptly and, near the bottom, the water becomes anoxic, remarkably tepid (25 °C) and three times saltier than seawater. Concentrations of major and trace elements in the water column show low values in the upper layer and maxima at depth (e.g. Masuda et al. 1982; Canfield et al. 1995; Green et al. 1998; Table 8).

Density, pH and redox conditions in the water column, together with the solution chemistry of elements such as the Mn oxide phase, are usually assumed to play a prominent role in the behaviour of trace elements in Antarctic lakes. Like Lake Vanda, Shield Lake, a coastal saline lake in the Vestfold Hills, shows a vertical distribution of Mn and other trace elements closely related to oxic and anoxic (below 20 m) conditions (Masuda et al. 1988). Concentrations of elements in unfiltered waters from Lake Hoare (lower Taylor Valley; McMurdo Dry Valleys; Green et al. 1986) were lower than those in Lake Vanda and Shield Lake. Moreover, the vertical distribution of trace metals in Lake Hoare showed two distinct patterns: Mn, Fe, Co and Ni concentrations increased in the suboxic and anoxic region below 24 m, whereas Cd, Cu, Pb and Zn tended to remain constant or to decrease below 24 m.

Comparisons between values measured in filtered (0.45 µm; Table 8) surface water samples from Lake Hoare (except Pb and Cd) are in the same range of average values in more than 100 samples of filtered meltwater, stream water and surface lake water from 45 different sites in northern Victoria Land (Cremisini et al. 1990). In general, these values are intermediate with respect to typical background concentrations reported in the literature for freshwater and seawater (e.g. Förstner and Wittmann 1983; Table 8).

Concentrations of dissolved Hg in waters from lakes and streams in the McMurdo Dry Valleys ranged from 0.10–0.44 ng l^{-1} (Lyons et al. 1999), i.e. values rather similar to that in Antarctic snow. Vandal et al. (1998) studied the chemical speciation of Hg in Lake Hoare and found that the percentage of methylmercury with respect to the total Hg content was about 10 % at depths of 4–20 m, and about 30 % below 22 m. This finding led Vandal and co-workers to suppose "in-situ" methylation of Hg at the oxic–anoxic boundary, within the sediment or at the sediment–water interface, followed by diffusion into the overlying waters of Lake Hoare.

Table 8. Average concentrations (μg g^{-1} dry wt.) of trace elements in filtered (0.45 μm) and unfiltered samples of water from Antarctic lakes and streams, compared with global average background values for freshwater and seawater (* unfiltered samples)

Site	Depth range (cm)	Al	Cd	Co	Cu	Fe	Mn	Ni	Pb	Zn
Lake Vanda*[a]	4–45	8.0±1.5	-	0.2±0.1	25±9	16±6	3.0±0.7	10±6	-	36±15
	50–64	10±2	-	1.1±0.7	631±340	179±169	1,300±900	73±62	-	400±19
Lake Shield*[b]	1–16	87±26	-	1.3±0.4	9.0±3.2	207±88	76±38	-	-	25±9
	20–37	100±26	-	2.4±0.7	11±9	664±239	131±84	-	-	54±29
Lake Hoare*[c]	3–15	-	-	-	-	10.4±5.7	0.7±0.2	2.0±0.5	0.8±1.1	0.5±0.5
	18–29	-	-	-	-	285±385	112±152	3.0±2.2	0.6±0.8	4±0.2
Lake Hoare[c]	3–15	-	0.6±0.2	0.04±0.01	1.5±0.4	5.4±4.4	0.7±0.2	1.5±1.2	1.7±1.5	<0.3
	18–29	-	0.9±0.1	0.06±0.04	0.8±0.6	185±141	106±141	2.3±2.0	0.2±0.1	0.2±0.1
Northern Victoria Land (streams, lakes and meltwaters)[d]	Surface waters	-	0.008±0.002	0.05±0.01	0.3±0.2	6.5±1.3	3.1±0.6	0.35±0.06	0.06±0.001	0.6±0.1
Freshwater[e]		2	-	0.05	0.5	4.0	4	0.4	0.02	0.5
Seawater[e]		1	-	0.04	0.1	1.3	0.2	0.2	0.005	0.01

[a] Data extracted from Masuda et al. (1982)
[b] Data extracted from Masuda et al. (1988)
[c] Data extracted from Green et al. (1986)
[d] Data extracted from Cremisini et al. (1990)
[e] Data extracted from Förstner amd Wittmann (1983)

Human impacts on lake systems in the Larsermann Hills region were discussed by Burgess and Kaup (1997). In this region Ellis-Evans et al. (1997) found remarkable changes in microbial communities in a lake located near a scientific station, built only four years earlier. Although concentrations of persistent contaminants in these and other Antarctic lakes affected by human activities have not yet been determined, there is evidence that lake sediments in northern Victoria Land accumulate higher concentrations of PCBs and ^{137}Cs than soil or marine sediments (Triulzi et al. 1990; Fuoco et al. 1996). Moreover, shallow waters in most Antarctic lakes and ponds are characterised by benthic algal mats which, together with sediments, could provide valuable material for assessing environmental contamination and temporal changes in the biogeochemistry of trace metals in Antarctic lakes and their watersheds (Bargagli 2001).

5.8 Summary

Compared to the rest of the world, the Antarctic environment is less affected by anthropogenic contaminants, but it is not pristine. The deposition of artificial radionuclides on the continent was first detected in the 1950s, and that of pesticides a few years later. Although the first reports of DDT contamination in Antarctica were questioned in the belief that the residues reached the continent on ships and planes serving research stations, in the early 1970s it became clear that other continents of the Southern Hemisphere were the source of pesticides and other persistent POPs. Antarctic snow constitutes an immense, natural receptacle for atmospheric contaminants and, with respect to other depositional environments such as peat bogs or lake sediments, permanently frozen snow and ice have a chemical composition more directly related to that of the overlying atmosphere and also contain more detailed records, including individual precipitation events. However, the atmosphere/snow transfer of contaminants and post-depositional processes affecting their incorporation into snow are complex and still largely unknown. There is evidence that snow is chemically active and produces a number of chemical compounds which affect local atmospheric photochemistry. A better knowledge of these processes and the post-depositional fate of contaminants are a prerequisite for reliable quantitative estimates of chemical concentrations in recent and past atmospheres based on contaminant concentrations measured in snow and ice. Despite these limitations, the recent introduction of suitable procedures to prevent contamination of samples, and of analytical techniques for the direct determination of chemicals in snow at the sub-picogram per gram level has promoted a number of studies in Antarctica and the Arctic, Alps and Andes. These studies aim to reconstruct pathways of airborne contaminants and temporal changes in their deposition pattern. Research performed in Antarctic snow and firn during the last decade shows that, especially in the period between the 1950s and 1980s, the continent received ^{137}Cs and other artificial radionuclides, DDTs, PCBs, HCHs, and some metals such as Pb, Cu and, probably, Ag, Cd, U, Cr, Bi and Zn, derived from human activities in the Southern Hemisphere. A large amount of airborne metals was probably released in South America, even if the relative metal contribution from natural sources such as volcanoes, rock and soil particles, and the marine environment is still uncertain, and for some metals it may be greater than previously supposed.

Environmental contamination from local human activities in Antarctica may compromise studies which aim to assess biogeochemical cycles of elements or large-scale processes in the transport and deposition of persistent atmospheric contaminants. In general, snow data show that the impact of human activities is relatively small and highly localised, usually confined to within a few kilometres or few hundred metres of human settlements. The main exception was probably the release of Pb by aircraft, and other man-made emissions from the 1960s to the 1980s; this impact has been significantly reduced in recent years through the progressive phasing-out of leaded gasoline.

Many Antarctic scientific stations are located in coastal ice-free areas where soils, lichens and mosses are among the most suitable environmental matrices to assess the impact of human activities. Environmental impact assessment is among the actions required under the Protocol on Environmental Protection to the Antarctic Treaty. In the past, wastes were often dumped or open-burned close to stations; through the analysis of soils, the nature and extent of pollution in some degraded Antarctic sites was detected and measures to remediate local environmental pollution were undertaken.

At present, trace metals and PAHs from combustion processes and fuel and oil spill are among the most widespread contaminants around scientific stations. In general, most contaminants have a low mobility in dry Antarctic soils; however, climate change may affect soil leaching and drainage processes. Antarctic organisms can play an important role in the assessment of anthropogenic impacts. Changes in the biodiversity of indigenous flora and fauna, for instance, can be valuable indicators in Antarctic ice-free areas, and some species of Antarctic macrolichens and mosses, which are long-term integrators of persistent atmospheric pollutants, can be used as reliable biomonitors of atmospheric deposition of trace metals, PCBs and other chemicals. These organisms could be used to establish long-term biomonitoring networks based on the analysis of indigenous cryptogams or of mosses and lichens exposed near stations in small nylon bags.

Although this chapter has discussed environmental monitoring assessment and recovery around several scientific stations, many other abandoned stations and dumping sites exist throughout Antarctica, not only in coastal ice-free areas but also on ice. As for the latter sites, concern arises from the fact that small spills of drilling fluids and all forms of waste, debris and contaminants from human activities on the Antarctic plateau will remain in the ice for thousands of years and will be gradually transported towards the coast within the ice sheet.

6 Contaminants in Antarctic Seawater and Sediments

6.1 Introduction

Oceanographic research usually involves repeat sampling at least at monthly intervals. As access to and prolonged stays in the Southern Ocean are difficult due to its remoteness and the ice cover, compared to other oceans, very few data are available on its water and sediment chemistry. Most available data refer to samples collected during summer surveys in the Weddell and Ross Seas. Such data show that iron, manganese and other elements essential to phytoplankton metabolism may here occur in lower concentrations than in any other sea. As there are no rivers in Antarctica, the input of soluble ions from the continent is negligible; furthermore, many elements with prevailing terrestrial sources are highly reactive in seawater and are easily scavenged by particulate matter. These features, together with the water circulation pattern, contribute to the formation in the Southern Ocean of the world's largest areas of high nutrient, low chlorophyll (HNLC) waters. The fact that a large proportion of nutrients in Southern Ocean surface waters are returned to depth unused suggests that the Southern Ocean may be pivotal in partitioning CO_2 between the atmosphere and ocean. At present, some numerical simulations suggest that if the daily input of nutrients were fully utilised by Antarctic phytoplankton, then atmospheric CO_2 concentrations could be reduced by an amount roughly corresponding to glacial–interglacial CO_2 variations during the last four glacial cycles. However, some circulation and mixing models predict that the net effect of a more efficient Southern Ocean biological pump alone cannot explain the lower CO_2 concentrations during ice ages. Ongoing scientific debate highlights the need to gain a better understanding of major and trace element cycling, and of other factors limiting primary productivity in the Southern Ocean. The effective role of this ocean in global biogeochemical cycles is largely unknown, and linking the physico-chemical and biological environments of the Southern Ocean to those of other oceans and to global processes will be a major research objective for the next decades.

This chapter summarises the current state of knowledge on sources, levels, internal recycling, and sedimentary deposition and burial of trace elements and persistent organic contaminants in the Southern Ocean. The main sources of elements and chemical compounds to this ocean are the atmosphere, especially volcanic gases and aerosols, and hydrothermal vents in the oceanic crust. Elements are mainly lost through biological processes such as the formation of siliceous frustules or calcareous shells by organisms, through the adsorption of ions to clay and organic particles, and through the formation of sea spray and of new minerals in sediments. Seasonal formation and melting of sea ice, and the presence of several polynyas and of areas of enhanced upwelling determine the relative instability of the water column. The distribution of elements in the Southern Ocean is therefore usually less "structured" than in oceans at lower latitudes. Thus, as discussed in the previous chapter for Antarctic snow and ice cores, most literature data on trace element concentrations in the Southern Ocean have limited spatio-temporal significance, and comparisons with the composition of waters or sediments in other seas are only indicative.

The impact of trace elements from local or remote anthropogenic sources can hardly be detected in Southern Ocean waters or sediments. However, a better knowledge of their distribution and cycling is important to evaluate the possible role of phytoplankton in the global CO_2 budget, and to understand why in this remote and pristine ocean environment many organisms, particularly crustaceans, seabirds and marine mammals, accumulate very high concentrations of potentially toxic metals such as Cd and Hg. As Antarctic marine species have probably evolved in an environment characterised by enhanced bioavailability of potentially toxic metals, it seems unlikely that they can be adversely affected by chronic exposure to naturally high metal concentrations. However, the physico-chemical characteristics of Southern Ocean waters vary greatly in space and time, and the detoxification mechanisms of different species of marine vertebrates and invertebrates are extremely varied. This chapter will therefore discuss the biogeochemical cycles of Cd, Hg and other metals in different regions of the Southern Ocean, and possible processes responsible for their enhanced bioavailability.

On a global scale, coastal marine areas are among the ecosystems most affected by the impact of nutrients, trace metals and POPs released by human activities. Some Antarctic coastal ecosystems near scientific stations are contaminated by wastewaters, dumping of solid wastes, oil spills and chemicals released by human activities. Available literature data on environmental pollution near Antarctic stations will be reviewed, and the possible role of benthic organism assemblages in surveillance and trend monitoring surveys will be discussed.

6.2 Trace Elements in Antarctic Marine Waters

Water mixing and biogeochemical cycles which balance input and output of elements help maintain the composition of seawater throughout the planet rather constant. Although the Southern Ocean is not separated from other oceans by continental masses, its waters have distinctive physico-chemical features. The circulation pattern around the continent, which is covered by ice and lacks river drainage systems, and the seasonal distribution of sea ice and primary productivity are among the factors determining the distribution and different concentrations of trace elements in Antarctic waters.

Trace elements have different structural and thermodynamic characteristics which give them a wide range of chemical properties and affect their environmental fate (i.e. uptake by organisms, adsorption and/or co-precipitation with particulate matter). The distribution of different chemical forms in the water column must be known in order to understand the biogeochemical cycling of trace elements. This is a difficult task because seawater, like snow, contains very low trace element concentrations, especially when considering single chemical species. As it is very difficult to avoid contamination during water sampling and analysis, the reliability of many early published data is questionable. Improved techniques for collection, handling and analysis of seawater samples introduced in the 1970s improved the quality of results, and average trace metal concentrations reported for seawater decreased significantly with respect to earlier results (e.g. Patterson 1974).

Very few data on the chemistry of Southern Ocean waters were available before the 1990s. However, earlier studies (e.g. Boyle and Edmond 1975; Boyle et al. 1976; Harris and Fabris 1979; Orren and Monteiro 1985; Bordin et al. 1987) found that, in contrast to oligotrophic surface waters in other oceans, those in the Southern Ocean were often not depleted in nutrients. As discussed in Chapter 3, regions of intense blooms (>3 g C m^{-3} day^{-1}; 5–10 µg chl l^{-1}) in Southern Ocean waters only occur in shallow shelf areas, near receding pack-ice margins and upwelling regions along the Antarctic Polar Front Zone (APFZ; de Baar et al. 1995; Arrigo et al. 1999). Southern Ocean offshore waters, along with North Pacific and equatorial Pacific waters, constitute a major ocean region with high concentrations of macronutrients, low productivity rates and low biomass (0.1 g C m^{-3} day^{-1}; 0.1–0.2 µg chl l^{-1}). These paradoxical regions (comprising about 25 % of the World Ocean; de Baar et al. 1999) have long been investigated by oceanographers (e.g. Gran 1931). As a rule, besides the usual limiting factors for primary productivity, such as light, grazing rate and water-column stability, the low bioavailability of some essential micronutrients such as Fe was considered a possible explanation for the so-called HNLC waters (Hart 1934). During the last 15 years, HNLC regions have been the focus of large research efforts. After laboratory culture and/or shipboard Fe enrichments (e.g. Martin and Fitzwater 1988;

Morel et al. 1991), several large-scale "in-situ" fertilisation experiments were performed in the equatorial Pacific (e.g. Martin et al. 1994) and Southern Ocean (e.g. Boyd et al. 2000). These experiments essentially show that the lack of Fe and/or co-limitation by Si are responsible for offshore HNLC areas. The addition of Fe induces algal blooms, draw-down in macronutrients (in ratios consistent with the growth of phytoplankton), and shifts in algal community structure. Comparisons between iron-mediated responses in polar and tropical areas (Boyd 2002) show that main differences are essentially due to the temperature-dependence of biological processes.

Studies in different oceans (e.g. Bruland 1980; Danielsson et al. 1985) show that uptake by phytoplankton, assimilation and recycling by zooplankton and scavenging generally determine an increase in concentrations of major nutrients and trace elements such as Cd, Cu, Ni and Zn with depth (i.e. with increasing age of the water). In the Southern Ocean, bioutilised trace metals such as Zn, Cu and Ni do not usually show co-limitation with Fe and, under Fe-depleted conditions, their concentrations in waters are not decreased by algal uptake as expected (Frew et al. 2001). This chapter discusses the distribution and cycling of these trace elements, particularly of metals of environmental and toxicological concern such as Cd, Pb and Hg. The reader interested in general aspects of trace element biogeochemistry in the marine environment can refer to specific books such as those by Riley and Chester (1983), Libes (1992), Bidoglio and Stumm (1994), Salbu and Steiness (1995) and Stumm and Morgan (1996).

6.2.1 Element Input from Atmospheric Dust in the Southern Ocean

As discussed in Chapter 3, the Southern Ocean cannot be considered a single functional unit, but a mosaic of distinct subsystems with specific physical and biological processes. These processes affect the spatio-temporal distribution of major nutrients and trace elements, especially Fe, which is essential to respiratory electron transport chains, synthesis of chlorophyll and production of amino acids in phytoplankton organisms (Raven 1990). Dissolved Fe^{2+} was probably easily bioavailable during the first stages of biological evolution in the primordial ocean but, as photosynthesis and an oxygenated ocean developed, the metal began to precipitate into solid Fe^{3+} phases. Thus, although in marine organisms Fe is involved in many essential metabolic pathways, in some oceanic areas it has become a limiting factor for life. This generally occurs in open-ocean waters, especially in the Southern Ocean where there is negligible fluvial or aeolian input of Fe and other lithophilic elements from continents. Very low concentrations of total and dissolved Fe have been measured in surface waters of the Pacific region of the Southern Ocean, in the Drake Passage and in the Antarctic Circumpolar Current (Martin et al. 1990; Westerlund and Öhman 1991a; Löscher et al.

Table 9. Typical concentrations of dissolved trace metals (nmol l^{-1}; Co, pmol l^{-1}) at different depths in offshore waters of the Southern Ocean

Depth (m)	Fe	Mn	Cd	Co	Cu	Ni	Zn	Sampling period
South Drake Passage (60° 46′S, 63° 26′W)[a]								
30	0.16	0.08	0.28	25	0.97	–	0.63	March 1989
110	0.10	0.21	0.56	26	1.12	–	1.49	
300	0.26	0.25	0.81	29	1.52	–	4.74	
550	0.40	0.29	0.77	27	1.68	–	5.25	
1,420	0.76	0.31	0.66	21	2.07	–	5.67	
Weddell Sea (65° 20′S, 15° 27′W)[b]								
50	0.79	0.25	0.58	–	0.65	–	4.00	22 Dec. 1988
200	0.90	0.31	0.84	–	0.52	–	5.50	
600	1.05	0.25	0.81	–	0.70	–	5.80	
1,000	1.41	0.18	0.78	–	1.00	–	6.25	
73° 18′S, 39° 59′W (close to the Filchner Shelf)								
50	1.92	0.36	0.60	29	1.95	5.80	4.70	25 Jan. 1989
200	4.17	0.25	0.64	27	2.05	7.30	5.90	
400	2.63	0.25	0.65	32	2.15	7.20	6.95	
800	4.06	0.16	0.60	27	2.60	7.05	5.95	
Weddell Sea (ice–covered, 60° 59′S, 49° 03′W)[c]								
10	–	–	0.56	–	3.80	6.12	–	Dec. 1988
20	–	–	0.53	–	3.54	6.46	9.17	
80	–	–	0.60	–	3.20	5.91	4.43	
200	–	–	0.60	–	4.91	6.01	11.6	
300	–	–	0.55	–	4.01	–	10.6	
Scotia Sea (57° 03′S, 48° 51′W)								
10	–	–	0.17	–	2.44	8.52	6.27	Dec. 1988
30	–	–	0.23	–	1.53	6.47	1.68	
60	–	–	0.43	–	1.48	7.50	1.83	
100	–	–	0.57	–	1.81	6.13	3.21	
300	–	–	0.89	–	1.81	7.67	7.34	
Antarctic Confluence (59° 22′S, 48° 44′W)								
10	–	–	0.41	–	4.69	4.77	4.74	Dec. 1988
30	–	–	0.48	–	4.17	5.11	5.05	
60	–	–	0.55	–	2.80	–	5.20	
100	–	–	0.64	–	2.91	5.45	6.73	
300	–	–	0.61	–	4.47	4.60	9.48	

Table 9. (*Continued*)

Depth (m)	Fe	Mn	Cd	Co	Cu	Ni	Zn	Sampling period
Remote Pacific waters of the Southern Ocean (63° 59'S, 89° 31'W)[d]								
50	0.16	–	–	–	–	–	–	25 April 1995
100	0.24	–	–	–	–	–	–	
200	0.39	–	–	–	–	–	–	
400	0.20	–	–	–	–	–	–	
800	0.33	–	–	–	–	–	–	
Bellingshausen Sea (68° 16'S, 89° 24'W)								
50	0.21	–	–	–	–	–	–	18 April 1995
100	0.41	–	–	–	–	–	–	
200	0.86	–	–	–	–	–	–	
400	0.43	–	–	–	–	–	–	
800	0.47	–	–	–	–	–	–	

[a] Martin et al. (1990)
[b] Westerlund and Öhman (1991a, b)
[c] Nolting and de Baar (1994)
[d] de Baar et al. (1999)

1997; de Baar et al. 1999; Table 9). Only over or near continental and Antarctic island shelves are the concentrations of dissolved Fe (e.g. Martin et al. 1990; Westerlund and Öhman 1991a; Löscher et al. 1997; Fitzwater et al. 2000; Grotti et al. 2001; Table 10) in the same range as those usually reported for other seas (e.g. Martin and Gordon 1988). Studies on relationships between concentrations of bioavailable Fe and those of nutrients generally show that Fe influences phytoplankton uptake of N, P and Si. Waters with higher biomasses and statistically significant decreases of nutrient concentrations are usually more affected by Fe derived from continental sources. Tables 9 and 10 specify the water sampling period, as the effects of sea-ice melting and phytoplankton activity must be considered when comparing concentrations of dissolved Fe in surface waters. When the sea-ice cover disappears, possible atmospheric inputs of soluble lithophile elements are removed by phytoplankton and incorporated into organic particulate matter. In surface waters from the Terra Nova Bay shelf, for instance, Grotti et al. (2001) found that the dynamics of metal concentrations followed those of primary production, which was characterised by algal blooms in the second half of December and in late summer. In the absence of water-column stratification, the profile of dissolved Fe and Mn in intermediate and deep waters was rather constant until January; then, the progressive removal of these metals by sinking materials determined a decrease in concentrations.

Table 10. Typical concentrations of dissolved trace metals (nmol l^{-1}; Co, pmol l^{-1}) at different depths in neritic waters of the Southern Ocean

Depth (m)	Fe	Mn	Cd	Co	Cu	Ni	Zn	Sampling period
Gerlache Strait (64° 55'S, 63° 19'W)[a]								
15	7.40	5.05	0.56	82	2.24	–	4.90	1 April 1989
50	4.70	4.19	0.60	59	2.09	–	5.10	
200	6.85	3.86	0.70	82	2.16	–	5.90	
Gerlache Inlet (Terra Nova Bay, 74° 43'S, 164° 11'E)[b]								
4	1.5	123	–	–	7.1	–	–	24 Nov. 1997
50	1.3	2.20	–	–	2.8	–	–	
100	1.1	0.86	–	–	2.7	–	–	
380	1.2	1.04	–	–	4.3	–	–	
2	1.0	0.31	–	–	1.1	–	–	7 Jan. 1998
30	0.6	0.41	–	–	1.0	–	–	
110	0.8	0.85	–	–	1.3	–	–	
380	0.6	0.74	–	–	0.8	–	–	
4	4.1	4.05	–	–	11.6	–	–	
25	1.6	2.02	–	–	5.0	–	–	
100	3.3	1.33	–	–	2.6	–	–	
380	1.8	1.15	–	–	1.1	–	–	
Weddell Sea (74° 33'S, 35° 54'W)[c]								
00	3.58	0.33	–	–	–	–	–	19 Jan. 1989
200	4.28	0.29	–	–	–	–	–	
400	4.59	0.29	–	–	–	–	–	
455	5.66	0.24	–	–	–	–	–	
76° 22'S, 29° 54'W								
50	2.74	0.22	–	–	–	–	–	23 Jan. 1989
100	3.46	0.27	–	–	–	–	–	
200	5.75	0.40	–	–	–	–	–	
300	7.86	0.47	–	–	–	–	–	
Ross Sea (76° 30'S, 179° 39'E)[d]								
20	0.09	–	0.18	12	1.26	5.42	0.71	Jan. 1990
60	0.16	–	0.70	28	2.15	6.75	4.92	
150	0.22	–	0.69	26	2.08	6.66	5.01	
250	1.13	–	0.70	38	2.16	6.78	5.10	
72° 31'S, 172° 31'E								
30	<0.05	–	0.55	26	1.92	6.23	2.95	Jan. 1990
60	0.12	–	0.61	26	1.91	6.30	4.00	
120	<0.05	–	0.73	25	2.10	6.08	5.20	
250	0.43	–	0.73	22	2.21	6.08	5.26	

[a] Martin et al. (1990)
[b] Grotti et al. (2001)
[c] Westerlund and Öhman (1991a)
[d] Fitzwater et al. (2000)

Although open-ocean waters receive very few lithogenic materials from melting ice and icebergs or remobilised marine sediments, de Baar et al. (1995) suggest that increased productivity along the Polar Front downcurrent of the Drake Passage was due to the upwelling of continentally derived Fe. Hiscock et al. (2003) suggest that a winter recharge of upwelled Fe-rich Upper Circumpolar Deep Water (UCDW) within the Antarctic and Southern Antarctic Circumpolar Current (ACC) zones provides enough Fe to support a diatom bloom. This bloom annually propagates polewards to the Southern Boundary of the ACC. However, other scientists such as Measures and Vink (1999) sustain that upwelled water usually has insufficient Fe to match upwelled N species, and external sources of Fe are required for the full utilisation of nutrients. Aeolian dust is the only external source of lithogenic elements in open-ocean waters, but the current atmospheric dust load over the Southern Ocean is one of the lowest in the world (Duce and Tindale 1991).

Aluminium is one of the most valuable tracers of soil and rock dust input in ocean surface waters because it is a relatively invariant component (about 7–8 %) of the Earth' crust. It has a relatively short residence time on the ocean surface (3–5 years), and is removed from the water column with little biological recycling. Average concentrations of dissolved Al in the Weddell Sea (about 3 nmol l^{-1}; Moran et al. 1992; Sañudo-Wilhelmy et al. 2002) are about 20 times lower than those measured in equatorial Atlantic waters by Vink and Measures (2001). Like those of Al and Fe, concentrations of dissolved Mn in surface waters from the open Southern Ocean (Table 9) are often lower than those usually reported for other seas, especially in the Northern Hemisphere where continents are much more extensive. Based on the low concentrations of dissolved Mn (0.08 nmol l^{-1}) measured in surface waters from the South Drake Passage, Martin et al. (1990) suggest that Mn deficiency may be another factor contributing to the limitation of phytoplankton growth in the Southern Ocean.

Concentrations of dissolved Al, Fe and Mn are usually much lower than those in suspended matter; Westerlund and Öhman (1991a) found that particulate Fe/Al and Fe/Mn ratios in the Weddell Sea, compared to ratios in the Pacific Ocean, were much closer to the terrestrial ratio. On the Ross Sea shelf, below a depth of 125 m, Fitzwater et al. (2000) found that particulate Fe/Al and Mn/Al ratios (0.40 and 0.020 respectively) were higher than average upper crustal ratios. Like in other oxygenated shelf areas, the enrichment of the two metals was probably due to the re-suspension of shelf sediments containing Fe and Mn oxides.

6.2.2 Biogeochemistry of Cobalt, Copper, Nickel and Zinc

In contrast to Fe and Mn, concentrations of other essential trace metals such as Co, Cu, Ni and Zn in Antarctic seawater are usually higher or in the same range as those reported for other oceans (Tables 9 and 10). Martin et al. (1990)

found that concentrations of dissolved Co and Cu in the Drake Passage showed a minimal surface depletion when compared to profiles of the two metals in the Gulf of Alaska. In February and March 1991, concentrations of dissolved Ag, Cd, Cu, Fe, Ni and Zn in surface waters of the Weddell Sea shelf were higher than in surface waters of other marine environments (Sañudo-Wilhelmy et al. 2002). The results of this study suggest that elements with a strong coastal component, such as Ag, Cd, Co, Cu, Ni and Zn, may be exported from the Antarctic Peninsula to the open Weddell Sea. Furthermore, the inverse relationship between chlorophyll *a* concentrations and trace metal residence times suggests the importance of biological activity in cycling bioactive metals. Westerlund and Öhman (1991b) found that in the Weddell Sea, in contrast to other oceans, concentrations of trace metals are high throughout the whole water column. In general, Cd, Cu and Zn concentrations were slightly lower in surface waters, Ni only showed very small variations while Co concentrations clearly increased in surface waters. The melting of ice was considered responsible for the increase in Co concentrations, as comparisons between the elemental composition of Antarctic snow and seawater revealed that snow was enriched in Co and Pb, concentrations of Zn were similar, while those of Cd and Ni were higher in seawater. Tables 9 and 10 show that, except for higher values in the Gerlache Strait, concentrations of Co in water samples from different depths in neritic and offshore areas of the Southern Ocean are rather constant (about 26 pmol l^{-1}). However, Sañudo-Wilhelmy et al. (2002) reported median Co levels of 51 pmol l^{-1} in surface waters from the Weddell Sea shelf. The higher values were attributed to input of Co from diagenetic processes occurring in shelf sediments, and to the mixing of upwelling waters with local shelf/upper slope waters.

Cobalt is essential for several species of phytoplankton, but its uptake is generally minimal and the element can be replaced by chemically similar metals such as Zn. Through laboratory experiments, Sunda and Huntsman (1995) found that the cyanobacterium *Synechococcus bacillaris* needs Co but not Zn for growth, the coccolithophore *Emiliana huxleyi* has a Co requirement which can be partly met by Zn, and diatoms of the genus *Thalassiosira* have Zn requirements which can be largely met by Co. The replacement of Co by Zn could explain why Antarctic surface waters are not usually remarkably depleted in Co. By reviewing literature data on Zn and Co concentrations in ocean surface waters, Sunda and Huntsman (1995) noted that Co depletion occurs only after Zn depletion (i.e. when Zn concentrations fall below 0.3 nmol l^{-1}).

Like Co concentrations, those of Cu and Ni in Southern Ocean surface waters (Table 9) are usually higher than typical ocean surface values (usually about 0.5 and 2 nmol l^{-1} respectively; Bruland 1980), and they do not show depletion near the surface. Copper is one of the most studied metals in oceans because it is essential to life. However, if environmental bioavailability of Cu increases (e.g. in recently upwelled waters), it can become toxic, because an excessive absorption of the metal may inhibit enzymes which require other metallic elements. The behaviour of Cu in the marine environment is interme- diate to that of nutrients (which are depleted in surface waters, and accumulate in deep and older waters) and scavenged elements (such as Al and Pb, which mostly derive from sources external to the ocean and show higher concentra- tions in surface and younger waters than in deep or older waters). The bioavail- ability and geochemical behaviour of Cu in seawater are largely affected by the formation of complexes with organic ligands. Capodaglio et al. (1994) studied Cu complexation in surface waters from Terra Nova Bay during three summer expeditions. Total Cu concentrations ranged between 0.5 and 4.8 nmol l^{-1}, and showed a uniform spatial distribution and scarce variations during the three campaigns. Like in many other marine areas, two classes of Cu ligands were found: a stronger one with considerable spatio-temporal variations, and a weaker one with a rather homogenous distribution in the study area.

Sañudo-Wilhelmy et al. (2002) found that the Ag/Cu ratio in the Weddell Sea corresponds to that reported for the Pacific Ocean, and hypothesised that Weddell Sea surface waters may influence trace metal compositions in Pacific subsurface waters. Furthermore, Westerlund and Öhman (1991b) found that the Weddell Sea and the Pacific Ocean have similar nutrient/trace metal ratios. Indeed, as discussed in Chapter 3, intermediate/deep waters of the world ocean mainly originate in the Weddell Sea.

As in the case of Cu, spatial and vertical variations in Ni concentrations in different regions of the Southern Ocean are rather small and these are gener- ally not significantly linked to nutrient concentrations. Nolting and de Baar (1994) measured higher Ni concentrations in surface waters from the Scotia Sea and Antarctic Confluence (Table 9). They hypothesised two possible sources of Ni: uptake from marginal sediments by waters flowing through the Gerlache Strait, and/or transport by deep Pacific waters, whose Ni contents are in the same range as those measured in the Scotia Sea.

Zinc is one of the most important essential micronutrients, because it is a component of nearly 300 enzymes and is involved in many metabolic processes (Vallee and Auld 1990). Morel et al. (1994) showed that carbon uptake by marine phytoplankton may be limited by Zn, which is a constituent of the metalloenzyme carbonic anhydrase. Because of such bioactivity, open- ocean surface waters typically have low Zn concentrations (0.1 nmol l^{-1}; Bru- land and Franks 1983), which are usually strongly correlated with those of Si and N (Bruland et al. 1991). It has therefore been hypothesised that Zn avail- ability, like that of Fe, may limit oceanic primary productivity (Morel et al. 1994). Furthermore, while much Fe limitation is probably Fe–N co-limitation, that of Zn is probably Zn–C co-limitation (i.e. Zn may have a larger impact than Fe on the global C cycle).

Phytoplankton organisms have adopted different strategies to grow in environments with limited Zn availability. Coccolithophores reduce their Zn requirement and carbonic anhydrase activity by using CO_2 formed from

HCO_3^- through calcification reaction. This makes the biological carbon pump particularly inefficient, because the export of $CaCO_3$ regenerates CO_2. On the contrary, other marine primary producers, such as some species of diatoms, can replace Zn in enzymatic sites with other chemically similar metals such as Cd and Co (Price and Morel 1990).

Biological uptake of Zn, depletion of its concentrations in surface waters at sub-nanomolar concentrations, and statistically significant relationships between Zn and Si concentrations have been reported in some highly productive regions of the Southern Ocean (e.g. Orren and Monteiro 1985; Nolting and de Baar 1994; Fitzwater et al. 2000). Zinc is known to be incorporated into diatom frustules (Ellwood and Hunter 1999), which carpet the seafloor of the Southern Ocean. The downward flux of diatoms, also called the "silicate pump", might be the main mechanism for transferring Zn and other trace metals from surface waters to bottom water. During the sedimentation of diatoms, trace metals can be scavenged on the outer surface of frustules; both trace metals and Si are released when the diatoms dissolve. However, relationships between trace metals and Si often show different slopes and intercepts in different regions and/or at different depths, indicating rather complex interactions and biogeochemical cycles. In some highly productive areas of the Ross Sea shelf, Fitzwater et al. (2000) found remarkable surface depletions of Zn (<1 nmol l^{-1}), Co and Cd, while Si concentrations remained >30 µmol l^{-1}. In areas where soluble Fe concentrations were also high, the relative depletion of Zn, Cd and Co in surface waters may indicate Zn stress or limitation.

6.2.3 The "Cadmium Anomaly" in the Southern Ocean

Among trace metals, Cd is one of the most studied in seawater profiles, and its global marine biogeochemistry is reasonably well understood. The distribution of Cd in oceans is commonly strongly correlated with phosphate concentrations. The Cd/PO_4 ratio generally exhibits a linear relationship, which is important in understanding biogeochemical processes occurring in modern and ancient waters. Boyle (1988) plotted available high-quality datasets and suggested that the relationship between Cd and PO_4 concentrations worldwide can be described through a general equation, consisting of two separate relations: one for phosphate concentrations<1.3 µmol l^{-1} (with virtually zero intercept, in all upper ocean waters and North Atlantic deep waters), and one for concentrations>1.3 µmol l^{-1} (mostly North Pacific and South Atlantic deep waters). According to the latter relationship, for a given value of PO_4 in deep open-ocean waters, Cd concentrations are consistent within about ±7%.

More recently, de Baar et al. (1994) compiled a larger dataset of selected high-quality Cd and phosphate values for deep waters, revealing two data clusters with a statistically significant bimodality: deep North Atlantic versus deep Antarctic/Indo/Pacific waters. The two distinct biogeochemical provinces for Cd cycling were attributed to the scarce input of Cd- and phosphate-rich waters from the Weddell Sea to the deep North Atlantic. Evidence supporting the two Cd/PO_4 relationships was also provided by Yeats et al. (1995) and Frew (1995). The latter author attributed the apparent global Cd–phosphate link to the formation of bottom waters with high Cd/PO_4 ratios around the Antarctic continent. However, the question of whether the global Cd/PO_4 ratio is better described by one or two different linear relationships is still a matter of debate. By adopting the same selection criterion and adding new data to the de Baar and co-workers selection, Löscher et al. (1997) found statistically significant differences between the North Atlantic and Indo-Pacific Oceans. However, when only two datasets with the smallest phosphate concentrations ($PO_4 < 1\ \mu mol\ l^{-1}$) were excluded, the differences were no longer statistically significant. It follows that more accurate Cd and phosphate data are needed, especially in areas with low concentrations.

An important feature of Cd marine biogeochemistry is that the metal is incorporated into $CaCO_3$ exoskeletons of marine organisms in concentrations reflecting those of the seawater in which the organism is growing (Boyle 1992). The Cd/Ca ratio in benthic shells indicates Cd concentrations in waters and those of phosphate, via the Cd/PO_4 relationship. The biogeochemistry of Cd has therefore been used to study labile nutrient concentrations in oceans and to infer the palaeochemistry and palaeocirculation of bottom waters (e.g. Shen et al. 1987; Boyle 1992). However, Mackensen and Douglas (1989) showed that when dealing with several species of foraminifers, only one truly epibenthic species should be used to study bottom water palaeochemistry. The interpretation of results is complicated by the decoupling of Cd and phosphate during early diagenesis (Saager et al. 1992). In the Southern Ocean, for instance, $\delta^{13}C$ data suggest that deep waters were nutrient-enriched in the glacial period, whereas Cd data indicate only minor changes (Boyle 1992). According to several researchers (e.g. Hutchins and Bruland 1998; Takeda 1998), the use of opal accumulation rates as proxies to estimate past levels of productivity in the Southern Ocean may be biased. Indeed, in this ocean there is no evidence of higher opal accumulation rates during glacial eras, when the enhanced deposition of dust increased phytoplankton productivity (Boyle 1998). A better knowledge of Cd and $\delta^{13}C$ biogeochemistry in surface and deep Antarctic waters, particularly in areas where deep waters form, is probably necessary to clarify discrepancies and to gain a better understanding of relationships between ocean chemistry and circulation, and of atmospheric CO_2 concentrations (Frew 1995).

Southern Ocean surface waters are exceptional because Cd (see Tables 9 and 10) and phosphate (about $2.0\ \mu mol\ l^{-1}$) concentrations in these waters are much higher than in other ocean surface waters (often strongly depleted to levels as low as 0.01 nmol l^{-1} for Cd and less than 0.04 $\mu mol\ kg^{-1}$ for phosphate;

e.g. Yeats and Campbell 1983). Furthermore, while maximum Cd concentrations in Pacific and North Atlantic waters are found at depths of about 1,000 m, maximum values in Antarctic waters are already reached at depths of 100–300 m (Table 9). The Cd/PO$_4$ ratio in the Antarctic Circumpolar Current and in the northern portion and western rim of the Weddell Sea ranges from 0.50 to 0.58 (nmol l^{-1}/μmol l^{-1}; Nolting et al. 1991; Löscher et al. 1998; Sañudo-Wilhelmy et al. 2002). However, Westerlund and Öhman (1991b) reported a much lower ratio (0.31) for the southern portion of the Weddell Sea. In the upper 300 m of a transect in the Southern Ocean, Nolting and de Baar (1994) found different Cd/PO$_4$ ratios in each of the studied areas: the Scotia Sea, Antarctic Confluence and Weddell Sea. These results suggest that within the Southern Ocean, which is a mosaic of marine subsystems, the Cd/PO$_4$ ratio is susceptible to spatio-temporal variations likely due to phytoplankton growth in surface waters. Phytoplankton assemblages may be dominated by different species of diatoms or by *Phaeocystis antarctica*, which have species-specific macro- and micronutrient requirements. As discussed by Arrigo et al. (1999) for the N/P ratio, spatio-temporal differences in phytoplankton composition may contribute to variations in the Cd/PO$_4$ ratio of Southern Ocean surface waters.

Scarponi et al. (2000) studied the summer evolution of Cd distribution in the water column at Wood Bay (Ross Sea) and found that in November, before ice melting and phytoplankton blooms, the vertical profile of Cd along the water column was characterised by uniform concentrations (mean 0.64 nmol l^{-1}). These relatively high and uniform Cd concentrations were ascribed to the upwelling or vertical diffusion of Cd-enriched deep waters and to the scarce uptake or scavenging of this metal by organisms. Later in the season, with the onset of ice melting and the development of phytoplankton, Cd was strongly depleted in surface water to a depth of 50 m (minimum concentration 0.056 nmol l^{-1} at –10 m, at the end of January). In February, as a consequence of reduced phytoplankton uptake and shallow regeneration cycles, Cd concentrations in the upper 50 m began to increase.

6.2.4 Natural and Anthropogenic Sources of Lead

Snow data reported in the previous chapter show that even remote Antarctica is affected by atmospheric deposition of Pb and other metals produced by human activities in continents of the Southern Hemisphere. Until two decades ago, most Pb in the atmosphere and in ocean surface waters derived from the combustion of leaded gasoline and other anthropogenic sources (e.g. Nriagu and Pacyna 1988; Volkening et al. 1988). Combustion of leaded, low-leaded and unleaded gasoline continued to be the major source of atmospheric Pb emissions during the mid-1990s (Pacyna and Pacyna 2001). Fluxes of anthropogenic Pb have characteristic, conservative isotopic compositions

(^{204}Pb:^{206}Pb:^{207}Pb:^{208}Pb), and oceanic dispersion of the metal can be traced through the determination of Pb isotopic ratios. The transient input of Pb from industrial sources complements those of other seawater tracers such as freons and tritium, and has been widely used to study circulation in the North Atlantic (e.g. Boyle et al. 1986; Hamelin et al. 1997). Although fewer studies have focused on the Southern Hemisphere, a recent survey of Pb isotopes in the South Atlantic atmosphere and seawater (Alleman et al. 2001) indicates that anthropogenic lead from the Northern Hemisphere does not directly influence the Southern Hemisphere. Antarctic Bottom Water has a relatively radiogenic Pb component ($^{206}Pb/^{207}Pb$=1.186±0.002) which probably reflects a mixture of natural and anthropogenic Pb sources within the Southern Hemisphere.

During periods of intense algal blooms, Flegal et al. (1993) measured very low concentrations of Pb (Table 11) in unfiltered surface waters from marginal ice zones of the Weddell and Scotia seas. The low Pb concentrations were attributed to efficient scavenging associated with intense primary productivity and to the relative isolation of the study area. The aeolian Pb flux (about 0.01 pmol cm^{-2}) was at least two orders of magnitude lower than that of any other sea. However, concentrations of Pb in Antarctic waters were positively correlated with those of Fe, which are generally not increased by anthropogenic inputs. It was therefore concluded that a significant amount of Pb in surface waters of the Weddell and Scotia seas derived from natural sources such as volcanic emissions, aeolian dust from South America or sediment transport by ice rafting. The isotopic composition of Pb from the Mt. Erebus volcanic province, from lavas and sediments in Marie Byrd Land, and from landmasses in the Southern Ocean, such as the South Sandwich Islands and Chile, nearly matched the radiogenic component of Pb in Antarctic surface waters. In these waters the relative amount of anthropogenic Pb was highly variable at different sites. It was estimated to range from 30 to 70% based on the isotopic composition of Mt. Erebus volcanics, and from 0 to 60% based on the isotopic composition of South American crustal sources. According to Flegal et al. (1993), although the variability of Pb concentrations and isotopic compositions may also reflect anthropogenic point source inputs in Antarctica itself (ships, aircraft, research operations), the isotopic composition of industrial Pb in Antarctic surface waters generally reflects that of urban aerosols emitted in Australia, South Africa and other countries of the Southern Hemisphere. The major sources of this Pb were the Broken Hill and Mount Isa ore deposits in Australia ($^{206}Pb/^{207}Pb$ ~1.04).

A recent study on surface (filtered) waters collected in February/March 1991 from the north-west Weddell Sea (Sañudo-Wilhelmy et al. 2002) confirms that Pb concentrations are low (about half those usually measured in other ocean environments; Table 11), and the anthropogenic component of this metal reflects the isotopic composition of aerosols from South America and Palmer Station.

Table 11. Typical concentrations (range and/or mean±SD; pmol l^{-1}) of Pb in Antarctic waters

Location	Depth (m)	n	Pb (range)	(mean)	Notes
Weddell Sea water column	From surface to 3,600 m	35	5.0–47	10	December 1988–February 1989, unfiltered samples (Westerlund and Öhman 1991b)
Weddell and Scotia Sea	Surface waters (marginal ice zone)	1	9.8–103	45±39	March and June 1988; June and August 1988, unfiltered (Flegal et al. 1993)
Weddell Sea, northeast	Surface waters<1 m	11	4.3–40	16±13	February 1991, filtered (Sañudo-Wilhelmy et al. 2002)
		7	22–87	38±	19 November 1993, filtered (Scarponi et al. 2000)
Wood Bay, Ross Sea	From 0.5 to 350 m	6	7.6–37	20±9	9 January 1994, filtered (Scarponi et al. 1999)
		5	13–27	20±6	10 February 1994, filtered (Scarponi et al. 1999)

Capodaglio et al. (1991) used Differential-Pulse Anodic Stripping Voltammetry (DPSAV) to study the complexation of Pb by natural organic ligands in surface waters from Terra Nova Bay (Ross Sea). Total Pb concentrations in the January–February 1988 period ranged from 25 to 114 pmol l^{-1}; open-sea water samples showed lower concentrations of total Pb (25–38 pmol l^{-1}) and ligands (about 44 % of total Pb was in the labile fraction). Total Pb concentrations in coastal water ranged from 61 to 114 pmol l^{-1} and, due to the higher content of ligands, only about 25 % of total Pb was in the labile fraction. As for the possible origin of Pb ligands, a statistically significant relationship was found between ligand and chlorophyll concentrations in open-sea waters. In the same marine region, Scarponi et al. (1999) studied the evolution of Pb profiles in the water column during the 1993–1994 austral summer. The vertical profile of soluble Pb concentrations was nearly uniform (31±6 pmol l^{-1}) in November (Table 11), whereas Pb concentrations in the water column decreased progressively in December and during summer to about half the initial value. Lead concentrations in the water column seemed to increase in February, beginning from the deepest layers. The summer depletion of soluble Pb in the water column was attributed to adsorption on suspended matter and interactions with quickly settling particles, especially during the senescent phase of algal blooms.

6.2.5 A Neglected Element: Mercury

High concentrations of Hg were reported in water samples from the Ross Sea during the 1970s (e.g. Williams et al. 1974). The data were probably affected by the lack of suitable procedures for clean sampling and handling of seawater. However, there appear to be no recent data on Hg concentrations in Southern Ocean waters. This is a drawback because there is evidence that monomethylmercury (MeHg) biomagnifies in Antarctic marine coastal food webs – its concentrations in feathers of Antarctic skua and in tissues of a Weddell seal from Terra Nova Bay (Ross Sea) were in the same range as those in related species of seabirds and marine mammals from the Northern Hemisphere (Bargagli et al. 1998a, 2000). The recent introduction of automated techniques for accurately measuring gaseous oxidised Hg species at typical atmospheric concentrations is providing very useful data on the biogeochemical cycle of this metal in the environment. As discussed in Chapter 4, these techniques have revealed that in polar regions after the polar dawn, atmospheric concentrations of Hg° diminish over a period of a few days to as low as 10–20 % of their typical value. Hg depletion is due to gas-phase oxidation of Hg, probably by halogen atoms or halogen-containing radicals. Soluble and insoluble forms of oxidised Hg are deposited in polar ecosystems; this is a cause for concern because terrestrial organisms resume biological activity this time of year. Until recently, the conversion of Hg° to Hg^{2+} was mainly attributed to the reaction of aqueous and gaseous phases with O_3. However, recent research (Hedgecock and Pirrone 2001) on the marine boundary layer (i.e. the air directly above the sea surface) suggests that the role of H_2O_2 in Hg° oxidation is probably as important as that of O_3. Concentrations of oxidised forms of Hg, such as HgO, $HgCl_2$ and $HgBr_2$, in the boundary layer of the Mediterranean Sea are as high as in the more industrial areas of northern Europe (Wangberg et al. 2001). Recent modelling studies (Hedgecock et al. 2003) suggest that deliquesced sea-salt aerosol in the marine boundary layer provides not only a scavenging phase for oxidised Hg compounds but also an almost unlimited supply of Cl$^-$, with which Hg^{++} can form aqueous-phase complexes.

Processes in the marine boundary layer reproduce on a lesser scale many of the reactions which determine Hg° depletion events in Arctic and Antarctic terrestrial ecosystems (UNEP 2002b). It seems likely that in spring and summer, when the surface of the Southern Ocean is ice-free, there is constant deposition of oxidised forms of Hg which are probably replenished in the boundary layer by Hg from the sea and free troposphere.

During the 1980s, Fitzgerald and co-workers (e.g. Gill and Fitzgerald 1988) largely contributed to the establishment of sampling and analytical protocols for producing reliable data on Hg concentrations in the ocean environment. More recently, Mason and Fitzgerald (1997) reviewed aspects of the biogeochemical cycle of Hg in oceans: typical concentrations in ocean waters are <5 pmol l^{-1}, and the estimated residence time for Hg is about 350 years. Unlike Zn, Cd and other trace metals, Hg does not generally show nutrient-like regeneration in the water column. Average Hg concentrations are lower in the Pacific than in the Atlantic due to differences in external sources and in the scavenging intensity of settling particles (i.e. biological productivity). Methylated Hg species (MeHg and dimethylmercury, DMHg) are found throughout the ocean water column. In contrast to freshwater ecosystems, DMHg has always been found in deep ocean waters; however, DMHg is usually lacking in open-ocean surface waters, perhaps due to decomposition in the presence of light or loss via evaporation. As this compound is relatively unstable, continual production is necessary to sustain measurable concentrations in marine waters (Fitzgerald and Mason 1997). Elemental Hg (Hg°) is another ubiquitous Hg species in surface and deep waters. Its formation in surface waters appears to be both an incidental result of primary productivity and a result of photochemical reduction (Amyot et al. 1997). The reduction rate depends on the availability of ionic Hg, which may be supplied by the marine boundary layer and processes such as upwelling, which also drive productivity in surface waters. The flux of particulate matter from the euphotic zone to deeper waters is the main source of Hg to sub-thermocline waters where net methylation occurs.

Dalziel (1995) measured the vertical distribution of reactive Hg (i.e. the metal volatilised from water after the addition of 10 % acidic stannous chloride) in the eastern North Atlantic and the south-eastern Atlantic. The latter station (Angola Basin) showed higher biological productivity and the lowest concentrations of reactive Hg. The depletion of Hg in surface waters was probably due to enhanced biological activity (i.e. to the biological reduction of Hg^{++} and release of Hg° to the atmosphere, and to particulate scavenging processes). Higher concentrations of reactive Hg (about 1.4 pmol l^{-1}) in water samples from depths of 35–200 m coincided with an intense nutrient gradient and O$_2$ depletion. The thermocline extended to a depth of about 500 m (average reactive Hg concentration=0.92±0.54 pmol l^{-1}). Below the thermocline, to a depth of 1,200 m, Antarctic Intermediate Water had a lower average concentration of Hg (0.51±0.04 pmol l^{-1}). The deep vertical profile did not show remarkable variations in reactive Hg concentrations. Values measured in North Atlantic Deep Water (from depths of 2,000–2,800 m) and Antarctic Bottom Water (at depths>4,000 m) ranged from 0.45–0.67 pmol l⁻¹. In two other sampling stations, at lower latitudes (over the Cape Verde Abyssal Plain and the Seine Abyssal Plain), concentrations of reactive Hg in Antarctic Bottom Water progressively increased (from 0.67–1.25 pmol l⁻¹). Dalziel (1995) hypothesised that mixing with North Atlantic Deep Water and/or release of Hg from bottom sediments were the main sources of this metal. In contrast to Antarctic water masses, those coming from the Northern Hemisphere, such as North Atlantic Deep Water, showed progressively lower Hg concentrations (from 1.34–0.48 pmol l⁻¹).

Mason and Sullivan (1999) measured concentrations of different species of Hg in surface and deep waters from the South and equatorial Atlantic. Concentrations of DMHg in Antarctic Intermediate Water (i.e. low-salinity surface water sinking at the Antarctic Polar front) decreased northwards (from 0.11 pmol l⁻¹ at 33° S, 40° W to 0.037 pmol l⁻¹ at 7.5° N, 25° W). The same trend was observed in Antarctic Bottom Water, and Mason and Sullivan (1999) concluded that net decomposition of DMHg occurs in these water masses with time. Concentrations of MeHg in all samples were at or below the detection limit (0.05 pmol l⁻¹), and it was hypothesised that in deep waters this compound is decomposed more rapidly than it is produced and/or that it is scavenged by particulate matter. However, total Hg concentrations in intermediate and bottom Antarctic water (total Hg=2 pmol l⁻¹) were higher than those of Hg° (about 1.0 pmol l⁻¹), suggesting the presence of different Hg species (Mason and Sullivan 1999). In conclusion, available data on Hg concentrations and speciation in Antarctic waters, collected in the Atlantic, seem to indicate that concentrations of reactive Hg in waters from the Southern Ocean are approximately less than or equal to 1 pmol l⁻¹. This fraction mostly consists of Hg° (from about 0.4–1.0 pmol l⁻¹), probably a product of MeHg decomposition. Another significant amount of Hg in Antarctic waters probably consists of DMHg.

6.3 Particle Fluxes and the Composition of Surface Sediments

Despite the recent introduction of ultrasensitive analytical techniques, spatio-temporal fluctuations of biological, climatic and environmental factors determine large variations in ocean water-column element concentrations. Sampling and analysis of waters from different sites and depths should therefore be repeated in space and time to reach definite conclusions on the distribution and cycling of trace elements in the sea. This is particularly difficult in the Southern Ocean, which is remote and covered with sea ice for several months of the year. On the contrary, metal concentrations in marine sediments are 1,000–100,000 times higher than those in the respective waters, and they usually show negligible variations over large marine areas and intervals of time. Dated sediment cores provide a record of changes in persistent contaminant input and enable reliable estimates of background values. Sediments are therefore a suitable environmental matrix to study the spatial distribution and cycling of trace elements and other persistent contaminants.

The seafloor around Antarctica is unique among the world's continental shelves because of its great width, the exceptional depth of continental shelves (about 500 m) and the widespread occurrence of deep basins which act as traps for sediment accumulation (DeMaster et al. 1996). Since the Antarctic shelves are in good communication with the open Southern Ocean, surficial sediments are usually characterised by extensive regeneration of organic C, and dissolved products may affect the deep and bottom chemistry of C and nutrients well to the north of the Southern Ocean (Sarmiento et al. 1998).

The main sources of sediments in the Southern Ocean are phyto- and zooplankton. As this ocean contains large areas of extensive phytoplankton blooms, such as shelf regions and marginal ice zones, international research projects such as the Southern Ocean Joint Global Ocean Flux Study (JGOFS, also known as Antarctic Environment and Southern Ocean Process Study – AESOPS) have been developed in recent years to evaluate the role of these productive environments in the global ocean C and Si cycles (e.g. JGOFS 1996; Smith et al. 2000b). Langone et al. (2000) used time-series sediment trap moorings to measure fluxes of biogenic material sedimenting in the Joides shelf basin (western Ross Sea) from December 1994 to January 1996. The composition of particles in the upper trap showed a seasonal pattern, which was much less evident at depth. Fluxes of 0–64 mg m^{-2} day^{-1} of biogenic silica and 0–7.8 mg m^{-2} day^{-1} of organic carbon were measured at a depth of 200 m. Near-bottom fluxes of Si and C were almost always much higher – 11–141 and 1.5–19 mg m^{-2} day^{-1} respectively. Collier et al. (2000), who recovered over 80 trap samples from moorings (November 1996–December 1997) in the Ross Sea, also found a significant increase in the flux of particles to deep traps. They supposed that the increase was mainly due to horizontal transport and possible focusing of particulate matter. Organic matter, biogenic silica and calcium carbonate contents ranged from 8 to 80 %, 3 to 67 %, and 6 to 79 % of the sample mass respectively.

Although biosiliceous and biocalcareous oozes dominate the vertical flux of materials in the Southern Ocean, there are also some areas such as the Scotia Sea where sediments include a high proportion of terrigenous mud. Diekmann et al. (2000) studied the geochemical and mineralogical compositions of modern and Late Quaternary sediments from the Scotia Sea and found that terrigenous materials mainly originate from nearby terrestrial sources or are introduced through interbasinal sediment transfer from adjacent seas. Collier et al. (2000) found that re-suspension and horizontal near-bottom transport on the continental shelf of the Ross Sea can increase lithogenic contributions to over 30 % of particle flux in near-bottom traps.

Very few data are available on the elemental composition of materials sedimenting in the Southern Ocean. Frache et al. (2001) determined Cd, Cu, Fe and Ni concentrations in samples collected during different expeditions in the Terra Nova Bay continental shelf. The melting of sea ice and biological activity significantly affected concentrations of metals and their distribution along the water column. Grotti et al. (2003) developed a system for in-situ filtration of suspended particulate matter specifically suited for trace element determination. The system was tested in Terra Nova Bay coastal waters, and average concentrations of Ca, Mg, Si and Fe in particulate matter were 86 ± 10, 398 ± 46, 344 ± 40, and 3.5 ± 0.4 nmol l^{-1} respectively. A sequential extraction procedure was used to evaluate the association of Cd, Fe, Mn and Zn with different phases of suspended matter. Preliminary results showed that Cd and Zn in the photic zone were mainly associated with labile organic matter (probably phytoplankton cells); at greater depths the percentage of Cd and Zn bound to Fe and Mn oxides increased, while at a depth of 300 m their concentrations were very low. In contrast, the labile fraction of Fe was quite low at all depths, and the metal was mainly bound to oxides and more refractory matrices. Manganese showed an analogous "scavenging-type" behaviour, although the labile fraction (i.e. that associated with phytoplankton, carbonates and labile oxides) was dominant at all depths.

The geochemistry of pelagic sediments in Antarctica was first studied by Angino (1966). His analysis of 14 cores from the Ross, Bellingshausen and Amundsen seas revealed marked geographical differences in trace element concentrations. Those of Mo, Pb and Zn were invariably less than 20, 50 and 200 $\mu g\,g^{-1}$ dry wt. respectively, while average Fe, Cu and V concentrations were higher in samples from the Amundsen Sea (41,000, 185 and 240 $\mu g\,g^{-1}$ dry wt. respectively). Values for sediments in the Ross Sea were intermediate to those for sediments in the Amundsen and Bellingshausen seas. Angino (1966) found that the elemental composition of glacio-marine sediments revealed their distinctly different chemistry with respect to that of shelf sediments, deep-sea carbonates, and deep-sea clay. He therefore suggested that trace element criteria could help detect the presence of glacio-marine sediments in cores from the Southern Ocean. In the following two decades, several papers were published on the geochemistry of Antarctic sediments (e.g. Angino and Andrews 1968; Eisma 1973; Glasby et al. 1975, 1985; Barrett et al. 1984; Hieke Merlin et al. 1989), but very few data were produced during the 1990s.

It is difficult to compare data reported by different authors because trace element concentrations in sediment samples depend on sample mineralogy, grain size, and content and nature of organic matter (e.g. Förstner and Wittmann 1983; Bargagli et al. 1998d). Hung et al. (1993), for instance, measured much higher total Cu and Zn concentrations in sediments from the Southern Ocean containing more than 99 % mud (128 and 458 $\mu g\,g^{-1}$ dry wt.

respectively) than in those with 51.7 % mud (Cu=84, Zn=288 $\mu g\ g^{-1}$ dry wt.). Moreover, available data are less easy to compare owing to the lack or misuse of standardised dissolution/extraction procedures and of reference materials to evaluate the accuracy of analytical determinations (Loring and Rantala 1988). Worldwide investigations of global-change phenomena, especially those in Antarctica, are very expensive and require the application of standardised procedures and very reliable data. The Italian National Programme for Research in Antarctica has promoted the certification of trace element concentrations in Antarctic seawater and surface sediments (Caroli et al. 1996; Pauwels et al. 2001). Table 12 summarises the results of analytical determinations performed by Mazzuccotelli et al. (1989), Hieke Merlin et al. (1989), Bargagli et al. (1998d) and Giordano et al. (1999) in the fine fraction (<63 μm) of surface marine sediments from the continental shelf and slope of the Ross Sea. Samples were digested with hydrofluoric acid (HF) and aqua regia (hydrochloric:nitric acid, 3:1) in high-pressure Teflon vessels in a microwave oven at about 150 °C. Although the same granulometric fraction

Table 12. Concentrations ($\mu g\ g^{-1}$, Al and Fe, %) of trace elements in the <63-μm fraction of surface sediments from the Ross Sea continental shelf

Element	n	Range	Mean±SD	References
Al (%)	10	3.1–5.1	4.3±0.6	Giordano et al. (1999)
Fe (%)	10	2.1–6.4	3.9±1.5	Giordano et al. (1999)
Be	10	1.2–2.2	1.7±0.3	Giordano et al. (1999)
Cd	98	0.04–0.72	0.35±0.21	Giordano et al. (1999); Mazzuccotelli et al. (1989); Hieke Merlin et al. (1989)
Co	12	0.10–13	1.9±3.4	Mazzuccotelli et al. (1989)
Cr	98	3.7–109	50±34	Giordano et al. (1999); Mazzuccotelli et al. (1989); Hieke Merlin et al. (1989)
Cu	88	2.0–63	23±14	Mazzuccotelli et al. (1989); Hieke Merlin et al. (1989)
Hg	20	0.006–0.027	0.012±0.007	Bargagli et al. (1998d)
Mn	10	478–1,497	907±351	Giordano et al. (1999)
Ni	98	2–46	20±5	Giordano et al. (1999); Mazzuccotelli et al. (1989); Hieke Merlin et al. (1989)
Pb	98	7–32	18±7	Giordano et al. (1999); Mazzuccotelli et al. (1989); Hieke Merlin et al. (1989)
Sn	10	3.3–6.9	4.5±1.3	Giordano et al. (1999)
V	76	10–75	49±17	Hieke Merlin et al. (1989)
Zn	99	25–331	67±42	Giordano et al. (1999); Mazzuccotelli et al. (1989); Hieke Merlin et al. (1989)

was analysed in each sample, data in Table 12 show high variability. Statistically significant differences in element concentrations were found between samples rich in igneous and metamorphic complexes and those in which volcanic clasts were dominant. In general, samples from the outer shelf and slope had higher V, Cr and Zn concentrations, intermediate Ni and Cu concentrations, and lower Co, Cd and Pb concentrations. Sequential leaching procedures (e.g. Hung et al. 1993; Frache et al. 2001) usually reveal that the bioexchangeable fraction of metals is low (0.01–10%), and that large proportions of these are incorporated into crystalline oxides or combined with sulphur and organic matter.

Comparisons between average concentrations of trace elements in surface sediments from the shelf and slope of the Ross Sea with those reported in large-scale surveys of marine sediments from Greenland (AMAP 1997) and northern Arctic Alaska (Naidu et al. 1997; Fig. 42), prepared and digested through similar procedures, show that values from the Arctic and Antarctica are in the same range. The main difference is the higher average concentra-

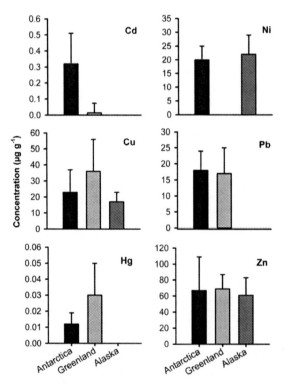

Fig. 42. Comparison between average trace element concentrations in marine surface sediments from the Ross Sea (Antarctica), Greenland (AMAP 1997), and Arctic Alaska. (Naidu et al. 1997)

tions of Cd in Antarctic sediments. Diatomaceous ooze in Antarctic upwelling areas is known to contain as much as 60 µg g^{-1} dry wt. of Cd (e.g. Fowler 1990).

6.4 Environmental Pollution in Marine Coastal Areas

Although the Southern Ocean environment is relatively unaffected by human activities, the development of fisheries, growth of tourism, and the construction and expansion of coastal stations and facilities to support research are changing the pristine conditions of Antarctic marine ecosystems. A number of studies have reported the occurrence of a contaminated halo around point sources on the Antarctic shoreline. In most cases concentrations of metals and other persistent contaminants rapidly decrease within a few hundred metres of sources, and measured concentrations are often much lower than those deemed to be toxic for marine organisms. In all scientific stations, fossil fuel hydrocarbons, oil and greases are widely used to generate electric power and for heating and transportation. Hydrocarbons, particularly polycyclic aromatic hydrocarbons (PAHs; persistent toxic substances which are produced through hydrocarbon fuel spillage and as unintended by-products of combustion), are the most widespread environmental contaminants in coastal marine areas near the larger and/or long-inhabited Antarctic stations. Hydrocarbons in the environment occur as complex mixtures, and it can be difficult to detect low-level anthropogenic contamination in Antarctica, because many compounds are also found in biogenic materials. Research by Cripps (1989) in Antarctic remote marine areas and in those affected by human coastal settlements revealed that methods used to assess the possible anthropogenic input of hydrocarbons were rather inconsistent. These methods assumed that organisms synthesise a range of n-alkanes dominated by compounds with odd numbers of carbon atoms, and/or that values greater than unity for the pristine/phytane ratio indicate the absence of petroleum. Even PAHs formed during the pyrolysis of organic materials cannot be considered reliable tracers of anthropogenic input, because the combustion of petroleum may result in high levels of alkanes and low levels of PAHs with three or more rings in their structure. Moreover, PAHs in the Southern Ocean are also produced by microorganisms (Cripps 1989). The author suggested that the identification of anthropogenic input in Antarctic coastal ecosystems should be based on the quantification of all compounds against a well-defined baseline; the source of pollution can then be estimated from hydrocarbon distribution patterns. In addition to the difficulty in discriminating between natural and anthropogenic hydrocarbons, data on PAHs and n-alkanes in the Southern Ocean environment are extremely variable, and this variability is at least in part due (Ehrhardt et al. 1991) to the wide variety of adopted analytical methods (e.g. spectrofluorometry, HPLC, capillary GC, GC/MS).

In addition to petroleum hydrocarbons, some studies have investigated the concentrations and distribution in coastal marine sediments of other lipid compounds such as sterols and fatty acids (e.g. Smith et al. 1986; Venkatesan and Kaplan 1987; Weber and Bigeco 1990; Green et al. 1992; Venkatesan and Mirsadeghi 1992; Green and Nichols 1995). In temperate regions, sterols such as coprostanol and epicoprostanol are often used as tracers of human faecal waste. However, in Antarctica these compounds are also produced by seals and other marine mammals (Venkatesan et al. 1986).

Despite difficulties in discriminating between natural and anthropogenic sources of hydrocarbons and sterols, clear evidence of local seawater and sediment contamination by human activities has been reported in areas located near long-inhabited research stations or land-based whale processing plants. Although the results of environmental studies in the Ross Sea Region, Palmer Peninsula, Davis Station and sub-Antarctic islands will be reported in this chapter, the reader should refer to Kennicutt and Champ (1992), COMNAP-AEON (2001), and UNEP (2002a) for more comprehensive reviews.

6.4.1 The Impact of Disused Whaling Stations in Peri-Antarctic Islands

South Georgia was the first land to be sighted south of the Antarctic Convergence when, in 1675, a London merchant rounding Cape Horn was blown off course during a storm. Later, the abundance of fur and elephant seals described in the account of Cook's voyage spurred the 1785 "gold rush" of sealers. Intense exploitation led to the virtual extermination of South Georgia fur seals by about 1810. Subsequent recovery of seal populations yielded lesser fur-sealing peaks in about 1820 and 1870. Elephant seals also were exploited for oil. Within the framework of the International Polar Year, a German scientific expedition operated a station at South Georgia in 1882 and 1883. The Norwegian whaling industry began to permanently settle the island in 1904, and six shore-based stations and eight floating factories operated up to 1917. Although sealing ceased in 1913, more than 175,000 whales were killed in this period. At this time, South Georgia also became "a Gateway to Antarctica"; two of Shackleton's expeditions, for instance, visited the island, where the explorer died and was buried. In 1925 the British Government established a scientific station on the island, but Norwegian and then Japanese whaling at South Georgia stopped only in 1965. On 3 April 1982 South Georgia become the only part of Antarctica ever directly involved in warfare (cf. the Argentinean attack of the UK scientific station at King Edward Point, which was recaptured by the Royal Navy on 25 April). For over a century South Georgia was the site of many events, as documented by the presence of a military base, sunken whalers along the seashore, derelict whaling stations, and newly introduced mammals such as reindeers, brown rats and mice.

Platt and Mackie (1979, 1980) analysed hydrocarbons in a sediment core from King Edward Cove (South Georgia), extending from the present-day to the early 1800s. They identified the temporal trend of hydrocarbon pollution associated with whaling activity. Cripps (1989) found that quantities of fuel oil, lubricant and whale oil were still present in South Georgia whaling stations, and that these contaminants leaked into terrestrial and marine environments. He collected samples of seawater from both control and more impacted sites, and found different n-alkane distribution patterns. Three sites were clearly contaminated (total concentrations of n-alkanes in seawater ranged from 7.5 to 10.1 $\mu g \, l^{-1}$) by weathered and degraded fuel oil or by tar. Several PAHs were detected, and phenanthrene, anthracene, fluoranthene and pyrene occurred in all seawater samples. The maximum total PAH concentration (1.2 $\mu g \, l^{-1}$) was measured in a sample from a cove adjacent to the Stromness abandoned whaling station.

Whaling at Signy Island (South Orkney Islands) commenced in 1907. Although a shore whaling station was built in 1921, it only operated for a few years; for about 20 years, most whaling activity was performed on floating factory ships anchored in Factory Cove. Since 1947 the British Antarctic Survey has maintained a research station at the site previously occupied by the whaling station. To evaluate the extent of contamination from the station, Cripps (1992a) measured concentrations of n-alkanes and PAHs in seawater and sediments. He found that total n-alkane concentrations in seawater decreased from 7.6–2.6 $\mu g \, l^{-1}$ within 500 m of the station. In contrast, although total PAH concentrations (from 110–216 ng l^{-1}) were slightly higher than in open-ocean waters (about 70 ng l^{-1}), they did not vary with distance from the station. Concentrations of n-alkanes and PAHs in surface sediments (top 2.5 cm) declined within 375 m of the station (from 1,731 and 280 $\mu g \, kg^{-1}$ dry wt. to 64 and 14 $\mu g \, kg^{-1}$ dry wt. respectively). Sediment core profiles revealed no consistent patterns of n-alkanes concentrations. The total PAH content near the station did not change with depth, while at a distance of 250 m it increased with depth. The latter trend was probably due to a small accidental spill in 1965.

6.4.2 Accidental Oil Spills

Unfortunately, local hydrocarbon pollution incidents have occurred in Antarctica (e.g. Jouventin et al. 1984; Cripps 1992b), and there are many reports of oiled or killed seabirds (e.g. Williams 1984; Croxall 1987). In general, the most significant oil spills have been caused by shipwrecks, collisions or accidents during bunker fuel transfer. Cripps and Shears (1997), for instance, studied the fate of 1,000 l of an accidentally spilled diesel fuel at

Faraday Station. Total PAH concentrations in seawater reached 222 µg l⁻¹ the day after the spill, and decreased to background values within 1 week. However, the worst incident in Antarctica was the sinking of the Argentine supply ship *Bahia Paraiso* on 28 January 1989. The shipwreck occurred at Arthur Harbor near Anvers Island, about 2 km from the small scientific US Palmer Station. The release of about 550 m³ of diesel fuel, together with the *Exxon Valdez* oil spill in Prince William Sound (Alaska) the same year, for the first time focused international attention on the possible environmental risk of human activities in polar environments. The US National Science Foundation undertook significant efforts to contain environmental pollution, and sponsored a Quick Response Team for environmental monitoring of the area affected by the grounding of *Bahia Paraiso* (Kennicutt et al. 1990). Within 4 days of the accident, 100 km² of sea surface was covered by an oil slick. Kennicutt et al. (1991, 1995) and Kennicutt and Sweet (1992) described and quantified the extent and duration of the environmental impact, which was compared to the low-level, long-term impact of Palmer Station. Over the first weeks, all intertidal areas within a few kilometres of the wreck were contaminated and populations of intertidal limpet *Nacella concinna* were reduced by 50%. Owing to the volatility of fuel and the generally high hydrodynamic energy of the marine environment, contaminated beaches were cleansed within days or weeks of the main phase of spillage; nevertheless, high concentrations of total PAHs (about 400 µg kg⁻¹ dry wt.) were detected on the beaches, and the intertidal environment was occasionally re-oiled or freshly oiled for about one year after the spill. As occasional releases from the wreck, calm weather conditions and marine currents enhanced the accumulation of hydrocarbons in relatively low-energy areas, beach sediment samples collected in 1991 showed unusually high PAH contamination with respect to those in 1990. However, PAH contamination in subtidal sediments from Arthur Harbor, except those within a few metres of the ship, was mainly attributed to other local inputs such as shipping, boating and station activities. After two years, in most intertidal zones affected by the spill, PAH concentrations in *N. concinna* tissues had decreased to values at or near the detection limit of the analytical method. An example of this temporal decrease is shown in Fig. 43, which reports average concentrations of total PAHs (Kennicutt et al. 1990; Kennicutt and Sweet 1992) for limpets, collected between February 1989 and April 1991 at different distances from the *Bahia Paraiso* wreckage.

McDonald et al. (1992) measured PAH concentrations in fish collected at Arthur Harbor and from remote sites of the Antarctic Peninsula. In general, higher total PAH values were detected in the stomach contents and liver of *Notothenia coriiceps neglecta* caught near the *Bahia Paraiso* wreck. Amphipods and the limpet *N. concinna* were the dominant identifiable materials in the stomach contents of fish, and phenanthrenes and dibenzothiophenes were the most abundant PAHs. Naphthalene was the major PAH in liver and mus-

Fig. 43. Indicative temporal trends (1988–1991) of total PAH concentrations ($\mu g\ kg^{-1}$ dry wt.; mean±SD) in composite tissues of the limpet *Nacella concinna* at different distances from the *Bahia Paraiso* wreckage. (Data from Kennicut et al. 1990, 1992)

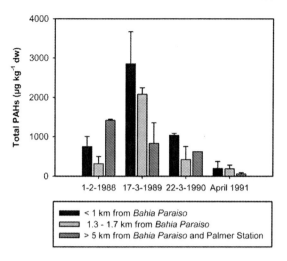

< 1 km from *Bahia Paraiso*
1.3 - 1.7 km from *Bahia Paraiso*
> 5 km from *Bahia Paraiso* and Palmer Station

cle tissue, suggesting a preferential bioaccumulation of more water-soluble PAHs. Total PAH concentrations in the stomach contents and muscle tissues of *N. coriiceps neglecta* captured near Palmer Station were significantly lower than those in fish caught near the wreck. The average content of naphthalene and phenanthrene in the bile of fish collected near *Bahia Paraiso*, Palmer Station and Low Island (a control site) was 69,000 and 9,000, 51,000 and 6,000, and 38,000 and 5,000 ng g^{-1} wet wt. respectively, with large standard deviations and considerable overlap of values. However, through gas chromatography with mass spectrometric detection (GC/MS), the metabolites of phenanthrenes and dibenzothiophenes were detected in the bile of some fish collected near the wreck and at Palmer Station, but not in samples from control sites. Likewise, hepatic EROD activity (ethoxyresorufin-O-deethylase, a biomarker of fish exposure to PAHs and to planar-halogenated and structurally similar compounds) was low to undetectable in *N. coriiceps neglecta* from remote sites of the Antarctic Peninsula, whereas values were variable in samples captured near the *Bahia Paraiso* wreck (30±29 pmol min^{-1} mg^{-1}), and highest (121±54 pmol min^{-1} mg^{-1}) near the pier at Palmer Station (Kennicutt et al. 1995).

The spill occurred at the middle to end of the seabird breeding season and potentially affected about 40,000 individuals (Kennicutt et al. 1990). The earliest evidence of the lethal exposure of seabirds to oil was recorded on 1 February 1989, when several dead, oiled Adélie penguins and blue-eyed shags were found in Biscoe Bay, adjacent to Palmer Station. Documented mortality from fouling and toxicity, in the 3-week period after the spill, was estimated to be less than 300 individuals. The direct effects of oil were not apparent in surface-feeding birds, although many of them fed on krill and limpets killed by the oil. However, mortality may have been higher due to other concomitant factors such as severe weather conditions, the efficiency of scavengers and predators, and the abandonment of breeding colonies during the period of seasonal dispersal of chicks. The effects of oil on reproduction were observed in blue-eyed shag nestlings which died of toxicity and abandonment. Mortality was exacerbated by the natural disappearance of prey. During the oil spill the local population of south polar skua (*Catharacta maccormicki*) suffered complete reproductive failure. As very few skuas (adults or chicks) died from direct contact with oil, it was hypothesised that sublethal oiling of adults temporarily reduced parental guarding (parents neglected their territories while cleansing themselves after feeding in slicks, thereby favouring the predation of chicks by other skuas; Eppley and Rubega 1989). However, other ornithologists working in the same region suggested that there was no relationship between the reproductive failure of skuas and the oil spill (Trivelpiece et al. 1990). In regions with extreme climatic and environmental conditions, natural events such as storms or food shortages in critical periods of the breeding season may occasionally result in reproductive failure. Without a better understanding of the natural variability of skua populations in the region, it is impossible to clear the controversy (e.g. Barinaga 1990; A. Anderson 1991) over the possible indirect effects of oil on seabird reproductive failure.

The impact of oil pollution on the metabolism of natural microbial communities is poorly understood. Crude oil contains thousands of hydrocarbons and related compounds, and these compounds can enhance (Bunch 1987), reduce (Griffiths et al. 1981) or have no effect (Carman et al. 1996) on the total abundance of sedimentary bacteria. It is well known that microorganisms play a critical role in the breakdown of hydrocarbons. Microbiological studies at Arthur Harbor showed that local densities of hydrocarbon-oxidising bacteria and degradation rates were extremely low (<100 g n-C_{16} cm^3 sediment year^{-1}) when compared with those in temperate regions (Kennicutt et al. 1990). On different sub-Antarctic intertidal beaches (Kerguelen Archipelago), Delille and Delille (2000) found that less than 5% of the total number of saprophytic bacteria consists of hydrocarbon-degrading microorganisms, but there are differences of one order of magnitude in the original richness of different sites. It has been suggested (Atlas 1991) that an initially high number of oil-degrading bacteria may indicate previous histories of oil contamination and/or chronic contamination. In the 3-year period following the *Exxon Valdes* oil spill in Alaska, for instance, Braddock et al. (1995) found significantly higher numbers of hydrocarbon-degrading microorganisms at sites affected by the oil slick (from 3.6×10^3 to 5.5×10^5 bacteria ml^{-1}) than at reference sites ($<10^2$ bacteria ml^{-1}). Delille and Delille (2000) performed oil-contamination experiments at nine intertidal beaches of the Kerguelen Archipelago; hydrocarbon-degrading microorganisms at some sites increased by several orders of magnitude within a few days, but no obvious effects were detected at two other sites. After 3 months of contamination, there was still a strong heterogeneity in the biodegradation capacity of different sites. It was hypothesised that this heterogeneity was due not only to differences in the grain size of sediments and in the availability of organic matter and nutrients, but also to previous contamination of some beaches by human activities, such as a whaling station operating at Port Jeanne d'Arc for more than 50 years.

George (2002) studied seasonal variations of anionic surfactant SDS (sodium dodecyl sulphate) biodegradation in Antarctic coastal waters at Rothera Station (Adelaide Island). He found that the half-lives of SDS in Antarctic coastal waters were generally far higher (160–460 h) than in temperate waters. Despite small seawater temperature differences (up to 2.45 °C), the persistence of surfactants in mid-winter was twice that measured in mid-summer. As at sites with higher hydrocarbon biodegradation capabilities, acclimatisation was taking place at the marine site receiving grey wastewater from the station. A large population of SDS-degrading bacteria developed in polluted waters during the summer months, and SDS half-lives were about 80 h shorter than in pristine Antarctic waters.

6.4.3 The Impact of Coastal Scientific Stations

The results of above-reported research at Arthur Harbor indicate that local anthropogenic sources released low concentrations of hydrocarbons and, except for the *Bahia Paraiso* spill, these sources are responsible for long-term contamination of the marine environment. In 1965 a station (Old Palmer Station) was built at Arthur Harbor, next to the British Base N Station. The present Palmer Station was built in 1968 at a distance of about 2 km. Near the old and the new station, there are abandoned dump sites where solid wastes were burned and some areas were contaminated by small fuel spills (from drains, leaks from fuel pipelines, overfilling of day tanks, and failure of fuel containers; Kennicutt et al. 1995). The marine environment is affected by wastewater, the transport of leachate from dump sites, the deposition of dust and particulates from station activity, and the impact of ships visiting the harbour. Most hydrocarbons used in Arthur Harbor are volatile and their transport to subtidal sediments mainly occurs through absorption onto sedimenting particles. Kennicutt and Sweet (1992) found elevated hydrocarbon concentrations in subtidal sediments directly adjacent to Old Palmer Station after cleanup activities at the site. They ascribed contamination to runoff and rapid sedimentation of soil particles containing hydrocarbons.

Concentrations of PCBs in sediments from Arthur Harbor were low (2.8–4.2 ng g^{-1}); however, intertidal limpets collected adjacent to the Old Palmer and Palmer Stations had rather high total PCB concentrations (28.5–75.7 ng g^{-1}). Molluscs from the vicinity of Palmer Station contained nearly equal amounts of hexa-, hepta- and pentachlorobiphenyls, and the profile of predominant congeners was similar to that of a mixture of Aroclor 1254 and 1260. PCB congeners with six or seven chlorines (i.e. those characteristic of Aroclor 1260) predominated in samples from the vicinity of Old Palmer Station. The predominance of high-molecular weight congeners indicated local sources of PCBs rather than long-distance atmospheric transport. Higher PCB concentrations in intertidal limpets than in subtidal ones suggested a runoff source. Except in samples collected directly adjacent to the station, PCB concentrations in limpets collected at Arthur Harbor in the 1989–1991 period were quite low and constant (8.7–28.5 ng g^{-1}; Kennicutt et al. 1995).

Although Ag, As, Cd, Cr, Cu, Pb and Zn concentrations were high in soil/waste samples from dumpsites at Palmer Station and Old Palmer Station, concentrations in subtidal sediment samples collected near the two stations were not significantly enriched above background values. This result was probably due to negligible transport of soil leachates from contaminated soil/waste to the marine environment. However, it could also be due to the fact that sediments in Arthur Harbor are quite mobile, and are regularly transported out of the bay by currents and ice scouring (Kennicutt et al. 1995).

At McMurdo Station (Ross Island) through the 1970s, wastes (drums, laboratory chemicals, obsolete vehicles, tyres, piping and other refuse) were routinely discharged along the eastern shoreline of Winter Quarters Bay (Manheim 1992). The bay provides docking facilities to visiting ships in an area adjacent to the former dumpsite. In 1988 the US National Science Foundation, in response to criticism by Greenpeace and other environmental groups, began a dumpsite cleanup and abatement programme. Inland and shoreline waste dumps were removed and waste was returned to the US. In 1989 a rigorous programme was initiated for reducing and recycling materials and for environmental monitoring. Within the framework of environmental studies aimed at evaluating the impact of human activities at specific Antarctic sites, Winter Quarters Bay became one of the most studied marine areas on the continent. Risebrough et al. (1990) determined the distribution of PCBs, Polychlorinated Terphenyls (PCTs) and DDT compounds in sediments collected at depths of 18–33 m from Winter Quarters Bay, from near the wastewater discharge site, and from locations (Cape Armitage, Turtle Rock and Cinder Cone) at increasing distances from the bay. PCBs were detected in all samples (except in one section of a sediment core, 13–18 cm deep), and values ranged from 0.01 ng g^{-1} dry wt. in deeper portions of the cores from Cinder Cone to 1,400 ng g^{-1} at the station nearest the former Winter Quarters Bay dumpsite. In general, values in the bay samples were three to four orders of magnitude higher than in Turtle Rock and Cinder Cone samples. Assuming a representative value of 500 ng g^{-1} in the top 10 cm of the Winter Quarters Bay sediment and an area of 80,000 m^2, Risebrough et al. (1990) estimated a total of about 4 kg of PCBs. The PCB profiles of all samples but one were identical to that of

Aroclor 1260, showing no degradation of any individual congeners. No trace was found of lower-chlorinated PCBs, which are assumed to have been present in capacitors and transformers.

Although the production of PCTs in the US ended at the beginning of the 1970s, these compounds were detected in the sediments of Winter Quarters Bay (concentrations ranged from 30 to 1,200 ng g^{-1} dry wt.). The bay was estimated to contain a total of about 2 kg of PCTs with composition similar to that of Aroclor 5460. Concentrations of DDE (0.4–4.7 ng g^{-1} dry wt.) and other DDT compounds in bay sediments were one to two orders of magnitude higher than at any other station. As global sources of POPs were insignificant, Risebrough et al. (1990) attributed environmental pollution in Winter Quarters Bay to past discharge of waste at the former dumpsite, and to the use of PCBs and PCTs as cutting and casting waves in machine shops on the ships or in local machine shops.

Kennicutt et al. (1995) determined PCB concentrations in sediments collected from Winter Quarters Bay and from other stations in McMurdo Sound. In general, results confirmed previous findings by Risebrough et al. (1990). PCBs were detected in all sediments collected near McMurdo Station, and the highest values were measured in samples from sewage outfall (350–690 ng g^{-1} dry wt.) and from Winter Quarters Bay (680–1,280 ng g^{-1}). Grid sampling of the latter area in December 1993 showed that total PCB concentrations ranged from 250 to 4,300 ng g^{-1}; in five samples concentrations exceeded 1,000 ng g^{-1} dry wt. Values decreased rapidly away from the bay: at Cape Armitage (less than 1 km away) they were more than one order of magnitude lower (18–28 ng g^{-1} dry wt.), and at the remotest sites they were less than 1 ng g^{-1} dry wt. Research by Kennicutt et al. (1995) confirmed that PCB compositions corresponded to Aroclor 1260, with little evidence of degradation. PAH concentrations in nine sediment samples collected from the Winter Quarters Bay in late 1993 ranged from 360 to 13,000 ng g^{-1} dry wt., with the highest values at the head of the bay. The predominance of naphthalenes and phenathrenes indicated that most environmental contamination was due to oil spills (at the site or transported to the bay by runoff from adjacent land areas). Pyrogenic PAHs were almost exclusively detected in sediments from Winter Quarters Bay, and their concentrations paralleled the increase in fuel PAHs. The abandoned dumpsite along the eastern shoreline of the bay was therefore considered the main source of pyrogenic PAH.

Subsamples of sediments analysed by Risebrough et al. (1990) for PCBs, PCTs and DDT compounds were also assessed by Lenihan et al. (1990) for trace metal concentrations. As in the case of POPs, concentrations of most metals (e.g. Ag, Cd, Cu, Hg, Pb and Zn) were higher (from two to ten times) in sediments from Winter Quarters Bay than in samples from outside the bay. The presence of a submarine sill was thought to prevent the spread of polluted sediments outside the bay area. However, the distribution pattern of metals was complex – Ni concentrations, for instance, were highest at reference sites

(Cinder Cones and Turtle Rock). Results by Lenihan et al. (1990) were substantially confirmed and extended by Kennicutt et al. (1995). They found that
sediments in Winter Quarters Bay and control sites were rich in basaltic materials characterised by high Ni and Cr concentrations, high Fe/Al ratios, and
rather low concentrations of Pb and Cd. It was rather difficult to reliably
assess the impact of human activities due to the geochemistry of McMurdo
Sound sediments, and the variable percentage of basaltic debris in analysed
samples. Furthermore, discarded metal scraps and ageing of structures had
probably enhanced Fe concentrations in marine sediments. Kennicut et al.
(1995) found evidence of anthropogenic Ag, As, Pb, Sb, Sn and Zn contamination in sediments from Winter Quarters Bay and the sewage outfall area, while
Cu, Ni and Cr concentrations (maximum values=92, 85 and 25.2 μg g^{-1} dry wt.
respectively) nearly corresponded to natural concentrations in basaltic
debris. In contrast to the results of Lenihan et al. (1990), Cd, Hg and Se concentrations were not found to be high. Figure 44 shows average trace metal
concentrations (μg g^{-1} dry wt.) in surface sediments from Palmer Station and
McMurdo Station (Winter Quarters Bay, outfall, and control sites) reported by
Kennicutt and McDonald (1996).

Davis Station was established in 1957 on the Vestfold Hills shoreline (Prydz
Bay, Princess Elizabeth Land) by the Australian National Antarctic Research
Expedition. The station is usually occupied by 20–30 people in winter, but
there may be a three- to fourfold increase in personnel during the austral
summer. A sewage treatment plant was installed in summer 1990/1991, and
since then all wastewater receives primary and secondary treatment before
discharge into the sea (Antarctic Division 1993). However, like in many other
Antarctic stations, the summer population at Davis is well above the optimum
population size served by the treatment plant (about 60 persons). Other
potential sources of contaminants are the fuel storage depot, refuelling stations and the power plant. Green and Nichols (1995) determined hydrocarbon
and sterol concentrations in marine and shoreline sediments adjacent to
Davis Station. The total sterol content (up to 119 μg g^{-1} dry wt.) in sediments
was much higher than concentrations previously measured in Antarctica,
such as at Bransfield Strait (0.15–0.42 μg g^{-1} dry wt.; Venkatesan et al. 1986)
and McMurdo Sound (0.40–8.0 μg g^{-1} dry wt.; Venkatesan 1988). Two distinct
sterol profiles were identified: one in sewage outfall sediments, shoreline sediments and wildlife faeces (dominated by coprostanol and/or cholesterol),
and another in Davis Bay sediments (dominated by algal sterols). The occurrence of faecal coprostanol (13.2 μg g^{-1} dry wt.; 60 % of total sterols) in sediments adjacent to the station sewage outfall and along the shoreline at Davis
Beach (up to 5 μg g^{-1} dry wt.) was due solely to human contamination,
because it was not detected in the faeces of elephant seals or Adélie penguins
(consisting almost entirely of cholesterol). Sewage contamination at Davis
Station extended at least 200 m offshore. Concentrations of hydrocarbons in
shoreline sediments near the station were rather low (up to 5.5 μg g^{-1} dry wt.)

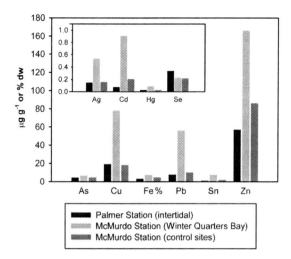

Fig. 44. Mean concentrations ($\mu g\ g^{-1}$ dry wt.) of trace metals in surface sediments from Palmer Station and McMurdo Station. (Data from Kennicutt and McDonald 1996)

and their profile indicated an anthropogenic origin, most likely from land runoff or small spills during ship-to-shore fuel transfer. PAHs, primarily naphthalenes, fluorenes and phenanthrenes, were present in all sediment samples in very low concentrations (parts per trillion). In a previous study of the same marine area, Green et al. (1992) measured very low PAH concentrations in sea-ice algae (mean=230 $\mu g\ m^{-2}$) and seawater particulates (mean=7.3 ng l^{-1}). The latter value was much lower than that measured by Desideri et al. (1989) at Terra Nova Bay (mean=0.15 $\mu g\ l^{-1}$) or by Cripps (1992a) in the Atlantic sector of the Southern Ocean (0.08 $\mu g\ l^{-1}$). In the Vestfold Hills region, Green et al. (1992) measured remarkably higher PAH concentrations (2.24–7.27 $\mu g\ g^{-1}$ dry wt.) in anoxic sediments from the Ellis Fjord. These "anomalous" values were attributed to anoxic conditions which can lead to indefinite preservation of PAHs. A novel bacterial source was also hypothesised for this particular environment, as two PAHs (1,3- and 1,6-dimethylnaphthalene) were overwhelmingly the dominant constituents, and their relative abundance was different from that in other sediment samples.

Desideri et al. (1989, 1990) determined concentrations of organic compounds in samples of seawater, particulate matter (mainly phytoplankton) and sediments from Terra Nova Bay. High concentrations of benzodicarboxylic acid esters, which are certainly anthropogenic, were detected in all matrices. The content of n-paraffins, aromatic hydrocarbons and phthalates was higher in sediments collected near the Italian Station. In the same marine area, Fuoco et al. (1994) reported a mean PCB concentration of 0.79± 0.36 ng l^{-1} in seawater, and of 0.11±0.08 ng g^{-1} dry wt. in surface sediments, with a quite homogenous distribution.

Poland (in 1961), Brazil (1984), and Peru (1988) established three stations at Admiralty Bay, on King George Island (South Shetland Islands). During four austral summers, Bicego et al. (1996) detected low concentrations of total aromatic hydrocarbons in surface seawater from Admiralty Bay (80 ng l^{-1}). Individual levels and types of PAHs were typical of marine environments with small oil-derived hydrocarbon input. In sediment samples from the same bay, Montone et al. (2001a) measured low PCB concentrations (total PCB congeners expressed as Aroclor 1254 ranged from 2.03 to 5.91 ng g^{-1}). In contrast to PCB profiles in Winter Quarters Bay sediments (Risebrough et al. 1990; Kennicutt et al. 1995), those in Admiralty Bay suggested a contribution from the commercial mixtures Aroclor 1242 and 1254, which have been widely used in electrical equipments. Owing to the absence of higher-chlorinated congeners, Montone et al. (2001a) concluded that there are no significant local sources of PCBs in Admiralty Bay, and long-range atmospheric transport is probably the main source of PCB contamination.

Alam and Sadiq (1993) roughly evaluated metal contamination near some scientific stations in the Antarctic Peninsula. They collected several sediment samples from beaches and intertidal sites and found evidence of contamination by Cd, Cr, V and other metals at Horse Shoe Island (near an old British station) and at Marsh Martin (King George Island, where several nations have established Antarctic stations).

6.5 Effects of Local Environmental Pollution on Benthic Communities

As benthic organisms have restricted mobility and live in contact with sediments, where organic matter and pollutants accumulate, they are well suited to environmental impact studies. The most common human impact on aquatic ecosystems is the deposition of organically enriched sediments around sewage discharges; the response of benthic communities usually consists in the loss of some resident species and an increase in organisms better adapted to reduced oxygen concentrations and changes in sediment redox potential. These opportunistic species are usually characterised by short generation times, high larval availability, and rapid colonising ability. The distribution of benthic assemblages is affected not only by the content and nature of organic matter but also by many other factors such as hydrodynamic conditions, particle size, microbial associations, food availability, and environmental contaminants (Pearson and Rosenberg 1978). The distribution of benthic assemblages is also affected by recruitment and succession processes, and the physico-chemical characteristics of the marine environment play an important role in the early mortality of larvae and in the settlement and metamorphosis of benthic organisms. It is well known, for instance, that some of the most common local Antarctic pollutants such as hydrocarbons and trace metals negatively affect the recruitment of benthic communities (e.g. Berge 1990; Watzin and Roscigno 1997; Olsgard 1999).

Relationships between the distribution of benthic fauna and the physico-chemical characteristics of the sedimentary environment can be evaluated through field surveys, manipulative field experiments or laboratory experiments, and mesocosms. In general, field surveys and experiments are rather complex and need to run for significant periods of time, with the consequent risk of organism loss and disturbance. These approaches are particularly difficult and risky in Antarctica, where benthic communities may be dramatically affected by physical disruption of the seafloor by iceberg grounding and anchor ice formation. Since the 1970s, attempts have been made to assess the impact of human activities on benthic communities at McMurdo and South Georgia (e.g. Dayton and Robilliard 1971; Platt 1978, 1980). Clear relationships between benthic community patterns and the impact of former whaling activity were not found at South Georgia, probably because of low residual hydrocarbon contamination in surface sediments. On the contrary, at McMurdo Station, the largest inhabited settlement in Antarctica, the availability of "historic" data on benthic communities and the presence of one of the most polluted sites in Antarctica (Winter Quarters Bay) allowed assessment of the responses of benthic communities to local anthropogenic pollution. Lenihan et al. (1990) found that species of sedentary infaunal invertebrates largely dominated Cape Armitage and McMurdo Sound, while two species of motile and highly opportunistic polychaete worms (*Capitella capitata antarcticum* and *Ophryotrocha claparedii*) dominated in polluted sediments at the back of Winter Quarters Bay. Closely related species of polychaetes live at temperate latitudes in organically enriched sediments. Dramatic changes along the pollution gradient were also observed in epifaunal invertebrate communities which, in McMurdo Sound, are usually characterised by abundant sea star (*Odontaster validus*), sea urchin (*Sterechinus neumayeri*) and large bivalve populations. The edge of the heavily contaminated bay (i.e. the transition area) contained several motile and less opportunistic polychaete species, while uncontaminated sediments were characterised by dense tube mats of infaunal animals dominated by polychaete worms, crustaceans and large suspension-feeding bivalves. A highly significant negative correlation was found between infaunal abundances and hydrocarbon concentrations in sediments. The results of laboratory bioassay experiments (Lenihan 1992; Lenihan et al. 1995) on local amphipod (*Heterophoxus videns*) and other crustacean species revealed that mortality was highest in sediments from Winter Quarters Bay, high in sewage outfall sediments, and progressively decreased in sediments collected at increasing distances from Winter Quarters Bay. Whole, intact infaunal communities were transplanted along the pollution gradient and sampled 1 year later. The community structure changed dramatically in Winter Quarters Bay, less so at the edge of the bay, and little at an uncontaminated site. Lenihan and Oliver (1995) found that community patterns along natural gradients of anchor ice disturbance or bottom areas gauged by icebergs were similar to community changes along anthropogenic disturbance gradients. A large number of the same motile and opportunistic polychaete worms were found in areas where anchor ice freezes and uplifts, and in the more recent ice gouges along the east side of McMurdo Sound.

Cleveland et al. (1997) studied the acute toxicity and genotoxicity of sediments from Winter Quarters Bay, the bioavailability of contaminants, and the influence of UV radiation on the bioavailability and toxicity of sediment-associated contaminants. Although the experiment was conducted with temperate species, such as the amphipod *Leptocheirus plumulosus*, and under temperate conditions, sediments from the bay (containing about 250 ng g^{-1} dry wt. of total PCBs and 20 µg g^{-1} dry wt. of total PAHs) elicited toxicity in the Microtox test, avoidance, and inhibited burrowing in the amphipod test. It was impossible to attribute toxic responses to specific sediment-associated pollutants; however, Cleveland et al. (1997) found that relatively unpolluted sediments from McMurdo Sound apparently contained some unidentified substance which was photolytically modified to a more toxic form.

Over the 1988–1998 period Conlan et al. (2000) studied the impact of McMurdo sewage outfall on benthic communities from a sampling site, located 62 m offshore, in 1988. In 1988 sewage was released at the shore's edge, but over the 1990–1992 period the sewage pipeline was extended 56 m off-shore (i.e. closer to the sampling site). This long-term study allowed assessment of spatio-temporal changes in benthic communities. Multivariate analyses of samples showed a marked reduction in the diversity of fauna after extension of the sewage outfall, while a promising increase in biodiversity was detected in 1997 and 1998. Further development of long-term monitoring projects at McMurdo (Kennicutt et al. 2003) will probably reveal a further increase in the biodiversity of this area, given the planned construction of a sewage abatement plant at McMurdo Station.

A number of studies have used Antarctic benthic communities as indicators of habitat conditions and to detect human-induced changes in the marine environment at Casey Station (e.g. Stark 2000; Stark et al. 2003; Thompson et al. 2003). Casey Station, situated on the Bailey Peninsula (Windmill Islands, Wilkes Land coast), is the third research station to be constructed in this area, with the Wilkes and "Old Casey" stations nearby. Wilkes Station was built by the US in 1958 on Clarke Peninsula for the International Geophysical Year; in 1959 it was taken over by the Australian National Antarctic Research Expedition (ANARE), while "Old Casey" was being built on Bailey Peninsula. The dilapidated buildings at Wilkes Station are surrounded by rusting drums of fuel, oil and lubricants, gas cylinders, and other types of refuse and environmentally hazardous substances (Deprez et al. 1999). "Old Casey" Station was completed in 1969 and "New" Casey Station, about 0.7 km away, was opened in 1989. The old station has undergone demolition and several disused buildings were removed; however, Deprez et al. (1999) found that an old waste disposal dump and other sites in the area were contaminated by petroleum hydrocarbons, PCBs and metals such as Cu, Pb and Zn, which partly leached into adjacent marine ecosystems. In addition to Brown Bay and Newcomb Bay, which are affected by old waste dumps near the Old Casey and Wilkes Station respectively, the current sewage outfalls in Shannon Bay and Casey Wharf are among the most potentially impacted marine areas adjacent to Casey Station (Snape et al. 2001). To determine whether human activities had affected marine benthic communities, Stark (2000) studied benthic assemblages at reference and potentially impacted sites. He determined the levels of stress and of correlations between assemblage patterns and environmental contamination. Through manipulative field experiments, Stark et al. (2003) then verified the causal link between observed patterns and contamination. It was found that the composition of soft-bottom benthic assemblages in the Casey region (dominated by crustaceans) was unlike that of other assemblages recorded in Antarctica (usually dominated by polychaetes; e.g. Richardson and Hedgpeth 1977; Knox 1994). Significant spatial variations at all spatial scales were found in both communities and populations of taxa, but the largest, most significant differences in diversity and species richness were those between reference and impacted locations. In agreement with findings by Lenihan and Oliver (1995) on polluted sediments at Winter Quarters Bay, some species of polychaetes were the most opportunistic and were very abundant at polluted sites but not at control sites. In general, cumaceans and the amphipod *Monoculodes* sp. were sensitive and extremely rare at polluted sites. However, some amphipod species were abundant at some impacted sites; although the species *Orchemenella franklini* was the most affected by hydrocarbon-treated sediments, it appeared to respond in an opportunistic manner to metal contamination. Among measured environmental variables (grain size, depth, total organic carbon and metal concentrations), the content of Cd, Cu, Pb, Sn and Zn showed the greatest correlation with biological patterns. In contrast, a 3-month field experiment on the recolonisation of sediments artificially contaminated by metals revealed no observable effect. As suggested by Stark and Riddle (2003), this result was probably due to the short duration of the experiment and/or to the scarce bioavailability of the metal added. Assemblages (especially crustacean abundance) in hydrocarbon-contaminated sediments were significantly different from those in control and metal-treated sediments at two locations. Polychaetes and gastropods tended to be more abundant in hydrocarbon- and metal-treated sediments than in controls.

As monitoring of human-induced changes in marine environments through benthic organism associations has been criticised as time-consuming and expensive (e.g. Ferraro et al. 1989), research was undertaken at Casey Station to improve the cost efficiency of marine pollution biomonitoring in Antarctica. Thompson et al. (2003) found that sieving of sediment samples through a 1.0-mm sieve instead of a 0.5-mm sieve can reduce sorting and identification time by 38 %, while identifying fauna to the family instead of species level could reduce sample processing by another 40 %. Although these are useful suggestions, standardised procedures are needed for environmental monitoring and management of areas near Antarctic stations through benthic community surveying. To this purpose, it is necessary to optimise the sampling (e.g. area and depth of sediment samples, and the number of replicates for each station) and verify if the proposed sieve mesh size and taxonomic resolution are applicable in other Antarctic marine environments.

6.6 Summary

The Southern Ocean is a mosaic of distinct subsystems with specific physico-chemical and biological characteristics. Owing to the low continental input, the bioavailability of lithophilic elements such as Fe and Mn, which are essential for phytoplankton metabolism, is lower in large areas of the Southern Ocean than in any other seas. Although concentrations of other essential trace elements such as Co, Cu, Ni and Zn are higher or in the same range as those reported for other ocean waters, the Southern Ocean water column is rather unstable and elements show a less structured vertical distribution. Their concentrations seldom show the surface depletion typical of all other seas, and it is hard to gain a better understanding of their biogeochemical cycles due to element recycling in the water column and to inter-replacement of elements in the enzymatic sites of phytoplankton cells. Considering the potential role of the Southern Ocean in global processes such as CO_2 partitioning between the atmosphere and oceans, understanding physico-chemical and biological processes in this ocean will be a major research trust for the next decades.

   One of the most astonishing features in Southern Ocean surface waters is the high Cd content. Antarctic marine organisms have probably evolved and adapted to the naturally enhanced bioavailability of this potentially toxic element, and there is evidence that some species of diatoms in Zn-depleted waters can replace Zn with Cd at enzymatic sites. Data on concentrations of Hg chemical species in Southern Ocean waters are completely lacking. However, there is evidence that Hg can reach high concentrations in the organs and tissues of Antarctic seabirds and marine mammals, through natural processes such as biomagnification of methylmercury along food chains. Lead fluxes from anthropogenic sources have characteristic, conservative isotopic compositions. Like in Antarctic snow, the isotopic composition of Pb in Southern Ocean waters indicates an anthropogenic contribution from sources probably located in other continents of the Southern Hemisphere and in Antarctica.

   Surface sediments are one of the most suitable environmental matrices to study the spatial distribution and cycling of trace metals and other persistent chemicals in the marine environment. Thus, although the geochemistry of Antarctic pelagic sediments has been scarcely studied, these sediments have been widely used for environmental studies in marine areas near scientific stations and disused whaling stations along the Antarctic or sub-Antarctic shoreline. Petroleum hydrocarbons, also released by accidental oil spills, are among the most widespread pollutants in Antarctic coastal environments. Notwithstanding the difficulties in discriminating between natural and anthropogenic hydrocarbons, high concentrations of persistent toxic compounds such as PAHs have been measured in sediments and organisms affected by the *Bahia Paraiso* wreck and near long-inhabited scientific stations. In all cases PAH, metal, PCB and PCT concentrations rapidly decrease within a few hundred metres of sources. Although further research is necessary to optimise and standardise procedures, surveys of benthic communities performed in McMurdo Sound and near Casey Station show that such assessments are one of the most suitable means of determining the impact of human activities on Antarctic coastal ecosystems.

Before concluding, the reader should note that, although most data on environmental monitoring and research near Antarctic stations reviewed in this chapter refer to American (Palmer, McMurdo) and Australian (Davis, Casey) scientific stations, this does not necessarily mean that they are the most polluted stations in Antarctica or the sub-Antarctic islands. Rather, these stations were selected because environmental pollutants in other disused (or operating) Antarctic stations and their biological effects were either not monitored, or results were not reported in widespread scientific journals.

7 Persistent Contaminants in Antarctic Marine Food Chains

7.1 Introduction

In previous chapters we saw that Antarctica and the Southern Ocean are scarcely affected by environmental contamination, except in localised areas adjacent to abandoned or inhabited human settlements. Nevertheless, this chapter will show that concentrations of persistent contaminants in Antarctic marine organisms are not necessarily low, and that they cannot always be taken as global reference values in biomonitoring studies. Although pesticides have never been produced nor applied in Antarctica, DDT and its derivatives have been detected in tissues of penguins and seals since the 1960s. Concentrations of these compounds in Antarctic biota increased up to the early 1980s and then apparently decreased, but less so than might be expected from global environmental data. A large body of data also exists on PCB concentrations in Antarctic marine organisms. The high variability of data does not allow the detection of clear temporal trends. However, the cold-condensation process and interactions between the atmosphere and the Southern Ocean suggest that Antarctic regions are a sink for the most volatile compounds, whose concentrations in Antarctic organisms are likely to increase.

In contrast to DDTs, PCBs and other xenobiotic compounds, trace elements occur naturally in the environment. The unique environmental features and biogeochemical cycles of major and trace elements in the Southern Ocean, together with the unique ecophysiological adaptations of marine organisms which underwent long evolutionary processes in isolation, affect the uptake of essential and potentially toxic metals. While it is possible to establish average background values of trace metals in sediments from different seas with the same grain size and organic matter content, there is probably no global reference value for trace metals in marine organisms. Although concentrations of persistent contaminants in organs and tissues of Antarctic organisms have often be used to demonstrate exposure, bioaccumulation data may not accurately reflect exposure. These organisms tend to be slow-grow-

ing and long-lived; as food availability (and the uptake of persistent contaminants) exhibits strong seasonal variability, concentrations of contaminants in
their organs and tissues may vary in relation to the sampling period and
metabolism. Thus, it is not surprising that some species of Antarctic marine
invertebrates, seabirds and mammals accumulate naturally the highest-ever
recorded concentrations of Cd and other potentially toxic elements. The
understanding of bioaccumulation processes is further complicated by
unique features of Antarctic food webs and spatio-temporal variations in the
physico-chemical characteristics of Southern Ocean water masses.

During the last three decades, a number of papers have been published on
concentrations of trace metals and other persistent contaminants in Antarctic
organisms, with the aim of using data from remote and uncontaminated areas
as background values for more contaminated marine areas. Unfortunately,
most data derive from one-off studies rather than repeated sampling; some
results seem unreliable, and background levels in Antarctic organisms are not
always clear. As in the case of water and sediment samples, there is a lack of
standard procedures for collecting, storing, preparing and analysing biological samples. It is often difficult to compare among values reported in different
studies for the same Antarctic species because results refer to the whole body
or to particular organs or tissues, and different measurement units are often
used (e.g. fresh or dry mass, total lipid content, etc.). To perform reliable comparisons between bioaccumulation data on Antarctic organisms and that on
related species from other seas, differences in age (i.e. exposure time), feeding
behaviour, growth, reproductive cycle, and species-specific detoxification and
excretion mechanisms should be considered.

As diet plays an important role in the bioaccumulation of environmental
contaminants, this chapter will survey the results of research on the feeding
behaviour of Antarctic marine organisms. Literature data on persistent contaminant concentrations will be reported according to the scheme adopted in
Chapter 3 – krill and pelagic food webs, benthic and epibenthic organisms in
the neritic province, and seabirds and seals breeding in Antarctica. The
pelagic environment in the oceanic province shows two different food webs:
one in permanently ice-free zones (dominated by copepods, salps and small
euphausiids) and another in the seasonal pack-ice zone, in which krill
(*Euphausia superba*) and the short food chain diatoms–krill–whales (or seals
or penguins) are the dominant element. As cephalopods and epipelagic fish
are important prey for a range of pelagic vertebrates such as large fish,
seabirds and marine mammals, they may occupy important niches in these
trophic chains. While primary producers include microscopic phytoplankton
with very short turnover times, krill and most of its predators (except
cephalopods) are typically long-lived organisms. The relatively high efficiency of short food chains in the transfer of organic matter to top predators
tends to reduce the amplification of pollutant loadings (biomagnification),
but this effect is partly counterbalanced by the long lifespan (i.e. exposure

time) of Antarctic marine organisms. Several species of seabirds and marine mammals migrate to lower latitudes (some even reach the Northern Hemisphere) during the austral winter; these seasonal migrations further complicate the interpretation of bioaccumulation data because migrating species reflect the integrated input of persistent contaminants (inside and outside the Southern Ocean).

The neritic province includes nearshore waters and ice shelves where phytoplankton may show intense and brief blooms. Due to the low biomass of zooplankton, however, most algal cells sink and become food for very rich benthic communities (sponges, hydroids, tunicates, polychaetes, molluscs, actinarians, echinoderms, amphipods and fish). Many invertebrates are suspension-feeders, grow slowly, have long lifecycles, and can survive long period of starvation at negligible metabolic cost. They constitute an important link between primary producers (phytoplankton and benthic macroalgae) and notothenioid fish, which are in turn eaten by birds and marine mammals. As penguins and seals may be eaten by leopard seals and killer whales, food webs on Antarctic continental shelves can become rather complex and long, thereby enhancing the biomagnification of methylmercury and other persistent contaminants.

Evolutionary adaptations to Antarctic conditions may have enhanced the sensitivity of marine organisms to POPs and trace metals (especially those with low natural bioavailability in the Southern Ocean). Many biological processes such as growth and reproduction are slower in Antarctic organisms than in related species from other seas, and the relatively slow metabolism may decrease their ability to actively detoxify or remove contaminants from their bodies. Some Antarctic stations have sophisticated laboratories and suitable facilities to investigate interactions between marine organisms and persistent pollutants. Besides reviewing data on concentrations of persistent contaminants in organs and tissues of Antarctic organisms, this chapter summarises the results of studies on the physiological response of Antarctic organisms to experimental exposure to environmental pollutants.

7.2 Trace Elements and POPs in Pelagic Plankton

A better understanding of the trophic transfer of persistent contaminants requires knowledge of their interactions with phytoplankton cells. In general, the uptake of metals depends on geochemical and biological processes such as metal speciation, metal–metal antagonistic interactions and environmental factors (e.g. temperature or CO_2 concentrations; Sunda 1994; Cullen et al. 1999). In two species of diatoms, for instance, the accumulation of Cd was found to depend on ambient nitrogen conditions (i.e. on the physiological status of phytoplankton cells; Wang et al. 2001). As large areas of the Southern

Ocean are characterised by high nutrient bioavailability, this could be a factor enhancing the uptake of Cd in phytoplankton and its transfer to marine food chains. Algae produce extracellular chelating compounds or porphyrin complexes, which play an important role in the sequestration of essential and nonessential elements and other persistent contaminants (Hutchins et al. 1999). Moreover, there is evidence that marine diatoms, like higher plants, respond very rapidly to Cd (and probably to other metal stress) by producing cysteine-rich polypeptides (phytochelatins). These compounds play a major role in the storage of metals and their detoxification. Diatom exposure to high inorganic Cd concentrations determines an efflux of phytochelatins and Cd from their cells. As the Cd–phytochelatin complex does not appear to be very stable in seawater once outside the cell, the exported metal remains available to diatoms and induces phytochelatin synthesis (Lee et al. 1996).

Relatively little effort has been put into determining concentrations of persistent contaminants in organisms from open-ocean waters, particularly in those from the Southern Ocean. In general, open-sea studies are more expensive and are often ancillary to coastal-oriented contamination surveys. Moreover, plankton samples usually include communities of mixed pelagic organisms representing many phyla; owing to different physiologies, life spans and bioaccumulative abilities, it is difficult to compare concentrations of persistent contaminants in samples of mixed plankton. In general, concentration factors of 10^2–10^5 relative to seawater concentrations have been reported for most environmental contaminants in phytoplankton. In the 1990s the Southern Ocean water was largely free of radionuclide contamination, but Wood et al. (1990) demonstrated that Antarctic microalgae were able to concentrate otherwise undetectable, artificial and naturally occurring radionuclides. Measured concentrations of ^{239}Pu, ^{95}Nb and ^{144}Ce, for instance, ranged from 0.1–106 fCi (femtocurie) g^{-1} wet wt.

Joiris and Overloop (1991) analysed organochlorine residues in samples of particulate matter (mainly phytoplankton) from the Indian sector of the Southern Ocean. They found that PCB concentrations (0.7 µg g^{-1} dry wt.) were similar to those in temperate zones but, when values were expressed per unit water volume rather than dry weight, Antarctic concentrations (1.2 µg m^{-3}) were seven times lower than North Sea ones (8.8 µg m^{-3}). The low phytoplankton density in the sampled marine area had determined high PCB levels per unit of biomass. Thus, the authors suggested the need to adopt a different system of units in order to correctly express the contamination level of a marine area and to identify mechanisms responsible for persistent contaminant accumulation. In netplankton (mainly zooplankton) samples from the same zone, PCB concentrations were comparable to those in phytoplankton on a dry weight basis, lower on a lipid weight basis, and much higher when expressed per unit seawater volume (1.2 µg m^{-3}). Very low concentrations or traces of lindane, heptachlor epoxide, dieldrin, DDE and DDT were detected in various netplankton samples, while these compounds were undetectable in bulk phytoplankton. The high DDT/DDE ratio indicated that Antarctic organochlorines had originated recently in the Southern Hemisphere, and suggested a possible increase in Antarctic contamination by POPs.

Corsolini and Focardi (2000) measured mean PCB concentrations (the sum of 50 congeners) in Ross Sea phytoplankton (mainly diatoms) and mixed zooplankton (copepods, amphipods and krill) of 1 and 4.2 ng g^{-1} wet wt. respectively. Although the percentage of each congener was a small percentage of the total PCB content (<10%) in seawater, phytoplankton showed a remarkable increase in PCBs 153, 138, 180 and 195.

Fronts, convergences and divergences are permanent features of the Southern Ocean, separating water masses with different physico-chemical and biological characteristics. As individual plankters remain within a particular body of water, Hennig et al. (1985) analysed samples of zooplankton collected between New Zealand and McMurdo Sound to determine whether their elemental composition could be used to identify their provenance. Although the results of this study may have been affected by the use of formalin-fixed materials, the most widespread species (i.e. the amphipod *Themisto gaudichaudii*, the chaetognath *Eukronia hamata* and the euphausiid *Euphausia triacantha*) had very different metal concentrations, and amphipods generally showed the highest values. There were also striking differences among samples of the same species collected in different water masses. Large variations in metal concentrations were also found among samples collected in the same area but on different legs of the cruise (i.e. collected 2 months apart). In addition to the species-specific and spatio-temporal variability of metal concentrations in zooplankton, the study by Henning et al. (1985) showed that metal concentrations in Antarctic organisms are not necessarily low. Fe and Zn concentrations in *E. triacantha*, for instance, were higher in samples from the Ross Sea than in those from the deep-ocean basin and New Zealand continental shelf.

Table 13 reports average trace metal concentrations in some species of pelagic crustaceans from the Weddell Sea. Rainbow (1989) found geographical variations in metal concentrations, and a significant negative regression between the size of the amphipod *T. gaudichaudii* and its Fe and Zn contents. This relationship suggests that surface-adsorbed metals may constitute a significant proportion of the whole-body metal content in small individuals with a higher surface area/body weight ratio. Furthermore, smaller amphipods are faster growing and probably have greater physiological requirements for essential trace elements. In general, when compared with related species from other seas, crustaceans from the Southern Ocean have lower Pb concentrations, comparable Zn and Ni contents, and much higher Cu and especially Cd contents (particularly in caridean decapods and some species of hyperiid amphipods). The bioaccumulation of Cd has an ecological significance because amphipods of the genus *Themisto*, for instance, are an important component in the diet of petrels (Croxall et al. 1988) and squid which, in turn, are eaten by other seabirds and marine mammals.

Table 13. Mean trace metal concentrations ($\mu g\ g^{-1}$ dry wt.) in some species of pelagic crustaceans from the Southern Ocean

	Species	Region	Cd	Co	Cu	Ni	Pb	Zn
Amphipoda[a]	Themisto gaudichaudii	60° 30'S/56° 04'W	53	–	31	–	–	45–83
		53° 55'S/37° 57'W	19	–	28	–	–	32–77
	Eusirus properdentatus	60° 49'–65° 45'S 55° 24'–68° 15'W	95	–	107	–	0.44	49
Decapoda[b]	Notocrangon antarcticus	73° 21'–77° 31'S	13	–	67	–	0.78	46
	Chorismus antarcticus	21° 24'–42° 12'W	13	–	93	–	1.60	44
	Metridia gerlachei	66° 0'–74 ° 30'S	10	0.06	26	11.3	0.72	518
Copepoda[c]	Calanoides acutus	53° 55'E–38° 32'W	4.6	0.02	10	4.3	0.31	183
	Calanus propinquus		5.6	0.02	26	4.8	0.51	191

[a] Rainbow (1989)
[b] Petri and Zauke (1993)
[c] Kahle and Zauke (2003a)

As discussed in the preceding chapter, Cd concentrations in Southern Ocean surface waters are higher than in other ocean waters. The bioaccumulation of Cd in several species of Antarctic crustaceans could be due (e.g. Petri and Zauke 1993; Kahle and Zauke 2003a) to their slower growth rates, later sexual maturity and especially to their longer moult cycles (an efficient way of eliminating metals). Moreover, as the Southern Ocean is an environment with unique biological and geochemical features, it cannot be excluded that the poor bioavailability of some essential trace elements may promote the accumulation of Cd and other potentially toxic trace elements.

Although seawater samples from polar seas have similar Cu concentrations, Kahle and Zauke (2003a) found that Cu contents in Antarctic copepods are higher than in related species from Arctic seas. Since Cu is an essential element involved in several enzymatic activities, it was supposed that differences in Cu contents were mainly due to different life stages (i.e. samples from the Greenland Sea were caught at depths below 500 m at the end of the diapause, while those from the Weddell Sea were caught in the upper water layer during the austral summer). Toxicokinetic studies on the uptake of water-borne metals revealed that the Antarctic copepod *Metridia gerlachei* and the copepod *Orchomene plebs* accumulate Co, Cu, Ni, Pb and Zn during exposure and depurate them in uncontaminated seawater (Kahle and Zauke 2002, 2003b). The two species turned out to be very sensitive biomonitors for these metals, even at minimal increments in ambient exposure concentrations (from 0.2–0.8 μg l^{-1}). However, Cd bioaccumulation was not observed, suggesting that the uptake of this metal does not occur in the soluble phase. These results seem to confirm the prominent role of phytoplankton in the uptake of Cd and its transfer to Antarctic marine food chains.

7.2.1 Bioaccumulation of Persistent Contaminants in *Euphausia superba* (Krill)

In the Southern Ocean, the marginal ice zone is the nursery ground for *E. superba* (by far the dominant herbivore), which channels organic matter produced by nannoplankton and diatoms to cephalopods, fish, seabirds, seals and baleen whales. As discussed in Chapter 3, krill has a number of unique characteristics – its life cycle closely matches seasonal cycles of primary productivity, adults are larger than other pelagic herbivorous crustaceans (up to 6.2 cm long), they have an unusually high metabolic rate and swimming speed, adapt to whatever food is available, and survive notwithstanding the low food supply in winter (Marschall 1988). A salient characteristic of euphausiids, especially of *E. superba*, is the formation of dense aggregations which vary considerably in area (from a few to 1,000 m^2 and sometimes even many square kilometres), density (from 0.5 to several kg m^{-3}) and depth (usually in the upper 100–200 m of the water column; Knox 1994). The different types of aggregations, indicated as patches, shoals, schools, swarms or super-swarms, comprise individuals of different size and gonad development, and they may be transient, lasting for hours or days, or may persist for weeks or a lifetime. In view of these characteristics and its central position in the Antarctic pelagic food chain, krill has been of extreme interest to man since the onset of sealing in the 18th century. Despite intensive research, however, secondary productivity by krill and its total biomass in the Southern Ocean are still unknown. Estimates based on its annual consumption by predators (40, 100, 40, 25 and 15×10^6 tonnes (t) by reduced whale stocks, seals, squid, birds and fish respectively; Knox 1994) give an idea of annual krill production and of its role in transferring metals and POPs to higher levels of the Antarctic pelagic food chain.

In the 1960s there was considerable speculation about the potential role which Antarctic krill harvesting might have in the world's projected protein shortage. Large-scale commercial capture commenced in the 1970s and reached a peak (528,201 t) in austral summer 1982–1983. Since then catches have declined, stabilising to around 100,000 t year^{-1} in the last decade (CCAMLR 2003). Although it has been shown that krill can be caught in quantity and processed into food for humans, the world krill harvest is currently limited above all by the lack of demand (Nicol and Endo 1999). One of the most unexpected setbacks in the use of krill as a food product was the very high F contents in its exoskeleton (up to 2,440 µg g^{-1} dry wt.; Soevik and Braekhan 1979). Procedures such as lowering the temperature to –40 °C, separating body fluids, and boiling the crustaceans were adopted to reduce the migration of F into the muscle tissue of frozen krill (Knox 1994). Despite considerable efforts to develop Antarctic krill products for human consumption, in recent years most of the harvest has been used as aquaculture feed, in which high F levels are not a problem.

The present level of krill exploitation in Antarctica in relation to the total stock and to natural predator needs does not seem excessive. However, it cannot be excluded that localised stocks at the northern limit of pack ice are already overexploited (Everson and Goss 1991). Indeed, as discussed by Hewitt and Low (2000), krill fishing mostly concentrates near colonies of land-breeding krill predators.

Several authors (e.g. Stoeppler and Brandt 1979; Yamamoto et al. 1990; Palmer Locarnini and Presley 1995; Barbante et al. 2000) determined concentrations of trace elements in E. superba samples from different zones of the Southern Ocean. Although samples were captured at different locations and in different seasons, and although there is evidence (Rainbow 1989) that concentrations of Cd, Zn, Fe and Mn in euphausiids are size-dependent and that the Cu content decreases following ecdysis (Nicol et al. 1992), available data do not show considerable variations. The main differences with respect to the elemental composition of another euphausiid species (Meganyctiphanes norvegica) collected in the NE Atlantic (Ridout et al. 1989) were the lower Zn

Fig. 45. Mean trace metal concentrations (µg g⁻¹ dry wt.) in *Euphausia superba* from different regions of the Southern Ocean compared with mean values in euphausiids from the NE Atlantic Ocean and central Mediterranean Sea (for references see the text)

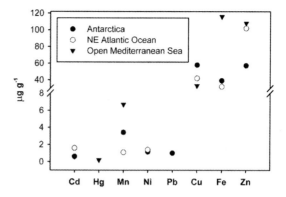

and Cd contents (Fig. 45). Mean Fe, Hg, Mn and Zn concentrations in euphausiids from the central Mediterranean Sea (Fowler 1986) were higher than in *E. superba*. The latter difference may reflect the low bioavailability of lithophilic elements in the Southern Ocean. Barbante et al. (2000) also determined average Cr, Co, Sn, Se and V concentrations in Antarctic krill, and the values (0.5, 0.07, 0.2, 7.8 and 1.2 µg g⁻¹ dry wt. respectively) were in the same range as those reported by Fowler (1986) for euphausiids collected in the Mediterranean Sea.

Comparisons between data in Table 13 and Fig. 45 reveal that *E. superba* accumulates lower Cd concentrations than other species of crustaceans in Antarctic zooplankton. However, such comparisons only provide an indication and are not completely reliable due to the relatively high growth rate and large size of *E. superba* with respect to other crustaceans. Reid (2001), for instance, found that at South Georgia *E. superba* increased in size from ca. 42 to ca. 54 mm in the period October–March, and 2-year-old individuals attained the same size as 3-year-old krill in the Antarctic Peninsula. As the size of organisms increases, their surface area/body weight ratio and the proportion of surface-adsorbed metals decrease. Moreover, Cd is known to mainly accumulate in the digestive gland (or liver in vertebrates) and kidney (e.g. Bargagli et al. 1996b), and the proportion of these organs with respect to that of muscle, fat and gonads is lower in adult *E. superba* than in other Antarctic crustaceans. Therefore, if concentrations of persistent contaminants in krill are measured in the whole individual and are expressed in relation to weight, they may change continuously during the growth of individuals, even when contaminant concentrations in the environment or in a single organ remain unchanged.

In a krill sample collected in 1975 along the Antarctic Peninsula, Risebrough et al. (1976) measured *p,p'*-DDE, *p,p'*-DDT and total PCBs of 14, 19 and 3 ng g⁻¹ lipid mass. The DDE:DDT ratio indicated a high proportion of unmetabolised DDT. Overall, data for birds and mammals endemic to the Antarctic region showed an increase in DDT and derivatives from the early

1760s to the early 1980s (UNEP 2002a). However, Bidleman et al. (1993) reported that significant input of fresh DDT was still evident in 1990 samples.

Sen Gupta et al. (1996) measured higher average concentrations of t-HCH ($\alpha+\gamma$ isomers=0.154 ng g^{-1} dry wt.) than of total PCBs (0.152 ng g^{-1} dry wt.) and total DDT (p,p'-DDT+o,p'-DDT+p,p'-DDE+o,p'-DDE=0.037 ng g^{-1} dry wt.) in samples of E. superba collected in December 1987–January 1988 at about 70° S and 12° E. The dominant isomers and metabolites of DDT, p,p'-DDT and p,p'-DDE occurred in almost equal percentages in krill. The dominant PCB congeners were PCB-138 and PCB-136. Corsolini et al. (2002) measured average p,p'-DDE and HCB concentrations (0.86±0.98 and 0.37±0.17 ng g^{-1} wet wt. respectively) in krill samples collected from the Ross Sea (71° 20'S-72° 331'S/170° 22'E-178° 04'E) in January 2000. Average PCB concentrations in these samples were much higher (167±85 ng g^{-1} wet wt.), and congener-specific PCB profiles showed a prevalence of low-chlorinated isomers (tetra-PCBs accounted for most of the residue). This pattern differed from that usually detected in organisms from low and mid latitudes, and it was likely due to global fractionation at high latitudes. Isomer patterns in organisms from the Ross Sea were similar to those of Kanechlor, a technical mixture mostly used in Japan and other eastern Asian countries, roughly located at the longitude of the Ross Sea (Corsolini et al. 2002). On a lipid weight basis, concentrations of total polychlorinated dibenzo-p-dioxins (PCDDs) and dibenzofurans (DFs) in the same krill samples (Kumar et al. 2002) were 27 pg g^{-1}, and those of non- and mono-ortho-substituted polychlorinated biphenyls (dioxin-like PCBs) were 0.9 ng g^{-1}.

An E. superba sample collected from the Ross Sea in November 1994 had a very low artificial radionuclide content (<0.11 Bq kg^{-1} dry wt. of ^{137}Cs), while ^{90}Sr, [239]Pu, [238]Pu and [241]Am were undetectable (Nonnis Marzano et al. 2000).

In order to evaluate the potential use of several biomarkers (esterases, mixed function oxidases, porphyrins), Minutoli et al. (2002) analysed acetylcholinesterase (AChE) activity in homogenates of whole zooplanktonic crustaceans from the Mediterranean and Ross seas. Among euphausiid species, M. norvegica from the Mediterranean Sea showed higher AChE activity (13.3 µmol min^{-1} g^{-1}) than E. crystallorophias and E. superba (4.5 and 1.7 µmol min^{-1} g^{-1} respectively). This difference was attributed to the lower basal metabolic activity of Antarctic organisms, while the lower activity in E. superba with respect to E. crystallorophias was probably due to their different dimensions (62 and 34 mm respectively). Indeed, AChE activity is known to be inversely proportional to organism dimensions (Fossi et al. 1996).

7.3 Transfer of Contaminants in Pelagic Food Chains

In ice-free zones of the Southern Ocean, except in a few areas such as South Georgia, krill is generally absent, and the most common organisms in zooplankton and nekton are herbivorous copepods, salps, small euphausiids, squid and myctophid fish. Antarctic waters lack mid-water crustacean fauna such as Penaeidae which are very common in boreal, temperate and tropical seas. In general, there is scarce knowledge of Antarctic nekton fauna, and very few data are available on their chemical composition. Table 14 reports mean concentrations of trace metals in the liver, kidney and muscle of pelagic myctophid fish and two species of octopuses from the Kerguelen Islands (Bustamante et al. 1998, 2003), and in myctophid fish and cephalopods from the Atlantic Ocean (Schulz-Baldes 1992) and Mediterranean Sea (Miramand and Guary 1980; Fowler 1986). The liver of myctophid fish and the digestive gland of cephalopods are two partly analogous organs which accumulate the highest concentrations of Cd and other trace metals, except Hg. The high Cd content in the liver and kidney (but not the muscle) of *G. piabilis* indicates that this metal is probably taken in through diet, which mainly consists of very effective Cd accumulators such as hyperiid amphipods and other planktonic crustaceans (see Table 13). Octopus prey includes crustaceans, molluscs and fish. Bustamente et al. (1998) hypothesised that the enhanced accumulation of Cd in Antarctic cephalopods could also be due to the disruption of Cu homeostatic mechanisms. Cephalopods from the Kerguelen Islands had much higher Cd/Cu ratios and lower Cu contents than samples from other seas. Cd and Cu are known to bind to the same type of metallothioneins in the digestive gland of molluscs. Thus, in marine environments characterised by enhanced Cd bioavailability, it cannot be excluded that this metal can compete and/or substitute Cu in the organ most involved in metal detoxification processes. Very high concentrations of Cd (up to 782 μg g^{-1} dry wt.) and other metals have also been measured in the digestive gland of cephalopods from the Pacific Ocean (Martin and Flegal 1975).

Data summarised in Table 14 have important ecological implications because cephalopods and myctophid fish occupy predominant niches in food chains of ice-free zones of the Southern Ocean. Although squid are the principal group of cephalopods in the Southern Ocean, data on their elemental composition are not available; as their diet mainly consists of myctophid fish, euphausiids and other crustaceans (Kear 1992), their composition could be quite similar to that of benthic octopuses.

Table 14. Mean trace metal concentrations (μg g^{-1} dry wt.) in organs and tissues of myctophid fish and cephalopods from Kerguelen Islands, and related species from the Atlantic Ocean and the Mediterranean Sea

Region	Species	Organ/tissue	Cd	Cu	Hg	Zn
Kerguelen Islands (myctophid fish)[a]	*Gymnoscopelus nicholsi*	Liver	4.2±0.3	5.8±1.3	–	93±18
		Kidney	2.7±1.0	4.8±1.3	–	86±7
		Muscle	<0.1	2.5±0.7	0.21±0.13	9.2±4.0
	Gymnoscopelus piabilis	Liver	28±17	10±3	–	142±31
		Kidney	16±8	11±5	–	113±23
		Muscle	<0.1	1.2±0.4	0.31±0.13	9.9±1.2
Atlantic Ocean (myctophid fish)[b]	–	Whole	2.0±1.6	7.6±5.6	–	
Mediterranean Sea (myctophid fish)[c]	*Myctophum glaciale*	Whole	0.2±0.1	3.6±2.0	0.21±0.2	69±18
Kerguelen Islands (cephalopods)[d]	*Graneledone* sp.	Digestive gland	369	1,092	–	102
		Whole	39±5	68±30	–	131±19
	Benthoctopus thielei	Digestive gland	215	306	–	416
		Whole	38±8	68±29	–	166±39
Atlantic Ocean (cephalopods)[b]	–	Whole	17±8	45±21	–	–
Mediterranean Sea[e]	*Octopus vulgaris*	Digestive gland	50±10	2,500±700	–	1,450±400
		Whole	1.2±0.1	260±70	–	150±500

[a] Bustamante et al. (2003)
[b] Fowler (1986)
[c] Bustamante et al. (1998)
[d] Schulz-Baldes (1992)
[e] Miramand and Guary (1980)

7.3.1 The (Hyper)Accumulation of Cd and Hg in Pelagic Seabirds

Albatrosses and petrels are the predominant flying seabirds in the Southern Ocean (i.e. about 20 species of truly oceanic Procellariiformes, which avoid land except when breeding). Most albatrosses, such as the black-browed (*Diomedea melanophris*) and light-mantled sooty (*Phoebetria palpebrata*), prevail in ice-free marine areas whereas petrels (e.g. the southern fulmar, Antarctic cape pigeon and Wilson's storm petrel; Fraser and Ainley 1986) prevail close to the edge of the seasonal pack-ice zone, where krill is the dominant element of the food chain. In ice-free waters, albatrosses (except *D. melanophris*) and some petrels of the genera *Fulmarus*, *Pterodroma* and *Procellaria* mainly feed on squid and fish (Croxall 1984).

Given their extreme mobility, these seabirds behave as spatial and temporal integrators of persistent contaminants over large areas of the Southern Ocean, and a number of studies (e.g. Anderlini et al. 1972; Norheim 1987; Thompson et al. 1990; Lock et al. 1992; Furness 1993; Hindell et al. 1999; Gonzáles-Solís et al. 2002) show that they accumulate some of the highest Cd and Hg concentrations reported for any vertebrate. As an example, Table 15 reports data from a comprehensive survey performed by Lock et al. (1992) on Cd, Cu, Pb, Zn and Hg concentrations in 64 taxa of tropical, subtropical, sub-Antarctic and Antarctic seabirds. Cadmium concentrations in the kidney are generally two- to fivefold those in the liver. Hindell et al. (1999) measured maximum Cd concentrations (>260 μg g^{-1} dry wt.) in the kidney of wandering and royal albatrosses, which have predominantly oceanic patterns and are largely confined to the Southern Ocean. Cadmium concentrations in the same range also occur in the kidney of Antarctic petrels and fulmars (Table 15), which feed on krill at the pack-ice edge. *Fulmarus glacialoides* is slightly larger and only a little paler than its counterpart in the Northern Hemisphere (*F. glacialis*). Cadmium concentrations in the liver and kidney of northern fulmars from northwest Greenland (Dietz et al. 1996) were lower (21 and 44% respectively) than those in the liver of Antarctic fulmars.

The fact that some species of long-lived albatrosses (e.g. wandering and royal albatrosses may live more than 50 years; Marchant and Higgins 1990) accumulate the same amount of Cd as petrels (mean life expectancy of 15–30 years, depending on species size; Croxall 1984) may be due to differences in diet and in Cd detoxification and elimination capabilities. Cadmium does not generally accumulate with age in seabirds, and concentrations in the liver and kidney of juvenile seabirds are sometimes in the same range or even higher than those in adults (Table 15). This could be due to different foraging or metabolic patterns; in any case, it indicates that seabirds can regulate Cd contents to some extent and that they are able to reduce the Cd load as they get older. As birds cannot eliminate Cd through eggs (e.g. Leonzio and Massi 1989), females and males which adopt similar foraging strategies usually show quite similar Cd concentrations in their organs and tissues.

Table 15. Mean concentrations (μg g^{-1} wet wt., ±SD) of Cd in the liver (*l*) and kidney (*k*), and of Hg in the liver (*l*) of adult (*a*) and juvenile (*j*) seabirds from the Southern Ocean. (Lock et al. 1992)

Species	Adult/juvenile	Organ	Cd	Hg
Diomedea exulans (wandering albatross)	a	l	14.4±5.3	295±173
	a	k	39.4±17.0	
Phoebetria palpebrata (light-mantled sooty albatross)	a	l	9.4±0.1	146±114
	a	k	46.5±2.1	
	j	l	6.1±1.3	9.5±2.9
	j	k	21.8±3.2	
Thalassoica antarctica (Antarctic petrel)	a	l	21.0	11.1
	a	k	43.2	
	j	l	14.9±2.5	9.2±4.2
	j	k	44.2±16.4	
Fulmarus glacialoides (Antarctic fulmar)	j	l	18.7±6.9	19.9±7.9
	j	k	43.1±18.8	
Daption capense (cape pigeon)	a	l	9.3±5.8	14.5±12.6
	a	k	32.8±9.4	

The behaviour of mercury is completely different from that of cadmium. Mercury, in particular methylmercury (MeHg), is biomagnified in aquatic food chains; its bioaccumulation is more similar to that of hydrophobic organic pollutants than to that of other metals (Bargagli 2001). All animals can generally eliminate inorganic Hg in several weeks, whereas the biological half-life of MeHg is months or years. Methylmercury therefore progressively accumulates in muscle and other tissues, reaching peak concentrations in long-lived animals at higher levels of the food web. The proportion of Hg and MeHg varies in different tissues and organs depending on the trophic status, age, and species-specific adaptive capacity for detoxification and elimination of the metal. Methylmercury concentrations in seabirds are usually much higher than in birds of terrestrial ecosystems because the former feed on crustaceans, fish and squid which "preconcentrate" MeHg by eating phyto- and zooplankton organisms. The highest concentrations of total Hg reported for seabirds and marine vertebrates throughout the world oceans probably occur in Procellariiformes (albatrosses and petrels) from the Southern Ocean. Hindell et al. (1999) measured 1,800 μg g^{-1} dry wt. (or 680 μg g^{-1} wet wt.) of Hg in the liver of a wandering albatross, and data in Table 15 show that although the Hg body burden increases with age, even in juvenile Antarctic seabirds average Hg concentrations are always >9 μg g^{-1} wet wt. This huge bioaccumulation is rather surprising.

As mentioned in the preceding chapter, although concentrations of Hg and MeHg in Southern Ocean waters are unknown, values are probably rather low. Low concentrations have also been reported for sediments, phytoplankton, macroalgae and zooplankton from the Ross Sea (Bargagli et al. 1998a), and concentrations in Antarctic krill are always <0.1 μg g^{-1} dry wt. (Fig. 45). As discussed by Hindell et al. (1999) and Bargagli et al. (2000), the bioaccumulation of Hg in Antarctic seabirds is a natural process mainly determined by species-specific life histories. Larger and long-lived species such as wandering albatrosses or grey petrels, which largely feed on squid, fish or carrion (Lock et al. 1992), have the highest Hg burden. Since wandering albatrosses have a low reproductive rate (one egg every 2 years) and replace feathers over a period of years rather than annually, they have very few ways of eliminating MeHg. Moulting is an important route for MeHg excretion, because seabirds deposit MeHg in their feathers during growth (Fimreite 1979). Part of the MeHg ingested with food is demethylated in the liver, where it is stored as inorganic molecules (Thompson and Furness 1989). In three species of albatrosses and a grey petrel, for instance, MeHg concentrations in the liver ranged from 4.8 to 16 μg g^{-1} wet wt., and inorganic Hg concentrations reached 280 μg g^{-1} wet wt. (Lock et al. 1992). Despite efficient demethylation mechanisms, levels of MeHg and inorganic Hg in the liver of Antarctic seabirds are higher than those usually found to be toxic for terrestrial and freshwater birds (Sheuhammer 1988); it has therefore been hypothesised that the latter are more sensitive than some Procellariiformes to Hg.

In addition to inorganic Hg, albatrosses and petrels also accumulate Se in the liver and kidney. Kim et al. (1966) measured a maximum value of 113 µg g^{-1} dry wt. in the liver of black-footed albatrosses. Ohlendorf et al. (1988) reported high incidences of bird embryo and adult mortality due to the toxic effect of Se. However, small quantities of this element are essential to the metabolism of birds and mammals, and there is evidence that Se acts as an antidote to the toxic effects of some heavy metals, particularly Hg (Koeman et al. 1973). Although Hg detoxification mechanisms in seabirds are not well known, co-accumulation of Hg and Se in the liver probably derives from the formation of inorganic Hg–Se complexes. Kim et al. (1996) found an equivalent molar ratio of 1:1 between total Hg and Se concentrations in grey petrel and light-mantled sooty albatrosses, which accumulated more than 100 µg g^{-1} dry wt. of Hg in the liver. On the contrary, this relationship did not emerge clearly in individuals with relatively low levels of Hg, and fluctuating molar ratios were ascribed to a different degree of exposure to Hg, species-specific demethylation capacity, and elimination through moulting. Dietz et al. (2000) reported a surplus of Se with respect to Hg (on a molar basis) in various tissues of Arctic seabirds and suggested that excess Se may reduce the potential threat of Hg poisoning.

In spite of the bioaccumulation of potentially toxic metals, populations of wandering albatrosses at South Georgia and in other sub-Antarctic areas are declining mainly because they are caught on hooks set by vessels which tow huge, heavily baited long-lines for Patagonian toothfish and other fish species. Seabird carcasses have sometimes been frozen to study their chemical composition. However, less destructive biomonitoring approaches such as the analysis of feathers can be used to evaluate interspecies and geographic variations in Hg bioavailability. In agreement with data on the liver and kidney, Fig. 46 shows that average Hg concentrations in feathers from Antarctic albatrosses and southern giant petrels (Thompson et al. 1993) are higher than those in feathers from seabirds at Foula Island (Shetland; northern Atlantic; Thompson and Furness 1989). By analysing museum feather samples of various species of seabirds from southwest Britain and Ireland, Thompson et al. (1992) found that anthropogenic emissions caused a threefold increase in Northern Hemisphere Hg concentrations over a 100-year period. On the contrary, the feathers of Southern Hemisphere Procellariiformes collected before and after the 1950s revealed no significant increase in Hg concentrations. These results led Thompson et al. (1993) to conclude that the bioaccumulation of Hg in albatrosses and petrels is mainly due to natural processes.

A number of papers have been published on POP concentrations in pelagic seabirds. However, these studies analysed eggs and different organs or tissues, and adopted different measurement units. Available data are highly variable, and it is difficult to interpret spatio-temporal variations in POP concentrations within Antarctic seabirds due to differences in trophic levels and to the wide

Fig. 46. Average concentrations ($\mu g\ g^{-1}$, ±SD) of total Hg in feathers of seabirds from South Georgia (Thompson et al. 1993) and Foula (Shetland Is.; Thompson and Furness 1989)

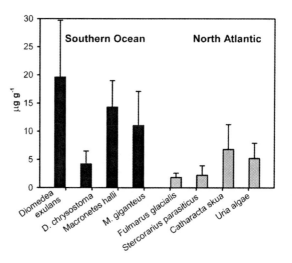

dispersion of some species outside the breeding area. Conroy and French (1974) found that pp'-DDE in the liver of giant petrels collected on Signy Island in 1968–1969 ranged from 20 ng g^{-1} wet wt. in younger birds to 30 ng g^{-1} wet wt. in 5–15 year old birds. Monod et al. (1992) detected DDE and PCBs in all seabirds sampled on the Kerguelen Islands in 1971 and 1975, but DDT and DDD were present only in penguins and albatrosses. Gardner (1983) measured high concentrations of chlorinated hydrocarbons in fat-storage tissues of feral cats on Marion Island (feeding mainly on seabirds). Gardner et al. (1985) subsequently measured concentrations of DDT, DDE, PCBs and dieldrin in eggs from 24 species of seabirds breeding on Marion and Gough islands. Relatively higher concentrations (expressed as ng g^{-1} of whole egg) were measured in eggs of wandering albatrosses and southern giant petrels: DDE values ranged from 56 to 4,242, DDT from 5 to 122, PCBs from 13 to 211, and dieldrin from 1 to 3. Eggs collected after the 1980s generally contained higher PCB levels than those of the same species collected in earlier periods. Luke et al. (1989) provided a comprehensive review of DDT residues in eggs of Antarctic seabirds, and measured the highest levels in species which breed in Antarctica and then migrate to regions well north of the Antarctic Convergence. Concentrations of HCB, DDE and PCB in the eggs of northern giant petrels were 0.11, 0.95 and 1.8 $\mu g\ g^{-1}$ wet wt. respectively. There was an apparently consistent increase in p,p'-DDE concentrations in eggs of southern and northern giant petrels collected in the period 1978–1983. Van den Brink (1997) determined PCB and HCB concentrations in preen oil from birds with different distribution ranges – those restricted to Antarctica and the Southern Ocean year round (e.g. snow petrels, Antarctic petrels, southern fulmars and Adélie penguins), those which spend summers in Antarctica and winters in sub-Antarctic regions (cape petrels) and, finally, a seabird (the common tern) of the Northern Hemisphere which spends summers in temperate regions and winters in subtropical regions. The first group of birds had low total PCB contents (mean=270 ng g^{-1} fat) and rather high HCB concentrations (mean=497 ng g^{-1} fat). PCB concentrations were significantly higher in cape petrels (mean=1,207 ng g^{-1} fat) than in birds restricted to the Southern Ocean. Cape petrels also showed the highest HCB concentrations (mean=2,096 ng g^{-1} fat). In the common tern, PCB concentrations were much higher (mean=13,095 ng g^{-1} fat) and HCB concentrations were much lower (mean=3 ng g^{-1} fat) than those in Antarctic seabirds. As HCB is relatively volatile, has a long environmental half-life and accumulates in species at higher trophic levels (Mackay et al. 1992), the much higher HCB content in Antarctic seabirds than in common terns from the Northern Hemisphere was attributed to the global distillation process. Based on HCB levels, seabirds from sub-Antarctic regions were considered at risk from the possible toxic effects of POPs.

In Antarctic seabirds the potential risk from exposure to POPs may be enhanced by the extreme variability of physiological conditions throughout the breeding season. During periods of starvation, the fat pool of birds decreases, and concentrations of POPs stored in fat consequently increase (Subramanian et al. 1986). To investigate the effects of breeding ecology on concentrations of organochlorines, van den Brink et al. (1998) collected blood and uropygial oil samples from southern fulmars. They found that fluctuations in HCB and p,p'-DDE concentrations were related to changes in body mass, while PCB and dieldrin contents did not vary significantly during the season. It was therefore hypothesised that the two groups of compounds were stored in different tissues, and that utilisation of stores in these different tissues at different moments of the breeding cycle could produce fluctuations in the content of various POPs during the season.

7.3.2 POPs and Heavy Metals in Pelagic Marine Mammals

While commercial hunting was once the biggest threat to the survival of several species of marine mammals, a range of other human-induced threats are now affecting their populations. Among the most widespread are fishing activities and by-catching of cetaceans, environmental pollution, and ocean noise pollution from active sonar systems. A "whale sanctuary" was established in the Southern Ocean by the IWC (International Whaling Commission), and several research programmes are currently being planned by IWC, CCAMLR and Southern Ocean GLOBEC to study interactions between cetaceans and their environment and to develop predictive models for major changes induced in Antarctic marine ecosystems. Besides endocrine disrupters and other persistent environmental contaminants, the possible effects of climate change are a major threat to marine mammals in the Southern Ocean. As discussed in previous chapters, warming would reduce the seasonal sea-ice cover, primary productivity and the availability of krill on which most Antarctic whales feed. Armstrong and Siegfried (1991), for instance, estimated that the minke whale (*Balaenoptera acutorostrata*) population in the Antarctic (60° S and higher) consumes 35.5×10^6 t krill annually (about 56 t krill for each average-sized female during a 120-day stay in Antarctic waters). The killer whale (*Orcinus orca*) is probably the main exception, as it feeds on squid, fish, cephalopods, birds, seals and other cetaceans. There is apparently no data on the accumulation of persistent contaminants in this top predator. Szefer et al. (1993, 1994) determined metal concentrations in the organs of leopard seals (*Hydrurga leptonyx*, another Antarctic top predator), and average values were in the same range as or lower than those measured in organs of other seal species. As discussed by Bargagli (2001), Cd contents in the liver and kidney of killer whales are probably similar to or lower than those in Antarctic seabirds and seals. The Cd content in the kidneys of Arctic belugas and polar bears is about half that in the kidneys of seabirds and seals (Dietz et al. 1996); in the liver of a killer whale stranded in the United Kingdom the Cd content was 15 μg g^{-1} dry wt., while the Hg content was very high (368 μg g^{-1} dry wt.; Law et al. 1997; Table 16).

The analysis of livers from a large number of southern minke whales (Honda et al. 1987) showed that concentrations of Fe, Cd and Hg were positively correlated with age. However, the increase in Hg and Cd ceased at the age of about 20 years and thereafter decreased year by year. This unusual age trend was linked to changes in the amount of food intake by minke whales as a result of disturbance to Antarctic ecosystems by commercial whaling. Data in Table 16 show that Fe concentrations in the liver of *B. acutorostrata* are lower in females than in males; Honda et al. (1987) detected a progressive decrease in Fe, Pb, Ni and Co contents during gestation. Hg contents in the liver of Antarctic marine mammals are much lower, whereas average Cd concentrations are in the same range or higher than those in the liver of seabirds (Table 15). Like seabirds, Antarctic marine mammals accumulate much higher Cd concentrations in the kidney than in the liver. In the kidney of some mammals such as the Ross seal, Cd concentrations may be much higher than the critical value of 200 μg g^{-1} wet wt. associated with kidney damage in mammals, including humans (WHO 1992). According to a review of the risk of Cd contamination in humans (Elinder and Järup 1996), elderly persons in environmentally exposed populations may display evidence of Cd-induced renal dysfunction at Cd concentrations of about 50 μg g^{-1} wet wt. However, Antarctic and Arctic seabirds and marine mammals are probably adapted to naturally high levels of Cd. In a pilot study on the seal *Phoca hispida* from northwest Greenland, Dietz et al. (1998) found no morphological differences between the kidneys of seals with low or high (up to 581 μg g^{-1} wet wt.) Cd concentrations.

Like seabirds, marine mammals intake MeHg above all with food. Hg and Se concentrations in the liver show a 1:1 stoichiometric relationship, and

Table 16. Average trace metal concentrations ($\mu g\,g^{-1}$ dry wt.; Fe, %) in the liver of marine mammals from the Southern Ocean and West Greenland

Region	Species	n	Cd	Cu	Fe	Hg	Mn	Ni	Pb	Zn
Southern Ocean	Balaenoptera acutorostrata[a]	96 Males	36±19	17±4	0.53±0.19	0.19±0.08	10±3	0.16±0.09	0.41±0.14	142±18
		39 Females	45±26	18±4	0.19±0.06	0.21±0.08	10±2	0.17±0.12	0.42±0.022	156±31
	Ommatophoca rossi[b]	3 Males	71±40	114±89	0.06±0.01	2.2±0.9	22±15	3.3±1.0	0.04±0.04	235±150
		17 Females	110±88	77±65	0.05±0.02	3.6±4.3	15±6	2.6±1.0	0.0±0.01	208±54
	Lobodon carcinophagus[c]	27	13±12	71±21	1.3±0.8	7.6±5.6	15±3	0.04±0.01	0.09±0.06	160±39
West Greenland	Balaenoptera acutorostrata[d]	–	3.6	–	–	1.56	–	–	–	138
	Orcinus orca[e]	–	15	35	–	368	–	–	–	201

[a] Honda et al. (1987)
[b] McClurg (1984)
[c] Szefer et al. (1993, 1994)
[d] Hansen et al. (1990)
[e] Law et al. (1997)

more than 90 % of Hg in this organ occurs in inorganic forms such as inert Hg–Se complexes, metallothioneins or non-degradable granules in hepato-cytes (André et al. 1990). As a rule, Hg concentrations increase in predatory species of marine mammals, and these species show a positive correlation between Hg concentrations in the liver and body length (Hansen et al. 1990; Meador et al. 1993; Wagemann et al. 1998). Minke whales in the Southern Ocean mainly feed on E. superba, which has a very low Hg content (Fig. 45); despite their long lifespan (about 50 years), the liver of these whales accumu-lates low concentrations of Hg (Table 16). West Greenland minke whales feed mostly on sand eels; with respect to Southern Ocean minke whales, their liver stores higher Hg concentrations and lower Cd concentrations. It seems likely that such differences between northern and southern minke whales are mostly due to their different diets.

During the austral spring, minke whale populations migrate southwards from low-latitude (10–20° S) breeding grounds to the Southern Ocean for intensive feeding. Contaminants accumulated in their tissues can be used as chemical tracers to understand their feeding, migratory and reproductive behaviour (e.g. Tanabe et al. 1987; Subramanian et al. 1988). The non-destructive sampling of skin biopsies from free-ranging cetaceans has recently been used to analyse DNA and contaminants in different whale pop-ulations (e.g. Monaci et al. 1998; Gauthier and Sears 1999). This sampling strategy causes limited behavioural disturbance. Kunito et al. (2002) com-pared trace element concentrations in skin biopsies and livers of whales col-lected from different zones of the Southern Ocean (from summer 1988/1989 to summer 1995/1996, within the framework of the Japanese Whale Research Program under Special Permit in the Antarctic) to evaluate the suitability of skin biopsies in detecting environmental contamination and the identity of different southern whale populations. Trace element concentrations in the skin were generally lower than those previously reported in the skin of cetaceans from other seas. Statistically significant positive correlations were found between Cd, Cr, Cs, Cu, Mn, Rb and Zn concentrations in the skin and liver of individual animals. Significant differences were also found in ele-ment concentrations in the skin of samples (especially males) collected in different Antarctic feeding areas. However, lipophilic POPs, which are exclu-sively released by human activity, may be more reliable tracers of whale migrations and reproductive behaviour.

Concentrations of p,p'-DDT (160 ng g^{-1} wet wt.), p,p'-DDE (from 267 to 456 ng g^{-1} wet wt.) and HCH (25 ng g^{-1} wet wt.) were much higher in the adipose tissue of leopard seals collected in 1979–1981 at King George Island than in that of four other species of seals (Karolewski et al. 1987). This may be due to the fact that the leopard seal is a top predator and that it disperses widely in the South-ern Ocean during the austral winter. In blubber samples from Ross seals, McClurg (1984) detected low DDT, PCB and dieldrin concentrations and found no relationships between residue levels and sex or size (age) of animals.

Baleen whales have often been used to biomonitor the distribution of POPs in the Northern and Southern hemispheres (e.g. Tanabe et al. 1983b; O'Shea and Brownell 1994). Although minke whales have a wide range of migratory areas, there is no evidence of hybridisation between populations in the two hemispheres. In general, POP concentrations in whales from the Southern Ocean are much lower than in those from the northern Pacific. This is likely due to lower environmental contamination in the Southern Hemisphere and, as discussed above for Hg, to the different feeding behaviour of northern and southern minke whales. However, data by Aono et al. (1997) show that DDT and HCB concentrations in the blubber of northern minke whales decreased in the period 1984–1994, while they increased in that of southern whales. The predominant DDT compound was p,p'-DDE; it tended to increase in Antarctic cetaceans, with a concomitant decrease in p,p'-DDT. This trend indicates that fresh inputs of p,p'-DDT were much lower than the degradation of this compound. The composition of HCH isomers in the diet and blubber of southern whales differed from that in the diet and blubber of northern whales; according to Aono et al. (1997), this suggests that lindane was being used to a greater extent in countries of the Southern Hemisphere than in those of the Northern Hemisphere.

7.4 Contaminants in Coastal Benthic Organisms

The distribution of benthic organisms on Antarctic shelves is largely affected by glacier transport of coarse materials into the sea, ploughing by floating icebergs, plucking by anchor ice and abrasion of shorelines by fast-, pack- and brash ice or by floes driven ashore and piled up by storms (push ice). In spite of these disturbances, most benthic assemblages of sessile and vagile organisms in Antarctic coastal ecosystems are characterised by high diversity and structural complexity, with most species having long life spans, low growth rates and lacking pelagic larvae (e.g. Dayton 1990; Gambi et al. 1994; Knox 1994). Although Antarctic seaweed does not have the trophic significance of phytoplankton, it contributes to primary productivity in coastal ecosystems and provides suitable habitats for a variety of animals. Some of the dominant species in Antarctica are *Iridaea cordata* in the infralittoral zone, *Leptosomia simplex* in the lower intertidal zone, *Phyllophora antarctica* at intermediate depths, and the largest Antarctic brown alga *Himantothallus grandifolius* down to depths of 25–30 m. The cell wall of algae consists of a variety of polysaccharides and proteins, some of them containing anionic carboxyl, sulphate or phosphate groups which are binding sites for metals. Owing to its abundance, limited mobility, and its ability to uptake metals and organic lipophilic compounds, seaweed has often been used to biomonitor persistent pollutants in the Arctic (e.g. Asmund et al. 1991; Johansen et al. 1991; Riget et al. 1997).

Very few data exist on trace metals and POP concentrations in Antarctic macroalgae. Average Cd, Cu, Fe, Hg, Pb and Zn concentrations in *I. cordata* samples from coastal sites in Terra Nova Bay were 2.8, 3.5, 47, 0.12, 0.05 and 42 µg g^{-1} dry wt. respectively (Bargagli 2001). When compared with values measured by Riget et al. (1997) in samples of *Fucus vesicolosus* from relatively unpolluted coastal areas of West Greenland, the main difference was the much lower Pb content in Antarctic seaweed. Algae of the genus *Desmarestia* are abundant in the intertidal zone of the Antarctic Peninsula and South Shetlands. Montone et al. (2001b) determined concentrations of PCB congeners in *Desmarestia* sp. samples collected from various locations at Admiralty Bay (King George Island). Total PCB concentrations ranged from 0.46 to 3.86 ng g^{-1} dry wt., and the predominance of low-molecular weight congeners indicated that there were no significant local sources of PCBs at Admiralty Bay.

Sponges are one of the most striking epibenthic sessile organisms of Antarctica. Besides their significance in terms of benthic biomass, they are important substrates for epizoic animals and are the main source of densely populated biogenic sediments. Capon et al. (1993) found that aqueous ethanol extracts from *Tedania carcoti* specimens collected in Prydz Bay contained some mg g^{-1} dry wt. of Cd and Zn. The potent antibacterial properties of the extract were ascribed to the two metals, and it was speculated that their accumulation by sponges serves as a natural antibiotic or an agent against predation and fouling. However, sponges from Terra Nova Bay (Bargagli et al. 1996b) had lower Cd concentrations (mean 26±15 µg g^{-1} dry wt.), and their Zn content was similar to or lower than that in other benthic invertebrates from the same marine environment (Table 17). Although the average Cd content in surface sediments from Terra Nova Bay was very low (0.22±0.12 µg g^{-1}) and roughly corresponded to that in Tyrrhenian Sea sediments of comparable grain size, Cd concentrations in benthic organisms from the Ross Sea were one order of magnitude higher than in related species from the Tyrrhenian Sea (Bargagli 1993). The enhanced bioavailability of Cd in Antarctic coastal marine environments is probably due to the rapid regeneration of the metal in the water column and/or to rapid mineralisation in surface sediments during early diagenesis. During the austral summer the upwelling of waters favours the ad/absorption of Cd on phytoplankton cells, and primary consumers uptake metal directly from water and food. Diatoms and benthic macroalgae at Terra Nova Bay have mean Cd concentrations of about 2.5 µg g^{-1} dry wt. Sponges draw in diatoms through their inhalant water system and exopinacocytes (which take in diatoms settling on the sponge surface and store them in the mesohyl matrix to strengthen the sponge cortex and as food reserve; Gaino et al. 1994). During the austral summer the amphipod *Paramoera walkeri* is a very important component of the fast-ice community and of the littoral benthos. Like sponges, it feeds mainly on diatoms (Gruzov 1977) and its Cd content is twice than of copepods (Bargagli et al. 1996b). Cad-

Table 17. Average concentrations (μg g⁻¹ dry wt.; mean±SD) of trace metals in coastal benthic organisms from Terra Nova Bay (Ross Sea; Bargagli 2001)

Invertebrate	Organ or tissue	Cd	Cu	Fe	Hg	Pb	Zn
Rhodophita (*Iridaea cordata*)	Whole	2.8±0.3	2.5±0.6	62±14	0.09±0.04	0.08±0.03	41±14
Ascidians (pooled across species)	Whole	26±15	11±3	307±97	0.08±0.05	0.46±0.23	48±13
Holothurians (pooled across species)	Whole	7.7±3.8	5.6±1.8	79±34	0.23±0.09	0.55±0.22	139±42
Polychaeta (*Marmothoe spinosa*)	Whole	6.8±2.8	7.5±3.1	318±49	0.07±0.02	0.11±0.07	76±24
Ophiuroids (pooled across species)	Arm	0.5±0.1	2.1±0.6	41±16	0.08±0.03	0.41±0.16	86±17
	Central disk	14±5	15±6	324±143	0.14±0.07	0.69±0.24	306±94
Asteroidea (*Odonaster validus*)	Arm	14±2	15±5	63±20	0.11±0.06	0.51±0.22	47±7
	Soft tissue	13±5	18±7	347±88	0.17±0.10	0.15±0.11	282±92
Echinoids (*Sterechinus neumayeri*)	Gonad	6.8±2.8	3.5±1.8	70±30	0.13±0.07	0.16±0.09	90±31
	Soft tissue	13±5	6.7±3.1	418±129	0.09±0.05	0.49±0.15	129±36
Bivalves (*Adamussium colbecki*)	Muscle	1.2±0.6	3.2±1.6	50±21	0.20±0.10	0.11±0.03	58±13
	Digestive gland	55±27	27±10	360±122	0.35±0.08	0.15±0.09	99±26
Prosobranches (*Neobuccinum eatoni*)	Muscle	6.1±3.5	8.5±3.3	105±24	0.28±0.15	0.08±0.02	71±22
	Digestive gland	227±65	10±4	278±39	0.24±0.10	0.37±0.11	251±112
Amphipoda (*Paramoera walkeri*)	Whole	4.8±0.6	14±4	66±13	0.07±0.03	0.08±0.03	78±19

mium concentrations in other herbivorous organisms such as the sea urchin *Sterechinus neumayeri* are three times higher than the total body content in Arctic sea urchins (Fallis 1982). A toxicity test was developed to evaluate the effects of Cd, Cu, Pb and Zn on *S. neumayeri* larvae and embryos (King and Riddle 2001). Results showed that the long-term test with two-arm pluteus-stage embryos was more sensitive (especially to Cu) than the short-term test on hatched blastulae. The embryonic development of *S. neumayeri* was relatively insensitive to Pb.

Asteroidea is one of the most important benthic groups in the Ross and Weddell seas (e.g. Jarre-Teichmann et al. 1997). These organisms are opportunistic feeders, and some species feed on sponges. Data in Table 17 show that Antarctic starfish may accumulate high Cd concentrations. However, it is difficult to estimate the flow of metals and POPs through the dominant groups of benthic organisms because the ecophysiology of many Antarctic species is practically unknown. Starfish are probably major predators of sponges, whereas sea cucumbers, brittle stars and sea urchins feed on detritus, polychaetes and other invertebrates. Sponges and echinoderms are mostly inedible or of little value as food; moreover, several species of Antarctic sponges, tunicates, nemerteans, coelenterates and echinoderms produce toxic compounds against predation (McClintock 1989). Vagile polychaetes and other "worms" are probably major prey of fish, gastropods, isopods and echinoderms, and are the main path for the transfer of persistent contaminants to higher levels of the benthic food web.

In general, Hg concentrations in benthic invertebrates from Terra Nova Bay are low and accumulation patterns differ from those of Cd (Table 17). Although values are highly variable, probably as a consequence of marked differences in the life span, habitat and feeding habit of animals, the total body burden of Hg increases from primary consumers such as sponges and *P. walkeri* to more omnivorous, opportunistic feeders and especially to scavengers and predators such as starfish. Data in Table 17 show that molluscs, particularly their digestive glands, accumulate the highest levels of Cd, Hg and other metals. The section below deals with the accumulation and detoxification of persistent contaminants in circumpolar species of Antarctic molluscs.

7.4.1 Metal Accumulation and Homeostasis in Antarctic Molluscs

The capability of marine molluscs to tolerate and accumulate metals and other persistent contaminants is well known. Since the 1970s, mussels and oysters have been used extensively around the world as bioaccumulators of contaminants for long-term monitoring of coastal marine waters within the context of "Mussel Watch" programmes (e.g. NAS 1980; Lauenstein et al. 1990). A number of papers have been published on metal accumulation in Antarctic molluscs (e.g. Mauri et al. 1990; Berkman and Nigro 1992; Ahn et al. 1996;

Lohan et al. 2001). These studies have generally revealed that several species of Antarctic molluscs meet most criteria defined in "Mussel Watch" programmes, and that the accumulation pattern of contaminants in their organs and tissues is similar to that in related species from other seas. The main difference between the elemental composition of Antarctic bivalves and that of related species from unpolluted environments in other seas is the huge accumulation of Cd in the digestive gland and kidney of Antarctic molluscs. Nigro et al. (1997) also found a remarkable accumulation of some elements, particularly As (up to 394 μg g^{-1} dry wt.), in the gills and digestive gland of another Antarctic bivalve (*Yoldia eightsi*). There are other reports on As accumulation in marine bivalves – Benson and Summons (1981), for instance, measured up to 1,025 μg g^{-1} dry wt. of As in giant clams from the Greet Barrier Reef.

Among Antarctic molluscs, circumpolar filter-feeding bivalves such as *Adamussium colbecki* and *Laternula elliptica* or the Antarctic limpet *Nacella concinna* (which occurs in intertidal and shallow subtidal zones around the Antarctic Peninsula and adjacent islands) are the species most frequently indicated as useful biomonitors of environmental contaminants in coastal ecosystems. *L. elliptica* is one of the most studied species (e.g. Ahn et al. 1996; Nigro et al. 1997; Vodopivez and Curtosi 1998; Lohan et al. 2001). In shallow, sheltered areas this deep-burrowing, large-sized bivalve (shell>10 cm long) may occur in dense patches (>100 individuals m^{-2}); it lives for more than 15 years and lays down distinct annual growth bands (Brey and Mackensen 1997). As a part of the biomonitoring campaign linked to the *Bahia Paraiso* sinking, Kennicutt et al. (1991) found that the total PAH content in *L. elliptica* samples collected near the wreck was 17,500 ng g^{-1} dry wt., while that in samples from central Arthur Harbor was 1,200 ng g^{-1} dry wt. In McMurdo Sound, Kennicutt et al. (1995) found that PCB concentrations in tissues of *L. elliptica* from Winter Quarters Bay and the sewage outfall (about 400 ng g^{-1}) were 20–80 times higher than in clams from more remote sites. Low concentrations of others POPs such as DDT and its derivatives, HCH, HCB and α-chlordane were also detected in molluscs from more polluted zones. Compared to tissue extracts from clams collected at remote sites of McMurdo Sound, those from *L. elliptica* in Winter Quarters Bay exhibited higher EROD (P4051A1-dependent ethoxyresorufin-O-deethylase) induction activity in rat hepatoma H4IIE cells (McDonald et al. 1994). Additionally, there was an excellent linear correlation between induced EROD activity versus total PCB levels. This result indicated the complementary nature of analytical and bio-analytical data. Furthermore, the assay provided a method for estimating TCDD toxic equivalents in extracts from marine organisms.

For its size, geographical distribution and biomass, the Antarctic scallop *A. colbecki* is one of the most conspicuous mollusc species in Antarctica. This suspension-feeder species has a wide bathymetric distribution and, in the inner shelf of the Ross Sea, it constitutes an important food source for benthic fish such as *Trematomus bernacchii* and invertebrate top predators (e.g. the gastropod *Neobuccinum eatoni* and the nemertine *Paraborlasia corrugatus*; Vacchi et al. 2000b). There are many studies on the ecophysiology (e.g. Dell 1972; Stockton 1984; Heilmayer et al. 2003) and elemental composition of *A. colbecki* (e.g. Mauri et al. 1990; Berkman and Nigro 1992; Viarengo et al. 1993; Nigro et al. 1997). Like in other Antarctic organisms, the increase in food availability during the austral summer determines the growth of its digestive gland and other organs, with a consequent "dilution" effect on trace element concentrations. However, there is evidence that Cd accumulates very efficiently in the digestive gland of *A. colbecki* and that its content does not vary during summer (Nigro et al. 1997). The huge bioaccumulation of Cd and other potentially toxic metals in the digestive gland of molluscs is made possible by several efficient detoxification mechanisms such as compartmentalisation within lysosomes, accumulation of metals into granules and/or membrane-bound vesicles, and binding of metals to specific ligands such as metallothioneins (Viarengo et al. 1993). Metal storage sites also occur in the kidney of *A. colbecki*. Nigro et al. (1992) found concretions containing Fe, Zn and Cu in the main vacuole of the renal cell; Ag/Se-rich particles and other electron-dense particles containing Fe, Cu and other metals were found in the basal membrane and in amoebocytes. These findings suggest transport of trace elements from uptake to excretion/storage sites.

Metallothioneins are soluble, low-molecular weight, cysteine-rich proteins. Metal ions within cells cause their rapid neosynthesis by enhancing metallothionein gene transcription (Squibb and Cousin 1977). Metallothioneins have been reported not only in molluscs but also in many other Antarctic organisms such as the sea urchin *S. neumayeri* and several species of fish (e.g. Scudiero et al. 1997; Carginale et al. 1998). The cysteine content in cadmium-induced metallothionein from the digestive gland of *A. colbecki* (Ponzano et al. 2001) is lower than that in metallothioneins purified from vertebrates and invertebrates in other seas. In addition to metallothioneins, many other biochemical markers can be used as biological indices of exposure to or of the effects of contaminants in molluscs. The enhancement of reactive oxygen species (ROS) has often served as a general pathway of toxicity induced by pollutants and oxidative stress (Winston 1991). Cells have evolved low-molecular scavengers and antioxidant enzymes for detoxification and removal of ROS and other oxidant molecules. Glutathione is one of the most important antioxidant agents, and variations in the content of glutathione and in the activity of glutathione-dependent and antioxidant enzymes are potential biomarkers of contaminant-mediated oxidative stress in several marine organisms (e.g. di Giulio et al. 1989; Viarengo et al. 1990; Regoli and Principato 1995). Cold Antarctic seawater has higher levels of dissolved oxygen, and research has shown that catalase activity and other antioxidants in *A. colbecki* are enhanced compared to those of the Mediterranean scallop *Pecten jacobaeus* (Regoli et al. 1997). Further comparative studies with the Arctic scallop *Chlamys islandicus* (which during summer experiences higher water temperatures than in the Southern Ocean) showed that *A. colbecki* has the highest scavenging capacity for peroxyl and hydroxyl radicals. These results indicate a possible biochemical adaptation of the Antarctic scallop to high levels of dissolved oxygen in Southern Ocean waters (Regoli et al. 2000).

7.4.2 Antarctic Fish and the Transfer of Contaminants to Higher Vertebrates

Unlike shelf waters in the other continents, the Antarctic shelf is dominated by a single suborder of fish (notothenioids), and most benthic and epibenthic species are notothenioids belonging to the genus *Trematomus* (e.g. *T. bernacchii, T. pennellii, T. scotti, T. hansoni*). These species are usually considered circum-Antarctic benthic feeders, and their diet consists of polychaetes, amphipods, fish eggs, molluscs and other epibenthic organisms, depending on availability (Kiest 1993; Hureau 1994; Vacchi et al. 1994). Epibenthic notothenioids such as *T. newnesi* and *T. loennbergii* mainly feed on zooplankton (hyperiids, copepods and euphausiids), fish, benthic polychaetes and gammarids (Gon and Heemstra 1990; Ekau 1991). Nototheniids also include a few pelagic species such as *Pleuragramma antarcticum* and species living on the undersurface of ice (e.g. *Pagothenia borchgrevinki*). Besides notothenioids, Antarctic waters contain plunderfish, dragonfish and icefish (about 50 species), which are mostly pelagic or semipelagic and prey on crustaceans and fish (Eastman 1993). Icefish, one of the most unusual groups of fish in the world (they lack haemoglobin and are white-blooded) have low concentrations of Fe; in the kidney and liver of the widespread species *Chionodraco hamatus*, Fe content is three to four times lower than that in the liver and kidney of red-blooded Antarctic fish (Bargagli 2001).

The feeding behaviour of notothenioids at Terra Nova Bay is somewhat reflected in the Cd and Hg concentrations in their organs and tissues. Benthic feeders (*T. bernacchii* and *T. hansoni*) have higher body burdens of heavy metals than epibenthic (*T. newnesi*) or semipelagic fish (*C. hamatus*). The latter two species mainly feed on copepods and euphausiids, which have lower Cd and Hg concentrations than many species of benthic invertebrates. At Terra Nova Bay *A. colbecki* and other molluscs are frequently included in the diet of *T. bernacchii* (Vacchi et al. 2000b), and this helps explain why this species, compared to other notothenioids, has the highest concentration of metals in the liver and accumulates 2–3 times more Hg in the muscle tissue. Minganti et al. (1995) reported Hg concentrations in *A. colbecki* and *T. bernacchii* which match those in Fig. 47, and showed that up to 96 % (median 80 %) of total Hg in the muscle of *T. bernacchii* was MeHg; in the liver, kidney and gonads this percentage was usually less than 50 %. Mercury concentrations in the fish muscle were significantly correlated with body weight. Carginale et al.

Fig. 47. Average trace metal concentrations ($\mu g\ g^{-1}$ dry wt., mean±SD) in different organs and tissues of benthic, epibenthic, and semipelagic fish from Terra Nova Bay. (Bargagli 2001)

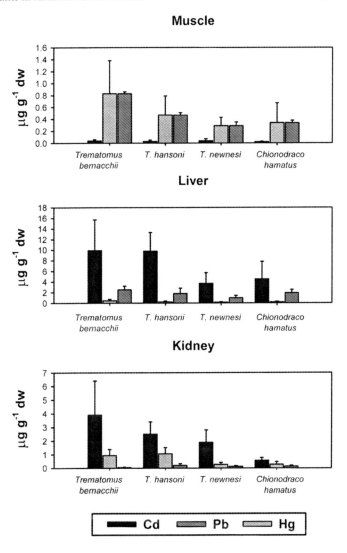

(1998) found that *T. bernacchi* contained higher levels of metallothionein than species of white-blooded fish collected in the same marine area. The only low-molecular weight, Zn-binding protein in the liver of icefish was a protein with a low cysteine content and rich in amino acid residues, whereas the predominant metal-binding protein in the red-blooded *T. bernacchii* was a metallothionein with molecular characteristics similar to those of rat metallothionein. Santovito et al. (2000) studied the bioaccumulation of Cd, Cu and Zn in different tissues of *T. bernacchii* and *C. hamatus* and found that the hepatic content of metallothioneins correlated positively only with Cd concentrations.

Trematomus bernacchii is an ideal bioindicator of local contamination because it not only has metal detoxification and accumulation capabilities, but it also has restricted home ranges and is ubiquitous. Indeed, this species spends much time in restricted areas; for example, some marked individuals released in February 1990 from a coastal site at Terra Nova Bay were recovered at the same site 2 years later. *Trematomus bernacchii* also has been used as a bioindicator of POPs in marine ecosystems adjacent to scientific stations such as McMurdo (Larsson et al. 1992; Kennicutt et al. 1995), Terra Nova Bay (Focardi et al. 1992a, 1995a; Bargagli et al. 1998d) and Syowa (Subramanian et al. 1983). Comparisons among the results of these studies are difficult because some refer to the whole body content, while others refer to different tissues and/or are expressed on different mass bases. In general, DDTs, PCBs, PAHs and other POPs have always been detected; average concentrations were in the same range as those previously reported for krill and were significantly lower than those usually detected in fish from other seas. Concentrations of three highly toxic non-*ortho* coplanar PCB congeners (77, 126 and 169) were always extremely low (<0.5 pg g^{-1} wet wt.) in the muscle tissue of *T. bernacchi* from Terra Nova Bay (Bargagli et al. 1998d). Microsomal CYP4501A-dependent monooxygenase activity was evaluated in the liver of the same samples by EROD, BROD (benzyloxyresorufin-*O*-deethylase) and BPMO (benzo-*a*-pyrene) assays; values were slightly higher in samples from a small cove receiving wastewater from the Italian station. Experiments to evaluate the induction of the cytochrome P450 monooxygenase system by PCBs and benzopyrene (Focardi et al. 1992b, 1995b) in *T. bernacchi* and other species of Antarctic fish showed that, owing to slow metabolic rates, the highest levels of induction occurred after 10 days (i.e. much later than in fish from temperate seas). Miller et al. (1999) measured directly CYP1A mRNA levels (a more sensitive and accurate biomarker than measurements of enzymatic activity) in the liver of *T. bernacchi* from Winter Quarters Bay and a remote site in McMurdo Sound. They found an average 37-fold increase in CYP1A expression, and two-fold higher levels of naphthalene and phenanthrene in bile samples from fish collected at the polluted site. The extent of CYP1A induction correlated positively with the content of aromatic compounds in bile samples from the same fish. High average concentrations of PAH metabolites were also measured in the bile of *Notothenia coriiceps* samples collected near Palmer Station and the *Bahia Paraiso* wreckage (McDonald et al. 1992). The latter species is one of the most abundant fish at depths of 0–450 m along the Antarctic Peninsula, the Scotia Arc, and the South Georgia, Bouvet and Pedro I islands (Kock 1989). Analysis of organochlorine compounds in fish species from the peninsula (Weber and Goerke 1996) showed that HCB concentrations (20 ng g^{-1} wet wt.) were as high as those in *Limanda limanda* from the North Sea, and bioaccumulation was attributed to the cold-condensation process. Levels of DDT and PCB congeners (153, 138 and 180) were from one to two orders of magnitude lower than in North Sea fish. However, POPs in samples from the Antarctic Peninsula generally increased between 1987 and 1991.

The Antarctic silverfish (*Pleuragramma antarcticum*) is the most important circum-Antarctic notothenioid species in terms of both number and biomass. It feeds almost exclusively on euphausiids, copepods and chaetognaths, and *P. antarcticum* is the only true pelagic species in the water column of most Southern Ocean shelf areas. *P. antarcticum* therefore plays a prominent role as the main source of food for several species of predatory vertebrates such as the Antarctic cod (*Dissostichus mawsoni*) and other fish, penguins (gentoo, Adélie and Emperor), South Polar skuas, Antarctic petrels, Weddell seals, crabeater seals and whales (Eastman 1993). Average PCB, HCB, and *p,p'*-DDE concentrations in adult *P. antarcticum* samples from the Ross Sea were 348, 4.85 and 2.01 ng g^{-1} wet wt. respectively (Corsolini et al. 2002). The higher PCB content in larvae than in adults was attributed to the affinity of PCBs for suspended particles and to the greater surface:volume ratio in larvae than in adult silverfish.

7.5 Contaminants in Birds and Seals Breeding in Antarctica

The biomass of euphausiids and that of zooplankton may be negligible in shelf waters around Antarctica (Jarre-Teichmann et al. 1997). At Terra Nova Bay, for instance, during summer the zooplankton biomass in shelf waters down to a depth of 200 m usually consists of a few hundred individuals per m^3 of water. Dominant species include copepods (above all *Metridia gerlachei*, *Euchaeta antarctica* and *Calanoides acutus*), *Euphausia crystallorophias* and the pteropod *Limacina helicina*. During summer most phytoplankton and organic detritus from the euphotic zone settles to the bottom where it supports rich benthic communities. The involvement of benthic organisms in the transfer of energy and persistent contaminants from phytoplankton and other autotrophic organisms to benthic invertebrates, notothenioid fish, seabirds, seals and top predators such as killer whales lengthens coastal food webs and makes them much more complex than the relatively simple krill food chain.

Birds breeding along Antarctic coasts are almost exclusively marine species, and each species is uniform across very large coastal regions. As they mainly feed on zooplankton and fish larvae, food availability is limited to the summer after the break-up of pack ice. Seabirds have a short and intense breeding season in this period, after which they must migrate northwards. About 90 % of the avian biomass in Antarctica consists of penguins. There are 18 species of penguins, of which seven breed south of the Antarctic Convergence and only four breed on the continent: Emperor, Adélie, chinstrap and gentoo penguins. Only Emperor (*Aptenodytes forsteri*) and Adélie (*Pygoscelis adeliae*) penguins are true continental birds, because the other two species are found exclusively in the northern Antarctic Peninsula. Other birds which breed successfully along Antarctic coasts include the South Polar skua (*Catharacta maccormicki*) and some species of procellariforms such as the snow petrel (*Pagodroma nivea*), cape pigeon (*Daption capensis*), Antarctic petrel (*Thalassoica antarctica*) and Wilson's storm petrel (*Oceanites oceanicus*).

7.5.1 Penguins as Biomonitor Organisms

During the breeding season Adélie and Emperor penguins concentrates in dense colonies along Antarctic coasts between 60° S and more than 77° S (e.g. the penguin rookeries on Ross Island). According to several population estimates (Knox 1994), there are about 2×10^6 breeding pairs of Adélie penguins in Antarctica, while the Emperor penguin (about 200,000 breeding pairs) is one of the rarest penguin species. Together with Weddell seals, they are the most important biotic components of the sea-ice zone, and the distribution of sea ice affects their travelling and foraging activity, with strong implications for chick production and population growth (e.g. Croxall et al. 1988; Fraser et al. 1992; Watanuki et al. 1997; Cherel and Kooyman 1998; Wienecke et al. 2000; Wilson et al. 2001). Although penguins mainly feed on crustaceans and fish (especially juvenile specimens of *P. antarcticum*), their diet shows large spatial variations. On the South Shetland Islands Adélie penguins feed almost exclusively on *E. superba*, while in the Ross Sea they ingest *P. antarcticum* during periods or years with little pack ice and *E. crystallorophias* during periods or years of heavy pack-ice cover (Ainley et al. 1998; Olmastroni et al. 2000). About 90 % by mass of the stomach content of Emperor penguins along the Ross Sea coast consists of fish (especially *P. antarcticum*; Cherel and Kooyman 1998). However, the diet of emperor penguins along the Mawson coast is dominated by squid (mainly *Psychroteuthis glacialis* and *Alluroteuthis antarcticus*; Robertson et al. 1994), while Emperor penguins in the Weddell Sea feed not only on fish and squid but also on significant amounts of krill (Klages 1989). Spatial variations in the diet of penguins from different colonies makes it difficult to compare between concentrations of persistent contaminants in their tissues. Annual and seasonal variations in sea-ice extent and prey availability can even determine temporal changes in the diet of birds from one and the same colony. Males and females show some differences in feeding behaviour (Clarke et al. 1998), and moreover very little is known about the distribution and feeding behaviour of penguins during non-breeding seasons.

In order to evaluate the potential input of xenobiotics through diet, Corsolini et al. (2003) determined POP concentrations in the stomach content of Adélie penguins at Edmonson Point (northern Victoria Land). They found that the mean concentration of HCB, *p,p'*-DDE and PCBs from one foraging trip were 1,412, 1,508 and 303 ng g^{-1} wet wt. respectively. Stomach contents richer in euphausiids had generally higher xenobiotic concentrations, and this was attributed to the effective release of particulate materials containing POPs from melting ice and their adsorption on the body surface of crustaceans. Although dioxin-like PCBs (i.e. those with either *meta* or *para* chlorine substitutions or with one chlorine in the *ortho* position) were detected in all stomach content samples, the estimated toxicity for Adélie penguins was very low. The amount of POPs in the diet of penguins was rather high in comparison to concentrations usually measured in these birds. A significant amount of ingested xenobiotics is probably metabolised and does not build up in organs and tissues. Besides variations in their diet, van den Brink et al. (1998) showed that during the breeding season penguins utilise different fat stores at different times, thereby contributing to seasonal fluctuations of organochlorine levels in blood and uropygial oil.

Despite all these factors causing strong variability in contaminant levels, Adélie and Emperor penguins are useful biomonitors of persistent contaminants in Antarctic marine ecosystems because of their distribution around the continents (exclusively within the seasonal pack-ice zone), their lifespan of more than ten years, and the occurrence year after year of many individuals in one and the same colony. Studies on POP accumulation in penguins began in the 1970s. Risebrough et al. (1976) assessed chlorinated hydrocarbon concentrations in penguin eggs from different zones of the Antarctic Peninsula and compared these with data from penguins on sub-Antarctic islands. The results proved, for the first time, that POPs in the Antarctic environment did not derive from local human activity and that the atmosphere (rather than oceanic water masses) is the main pathway of transport of PCB and DDT compounds to the continent. Tanabe et al. (1986) examined the mother-to-egg transfer in Adélie penguins of PCB isomers and congeners and of *p,p'*-DDE. Although the transfer rate was low (about 4% of the body burden of mothers), the pattern of individual PCB isomers and congeners in eggs was similar to that in mothers. In subsequent years several papers were published on POP and Hg concentrations in penguin eggs (e.g. Luke et al. 1989; Focardi et al. 1992 c; Court et al. 1997; Bargagli et al. 1998a; Kumar et al. 2002) and in various organs, tissues and feathers (e.g. Subramanian et al. 1986; Focardi et al. 1993, 1995a; Inomata et al. 1996; Sen Gupta et al. 1996; Corsolini and Focardi 2000). POP concentrations in penguins are generally lower than in birds from other seas and usually fall below recognised threshold levels for eliciting toxicological effects in birds. However, toxicity threshold levels for penguins are unknown, and there is evidence (e.g. Court et al. 1997; Wanwimolruk et al. 1999) that the liver of Adélie penguins has a low capacity to detoxify PCBs and chlorinated pesticides.

In 1981 Honda et al. (1986) collected organs, tissues and eggs of Adélie penguins at a breeding site about 18 km south of Syowa Station. A detailed survey on the elemental composition of many samples revealed that the liver and kidney accumulated the highest concentrations of Hg and Cd. The Cd content was much higher than in seabirds from other seas. The Fe content was much higher in the pectoral muscle of adult penguins, which is rich in myoglobin (associated with their ability to dive; e.g. Tamburrini et al. 1999), than in their femoral muscle or in the pectoral muscle of other bird species.

In order to evaluate the uptake and elimination rates of trace metals in Adélie penguins, Ancora et al. (2002) determined Cd, Hg and Pb concentrations in stomach contents, excreta and feathers. High Cd concentrations were measured in stomach content and excreta samples, while Hg contents were significantly higher in feathers, which in birds are an important route for the excretion of MeHg. Bargagli et al. (1998a) measured slightly higher total Hg concentrations in the feathers of Emperor penguins (0.98 ± 0.21 µg g^{-1} dry wt.) than in those of Adélie penguins (0.82 ± 0.13 µg g^{-1} dry wt.) from northern Victoria Land. The eggs and muscle tissue of Emperor penguins from certain populations also had slightly higher HCB, DDT and PCB concentrations (Focardi et al. 1993). Schneider et al. (1985) found that the Cd content in the liver and especially the kidney of *A. forsteri* (48 ± 21 and 382 ± 199 µg g^{-1} dry wt. respectively) from Atka Bay (Weddell Sea) were much higher than values measured in the liver and kidney of Adélie penguins, South Polar skuas and seals from the same bay. The higher bioaccumulation of Cd, Hg and organochlorines in Emperor penguins than in Adélie penguins is likely due to their different diet, lifespan, and detoxification and excretion capabilities.

7.5.2 Contaminants in Seals and in a Top Predator Bird: the South Polar Skua

Only three species of seals with very distinctive features are limited in distribution to the south of the Antarctic Convergence: the crabeater seal (*Lobodon carcinophagus*, which comprises $30–40\times10^6$ individuals and constitutes more than half of total world pinnipeds), Ross seal (*Ommatophoca rossii*), and Weddell seal (*Leptonychotes weddellii*). The latter species breeds in areas adjacent to the Antarctic continent and has the most southerly distribution of any mammal. In spring Weddell seals form pupping colonies on fast ice (near broken ice, tide cracks and hummocking), and during the year they move only locally to exchange breathing holes (Siniff et al. 1977). In contrast, Ross and crabeater seals, along with leopard seals, inhabit unstable areas of shifting pack ice. The Weddell seal is therefore a more reliable biomonitor of environmental contaminants in coastal marine ecosystems around Antarctica. There are many studies on the diet of *L. weddellii* around Antarctica (e.g. Dearborn 1965; Green and Burton 1987; Plötz et al. 1991; Casaux et al. 1997; Burns et al. 1998). These studies show that Weddell seals forage on both pelagic and benthic-demersal organisms to depths of 600 m, and large differences in diet may occur due to location, season, fluctuations in prey abundance, and age of seals.

Most seals in McMurdo Sound forage on pelagic fish (mostly *P. antarcticum*) and squid, while some juveniles concentrate on shallow benthic fish of the genus *Trematomus* spp. (Burns et al. 1998). Plötz et al. (1991) observed that, in years of low *P. antarcticum* biomass, seals in the eastern Weddell Sea foraged on benthic-demersal fish species, which usually have a more constant biomass. In contrast, in other Antarctic coastal regions such as the Vestfold Hills, Weddell seals move to open waters to forage on pelagic fish only in years of low local benthic resource abundance (Green and Burton 1987). On the South Shetland Islands, pelagic (mainly myctophids) and benthic-demersal fish were the most frequent (95.7 %) and numerous prey (46.2 %) of Weddell seals, but octopods constituted the bulk mass (63.1 %) of their diet (Casaux et al. 1997).

These studies seem to indicate that Weddell seals are generalised predators and that they adjust their foraging strategy according to age and local food availability. As in the case of penguins, the bioaccumulation of Cd, Hg and organochlorines in Weddell seals is therefore affected by spatio-temporal variations in diet and significant differences in the content of persistent contaminants within pelagic and benthic organisms.

Investigations on trace metal and organochlorine concentrations in organs and tissues of Antarctic seals began in the 1980s (e.g. Schneppenheim 1981; Kawano et al. 1984; McClurg 1984; Schneider et al. 1985; Yamamoto et al. 1987). Some available data on trace metals in *L. weddellii* samples from different Antarctic coastal regions are summarised in Table 18. Average Cd concentrations in the liver of Weddell seals are much lower than those in the liver of *O. rossii*, while the latter species and *L. carcinophagus* have much lower Hg contents. In contrast to two other species of Antarctic seals, *L. weddellii* does not feed exclusively on zooplankton and pelagic fish. As the Weddell seal includes benthic and epibenthic fish and molluscs in its diet, it occupies a higher trophic level in the marine food chain and is consequently exposed to enhanced biomagnification of MeHg and other persistent contaminants.

A number of papers have been published on trace metal concentrations in tissues and organs from different species of seals in the Northern Hemisphere (e.g. Dietz et al. 1996; Becker 2000; Fant et al. 2001). Mercury and Cd concentrations in the liver and kidney of Arctic seals are generally in the same range or higher than those in Antarctic seals. By reviewing literature data on Hg and Cd concentrations in the liver of pinnipeds, Watanabe et al. (2002) came to the conclusion that these mammals can be classified into two groups: the one (e.g. sea lions, grey seals, harbour seals and Caspian seals) accumulates more Hg than Cd, and the other (harp seals, ringed seals and fur seals) accumulates more Cd than Hg. The former group comprises pinnipeds feeding mainly on fish, while the latter group essentially includes pinnipeds feeding on crustaceans and molluscs. On the basis of the common assumption that most marine mammals in Antarctica feed on krill, Watanabe et al. (2002) included Weddell seals in the group which accumulates

Table 18. Average concentrations of trace metals (μg g^{-1} dry wt.; * wet wt.) in organs and tissues of Weddell seals from different Antarctic coastal regions

Location	No. of samples	Organ/tissue	Cd	Cu	Fe	Hg	Zn	Reference
Weddell Sea (Atka Bay)	7	Muscle	0.21±0.12	2.8±0.2	–	–	60±25	Schneider et al. (1985)
		Liver	23±10	45±21	–	–	105±41	
		Kidney	174±67	36±12	–	–	142±43	
Antarctic Peninsula, South Shetland Islands	2	Muscle	0.03±0.02	2.8±0.4	1,100±200	1.9±0.7	124±8	Szefer et al. (1993, 1994)
		Liver	3.1±2.1	57±27	2,200±1,020	35±16	167±15	
		Kidney	25±18	23±1	440±60	13±8	122±30	
Ross Sea (Terra Nova Bay)	1	Muscle	–	5	1,632	1.9	132	Bargagli (2001)
		Liver	–	12	2,639	44	152	
Syowa Station	2	Muscle*	0.15±0.15	1.0±0.1	252±21	0.14±0.02	37±3	Yamamoto et al. (1987)
		Liver*	1.1±0.3	21±8	664±389	5.8±3.8	44±4	
		Kidney*	6.4±5.0	8.1±4.1	388±324	0.67±0.40	29±3	

more Cd than Hg in the liver. On the contrary, data in Table 18 indicate that this seal, in contrast to other Antarctic seals and baleen whales, bioaccumulates more Hg than Cd.

Bacon et al. (1992) analysed concentrations of organochlorines and polychlorinated biphenyls in pinniped milk samples from the Arctic, Antarctic, California and Australia. Concentrations of p,p'-DDE and PCBs in the milk of Antarctic fur seals were about two orders of magnitude lower than those in pinniped milk samples from California. Although PCB levels were dramatically different in different geographical regions, Bacon et al. (1992) found a similar ratio pattern for PCB congeners 153, 138 and 180 in pinnipeds throughout the world. They concluded that it is not the levels which shape the profiles, but rather how the pollutants react in metabolic systems. Like in the milk of Antarctic fur seals, average DDT, PCB, HCB and HCH concentrations in samples of Weddell seal blubber from various Antarctic locations (e.g. Kawano et al. 1984; Schneider et al. 1985; Focardi et al. 1993) are from one to three orders of magnitude lower than average values in the blubber of Arctic seals (e.g. Becker 2000). Toxaphene and chlordane components in chromatograms of Weddell seal samples were found to exceed the intensity of PCB signals (Luckas et al. 1990), and the HCB/αHCH ratio was >1 while in Arctic samples it was always <1. Focardi et al. (1995a) found a relationship between PCB concentrations and trophic levels of marine organisms from Terra Nova Bay. Concentrations of PCBs and of three highly toxic non-*ortho* coplanar PCBs, like those of Hg, were much higher in the Weddell seal than in penguins or fish.

Among marine organisms breeding in Antarctica, the South Polar skua (*Catharacta maccormicki*) combines the traits of the species more exposed to the bioaccumulation of persistent contaminants. Skuas are opportunistic top predators and, during the breeding season in Antarctica, they adopt a wide range of feeding tactics which enable them to prey or scavenge on all profitable marine or terrestrial food resources according to their temporal and spatial availability (Furness 1987; Pezzo et al. 2001). Many skuas nest close to penguin rookeries, and Adélie penguins are an important component of their diet. However, in areas such as the Larsemann Hills (Princess Elizabeth Land) where there are no breeding penguins, the snow petrel (*Pagodroma nivea*), together with other seabirds, various marine foods and refuse (meat, fish and vegetable remains) from nearby stations (Zipan and Norman 1993), becomes the major dietary component of skuas. South Polar skuas have one of the longest migration flights of any bird, and they can range over huge areas of the ocean up to the north Atlantic and Greenland during the Antarctic winter (Fullager 1976). Their feeding behaviour in Antarctica and especially that in more polluted marine ecosystems of the Northern Hemisphere exposes South Polar skuas to enhanced uptake of persistent contaminants. Early biomonitoring surveys of Antarctic wildlife identified the South Polar skua and brown skua (*C. lonnbergi*) as the species with the highest concentrations of persistent contaminants (e.g. Risebrough and Carmignani 1972; Norheim et al.

1982; Gardner et al. 1985). Total concentrations of Hg in the feathers, eggs and chick plumage of South Polar skuas breeding near penguin rookeries along the northern Victoria Land coast were three to six times higher than in the same type of samples from Emperor and Adélie penguins (Bargagli et al. 1998a). Nygård et al. (2001) determined trace element concentrations in the eggs, muscle, liver, kidney and stomach content of nestling and adult South Polar skuas and Antarctic petrels (*Thalassoica antarctica*; i.e. the main prey of skuas at Svarthamaren, Dronning Maud Land). The latter species mainly feeds on krill, and Cd concentrations in the kidney and liver of its nestlings were about 40 times higher than those in the same organs of skua nestlings. Mercury and Se concentrations in all organs of the latter were in turn four to ten times higher than in the organs of petrel nestlings. The content and distribution of Cd in organs and tissues of adult birds of the two species was similar, while Hg and Se concentrations in South Polar skua organs and tissues were more than twice those in *T. antarctica*.

Although *C. maccormicki* has been sighted farthest south during summer than any other bird in the world, from late June to early October it forages at much higher latitudes, even reaching Greenland. The amount of Hg in its eggs (1.6 ± 1.2 μg g^{-1} dry wt.; Bargagli et al. 1998a) and nestlings (about 0.5 μg g^{-1} dry wt.; Nygård et al. 2001) is therefore partly determined by dietary Hg uptake in contaminated environments before egg deposition in Antarctica (usually in November). However, the high Hg content in chick plumage (1.9 μg g^{-1} dry wt.; Bargagli et al. 1998a) and that of Se in organs and tissues of nestlings (about 20 μg g^{-1} dry wt.; Nygård et al. 2001) indicate that South Polar skuas have effective Hg detoxification mechanisms already in the early stages of development.

Although differences in lifespan and moulting strategies make it difficult to compare Hg concentrations in feather samples from different bird species (Stewart et al. 1997), feathers from snow petrels and penguins breeding in northern Victoria Land had much lower Hg concentrations than feathers from petrel and penguin species breeding on the sub-Antarctic islands (Bargagli et al. 1998a). In contrast, Hg concentrations in feathers of adult South Polar skuas feeding in penguin rookeries from the same region correspond to those reported for the Arctic skua (*Stercorarius parasiticus*) from the Shetland Islands (60° 08′N, 2° 05′W; Stewart et al. 1997; Fig. 48).

Since the South Polar skua is a top predator and scavenger and has one of the longest migration paths in more polluted marine environments, average concentrations of DDTs (369 ng g^{-1} wet wt.) and PCBs (908 ng g^{-1} wet wt.) in its eggs at Cape Bird (Ross Island; Court et al. 1997) were 13 and 22 times higher respectively than those in eggs of Adélie penguins from the same coastal area. Concentrations of DDE, HCB and PCB in skua liver were 11, 3 and 4 times higher respectively than in the liver of penguins. However, studies by Focardi et al. (1992b, 1997) showed that values of MFO (Mixed Function Oxidases) activity were very low in penguins, while South Polar skua values

Fig. 48. Total Hg concentrations (µg g^{-1}, mean±SD) in
feathers of adult seabirds
from northern Victoria Land
(Bargagli et al. 1998a) and the
Arctic skua (Stewart et al.
1997)

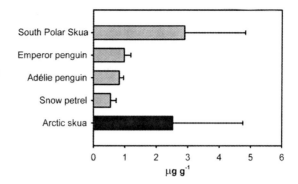

corresponded to those of certain Northern Hemisphere gulls. Although the
two species seem to have different metabolisms and although skuas accumulate most organochlorines in regions north of the Antarctic Convergence, the
prevalent PCB congener in penguin and skua samples was the same
(22′44′55′, a hexachlorobiphenyl; Court et al. 1997). Like in seals, levels of
chlorinated hydrocarbons in the eggs and tissue of South Polar skuas were
generally one to two orders of magnitude lower than those in the north
Atlantic great skua (*Catharacta skua*; Furness and Hutton 1979).

Kumar et al. (2002) determined concentrations of polychlorinated
dibenzo-*p*-dioxins (PCDDs), dibenzofurans (DFs) and dioxin-like PCBs in the
liver of polar bears from the Alaskan Arctic, in the liver of Weddell seals, and
in the eggs of Adélie penguins and South Polar skuas. Skua eggs had the highest concentrations of total PCDD/DFs (mean=181 pg g^{-1}), and estimated concentrations of 2378-tetrachlorodibenzo-*p*-dioxin equivalents (TEQs) of
PCDDs, PCDFs and dioxin-like PCBs in skua eggs (mean=344 pg g^{-1}) were
higher than in the liver of polar bears (mean=120 pg g^{-1}). These concentrations were close to values which may cause adverse health effects. The mean
value in skua eggs, for instance, was only two-fold less than the toxicity
threshold value reported for American kestrel eggs (Kemler et al. 2000).

7.6 Summary

In contrast to xenobiotic compounds, trace metals occur naturally in all terrestrial and marine environments. The unique physico-chemical characteristics of remote, pristine Southern Ocean water masses and the unique ecophysiological adaptations of Antarctic marine organisms which have
undergone long evolutionary processes in isolation enhance bioaccumulation
of Cd and other trace elements in some species of Antarctic invertebrates,
seabirds and marine mammals. Bioaccumulation is independent of human

activity or global processes and probably occurs naturally in marine organisms which grow and reproduce slowly, are long-lived, and have longer moult cycles than related species from other seas. Cadmium concentrations in Southern Ocean surface waters are higher than in other seas, and the efficient uptake of the metal by phytoplankton determines the transfer of Cd to primary consumers such as pelagic crustaceans or benthic invertebrates (in the neritic province). Seabirds, marine mammals and other organisms feeding almost exclusively on pelagic crustaceans or other primary consumers may accumulate very high concentrations of Cd in the liver and especially in the kidney. However, as the enhanced bioavailability of Cd is a natural characteristic of the Southern Ocean, Antarctic marine organisms likely adapted to this environment during their long evolutionary history. There is evidence that, beginning from phytoplankton cells, almost all Cd-accumulating marine species have efficient detoxification mechanisms. The uptake of Cd occurs mainly through diet, and the metal does not usually accumulate according to the age of organisms or their trophic level. As Cd cannot be eliminated through eggs, feathers or hairs, higher vertebrates release most of the metal through their urinary apparatus. The kidney of Antarctic (and Arctic) seabirds and marine mammals may thus contain impressive Cd concentrations; their kidneys can probably tolerate much higher concentrations of this metal than those associated with heavy damage in many species of terrestrial birds and mammals, including humans.

Mercury, like Cd, is one of the most toxic metals; in the environment, however, it behaves differently. The most toxic and bioaccumulating chemical form of Hg is MeHg, which is produced naturally in the environment by bacteria or abiotically. The chemical properties of this compound are more similar to those of POPs than to those of other metals. The uptake of MeHg by marine organisms mainly occurs through diet but in contrast to Cd, MeHg elimination processes are very slow; concentrations consequently increase in larger, long-lived species at the highest levels of trophic chains. There is no evidence of enhanced MeHg availability in Southern Ocean waters or sediments, and total Hg concentrations in phytoplankton, primary consumers and most Antarctic organisms are generally low. However, the highest-ever reported concentrations of Hg were measured in the liver of some species of albatrosses and petrels from the Southern Ocean. These are large, long-lived species which feed on squid, fish or carrion, have low reproductive rates and take years to replace feathers (i.e. compared to other bird species, they eliminate less MeHg through egg laying and feather moulting). The huge accumulation of Hg in the liver of some species of albatrosses and petrels seems a natural process determined by their life-history characteristics; it is therefore probable that, as in the case of Cd, these birds adapted to enhanced Hg accumulation during their evolution.

In contrast to metals, POPs are ubiquitous toxic compounds which are not natural in origin. They are released around the world and transported to polar regions by air masses according to global distillation and fractionation processes. In winter, Southern Ocean pack ice behaves as a sink for persistent contaminants, which are subsequently released during the spring and summer melt. Persistent airborne contaminants are then transferred to water and organisms, where they can accumulate in tissues and biomagnify in food chains. Antarctic seabirds and marine mammals are exposed to the potential toxic effects of some POPs because they have developed few metabolic detoxification pathways for xenobiotic compounds. The potential risk to organisms breeding in Antarctica from POP exposure could also be enhanced by the extreme variability of physiological conditions during the short, intensive breeding season.

Overall, available data for DDT and derivatives in Antarctic seabirds and mammals indicates that levels increased from the early 1960s to the end of the 1980s. Despite the reduction in the input of "new" DDT and DDE from outside Antarctica, the widespread occurrence of DDE in Antarctic organisms is indicative of the high persistence of this compound. Hexachlorobenzene is another POP with a long environmental half-life, and its concentrations in fish tissues from the Antarctic Peninsula are in the same range as those measured in fish from the North Sea. Concentrations of HCB in preen oil of seabirds such as the cape petrel wintering in sub-Antarctic regions were much higher than those measured in preen oil samples from European seabirds. With the exception of benthic organisms collected from a few polluted coastal ecosystems adjacent to scientific stations, concentrations of PCBs and other POPs in seabirds and marine mammals endemic to the Antarctic region are generally lower than those in comparatively widely dispersed sub-Antarctic species. However, available data indicate that the liver of Adélie penguins does not efficiently detoxify PCBs and chlorinated pesticides. Among seabirds nesting in Antarctica, the South Polar skua is a top predator and scavenger which migrates to more polluted marine areas of the Northern Hemisphere during winter. Mercury and POP concentrations in its eggs, feathers, internal organs and tissues are therefore among the highest recorded in Antarctic organisms. Although some detoxification systems such as the activity of mixed function oxidases or the co-accumulation of Hg and Se in the liver are more effective than in other Antarctic seabirds, there is evidence that concentrations of 2378-tertrachlorodibenzo-p-dioxin equivalents in its eggs are higher than in the liver of polar bears and are only twofold lower than toxicity threshold values reported for other birds.

8 Climate Change, Anthropogenic Impact and Environmental Research in Antarctica: a Synthesis and Perspectives

8.1 Introduction

One of the main values of Antarctica for science is the near-pristine environment where clean media (air, snow, ice, sediments) offer unique opportunities for many research activities such as astronomical and astrophysical observations, geophysical and ecophysiological studies, and investigation of the Earth's magnetosphere and ionosphere. Antarctic environmental matrices represent ideal archives of data on past and current trends in global processes and, in particular, the long-range transport of persistent atmospheric contaminants. The Protocol on Environmental Protection to the Antarctic Treaty provides strict guidelines for protection of the Antarctic environment, of its value for scientific research and of its wilderness and aesthetic values. The standards of protection set in these guidelines for Antarctica are generally higher than those required for other parts of the world. However, as discussed in this book, the effective conservation of the value of Antarctica for scientific research requires better knowledge of the structure and functioning of ecosystems, and the development of long-term monitoring programmes on a continental scale to detect changes induced by climatic or anthropogenic perturbations. All human activity in Antarctica (e.g. scientific research and associated logistic support, fisheries and tourism) causes localised environmental pollution and large-scale impacts (e.g. air pollution by aircraft and ships, accidental spillage of oil or fluids from ice drilling operations, waste materials, and thousands of released weather and research balloons which land on the continent or in the sea). The application of the Protocol on Environmental Protection will help minimise some of these impacts, especially near scientific stations. However, the data reported in the previous chapters show that most persistent airborne contaminants in Antarctica and the Southern Ocean mainly originate from anthropogenic sources in the Southern Hemisphere. Although current levels of environmental contamination do not represent a great threat to the sustainability of terrestrial and marine ecosystems, it cannot be excluded that climate change, together with increased inputs of some atmospheric contaminants, may compromise the scientific value of the Antarctic environment.

In the last century, most economical and industrial development took place in the Northern Hemisphere, where several countries grew rich while inadvertently burdening the global environment. However, the global pattern of anthropogenic emissions of persistent atmospheric contaminants, greenhouse gases and aerosols is changing rapidly. Most of the world's future population growth will occur in countries of the Southern Hemisphere, and this growth will necessarily determine an increase in the magnitude of production and resource exploitation. If rich countries in the Northern Hemisphere do not adequately help (through financial aid and transfer of technologies) developing nations to address new environmental threats, many of these nations will improve their overall standard of living following historical consumption and production practices. As a result, anthropogenic impacts on the climate and environment of the Southern Ocean and Antarctica may significantly increase in the near future. Although the international community has set these regions aside for protection as natural reserves and areas for the conduct of scientific research (through the Antarctic Treaty in 1961 and the signing of the Protocol on Environmental Protection to the Antarctic Treaty in 1991), wider international agreements will be necessary to address environmental challenges facing Antarctica in this century. Without the inclusion of new participants at all levels of the decision-making process for coordinated global action, and adequate financial support and technological innovation in developing countries, the Protocol to the Antarctic Treaty will probably not be adequate to effectively protect the Antarctic environment. However, problems such as the need for more intensive international cooperation among nations and people are beyond the scope of this book. Its aim was to give a general overview and a frame of reference of the distribution and cycling of natural elements and persistent contaminants between the atmosphere, snow, seawater, and organisms in Antarctica and the Southern Ocean. Reported values can be used as baselines to assess future variations in the environmental distribution of trace metals and persistent organic contaminants. This concluding chapter briefly discusses possible future trends in the environmental contamination of Antarctica, and suggests research activity and long-term monitoring approaches to improve the efficiency and cost effectiveness of programmes for the protection of the Antarctic environment and its scientific value.

The critical role of long-term, broad-scale studies and programmes to monitor changes and trends in global environmental systems has been acknowledged by major scientific initiatives such as the International Geosphere-Biosphere Program (IGBP) and the Long Term Ecological Research Network (LTER, implemented by the US National Science Foundation). The LTER network was established in 1981 in view of the fact that ecological processes must be studied for much longer periods than normally foreseen by most research grants. The network for long-term research on ecological systems has grown rapidly over the last two decades. In 2001 it included 24 sites ranging from the tropical rainforest of Puerto Rico to the northern Chihuahuan Desert of New Mexico. Access to data collected at each site, the diffusion of knowledge and of approaches to policy-making and management, and public understanding are integral parts of the LTER programme. Two polar sites are located in Antarctica – the one in the pelagic marine ecosystem at Palmer Station (Antarctic Peninsula), the other in the McMurdo Dry Valleys (southern Victoria Land). The Palmer LTER site essentially focuses on interactions among regional climate, hydrography and pack ice, and on the impact of variations in the extent of sea ice on the most representative marine pelagic species. The results of research on ecological processes in the western Antarctic Peninsula are also reported in a book by Ross et al. (1996). The McMurdo LTER site was established in 1994 with the aim of acquiring integrated knowledge of biological, chemical and physical factors involved in biogeochemical processes affecting the largest cold desert of continental Antarctica. Several publications on various aspects of this region have been cited in previous chapters. However, for a more comprehensive presentation of the results of ecological research in the McMurdo Dry Valleys, the reader can refer to the book edited by Priscu (1998). Other books have recently been published on ecosystem processes and protection of the Antarctic environment (e.g. Lyons et al. 1997; Hansom and Gordon 1998; Davison et al. 2000; Huiskes et al. 2003).

Having presented an overview of current knowledge on the occurrence and cycling of persistent contaminants in Antarctic ecosystems, it seems appropriate to conclude this book with possible future trends in environmental contamination and suggestions on possible approaches to monitoring of environmental change. Predicting how climate change and human activity in the Southern Hemisphere will modify contaminant transport and deposition in Antarctica and the Southern Ocean is an exceptional challenge. Although the biogeochemical cycles of trace metals and physico-chemical properties of most POPs are rather well known, knowledge of past and future trends in climate variability, of sources and pathways of persistent contaminants, and of environmental processes in Antarctica and the Southern Hemisphere is still very inadequate to delineate possible future scenarios for these regions. This chapter only attempts to outline observations and projected changes in the climate and atmospheric contamination of the Southern Hemisphere, with the aim of foreseeing the possible impact of persistent contaminants in Antarctic ecosystems and suggesting possible approaches for the early detection of environmental perturbations.

8.2 Climate Change and Pathways
of Persistent Contaminants

Chapter 1 described how one of the most important geological features of Antarctica is its location in an expanding lithospheric plate, which has determined the quite stable position of the continent with respect to the South Pole during the last 100 Ma. Climatic and environmental changes in Antarctica during this period are therefore mainly ascribed to global changes. The opening of the Drake Passage (about 25 Ma B.P.) and the consequent establishment of the Circumpolar Current (Early Miocene) increased the isolation of the continent and favoured the formation of the East and West Antarctic ice sheets (Early Pliocene). Isotope records from ice cores and deep-sea sediment cores (e.g. Denton et al. 1991; Petit et al. 1999) show that there have been periodic fluctuations in the extent of ice sheets during the last 500 ka, with a slow build-up over glacial periods (of about 100 ka) followed by enhanced deglaciation during warmer interglacial periods (about 10 ka). Like in many other areas of the world, air temperature in Antarctica fell markedly between 100 and 20 ka ago. Temperature data from the Vostok ice core (Fig. 13) show a marked warming trend during the last 10 ka, and this interglacial period (centred on 6 ka B.P.) saw the lowering of ice sheets, the recession of glaciers (especially in the Antarctic Peninsula and sub-Antarctic islands) and the emplacement of raised beaches (e.g. Robin 1983; Colhoun 1991). Most terrestrial life in Antarctica dates from this period, with colonisation of the continent by cryptogamic and microinvertebrate species able to cope with seasonal and climatic changes during the Holocene. Climatic amelioration during the late Quaternary also affected the marine environment through contraction of the sea-ice cover in the Southern Ocean and the southward penetration of warmer waters in the Southern Ocean (Sugden and Clapperton 1977; Hays 1978).

As the moisture-holding capacity of air increases rapidly with increasing temperatures, snow would probably accumulate over most of continental Antarctica if global warming were to occur. There is evidence (e.g. Giovinetto et al. 1990) that the spatial distribution of snow accumulation over the Antarctic ice sheet is related to geographical variations in air surface temperature or saturation vapour pressure immediately above the surface inversion. The mass balance of glaciers and ice sheets is affected by changes not only in temperature but also in snow precipitation, especially in regions constrained by atmospheric precipitation. In the McMurdo Dry Valleys, for instance, Holocene warmth determined the expansion of glaciers because the clearing of ice from the Ross Sea increased the atmospheric moisture content and enhanced snow precipitation (Denton et al. 1991). Glacier expansion due to increased precipitation has also been detected in several East Antarctic outlet glaciers (Domack et al. 1991). However, it is difficult to foresee how temperature changes will affect the pattern and amount of atmospheric precipitation in different Antarctic regions. King and Turner (1997) detected a significant warming trend and a 20% increase in precipitation at Faraday Station in the period 1956–1993, but little correlation was found between year-to-year variations in temperature and precipitation frequency. High snow precipitation in the region was above all associated with low mean annual surface pressures in the Bellingshausen Sea, indicating that interannual variability in precipitation was mainly due to variations in air mass circulation. Research on the influence of global change on contaminant pathways to and within the Arctic (AMAP 2003) indicates that most dramatic changes during the past decade were directly related to variations in atmospheric pressure fields (i.e. the Arctic Oscillation, which affects winds, precipitation, sea-ice drift and cover, marine currents and other components of environmental pathways; Serreze et al. 2000).

In continental Antarctica there is no evidence of a general warming trend and changes in precipitation are poorly understood; most future climate scenarios are uncertain, and estimates cover very large ranges of temperature and precipitation. However, all scenarios indicate that the climate of polar regions will be wetter and warmer, and some of the projected increases in temperature and precipitation are larger than those predicted for other parts of the world (IPCC 2001). Most available data from continental Antarctica show an increasing snow-accumulation rate (e.g. Morgan et al. 1991; Isaksson and Karlén 1994). A model by Smith et al. (1998) found that the accumulation rate in the East Antarctic ice sheet over the 1950–1991 period increased 1.9 mm year^{-1}. According to the same model, each degree of warming will result in a snow-accumulation rate of 12.5 mm year^{-1}. There is also indirect evidence of increased snow deposition in the Southern Ocean. By studying physico-chemical characteristics of water masses in the Indian Ocean, Bindoff and McDougall (2000) found that over the past 30 years Antarctic intermediate waters have become less saline and cooler, and they attributed such changes to increased precipitation in their Antarctic source regions. The increasing deposition and accumulation of snow in continental Antarctica will be scarcely affected by increasing temperature values because temperatures will remain well below freezing point, except in very limited, coastal ice-free areas. Warming of the Southern Ocean will be much lower, but precipitation is expected to increase by as much as 20% (IPCC 2001), with a likely increase in freshwater input to the ocean surface.

8.2.1 Future Trends in Trace Metal Deposition

Chapter 1 revealed that about 40% of water vapour falling as snow in Antarctica reaches the continent through the western sector (between the Ross Ice Shelf and the Antarctic Peninsula; Bromwich et al. 1995). Although this sector has the largest inter-annual variability in snow deposition, particularly in conjunction with the ENSO phenomenon (Trenberth and Hoar 1996), the annual accumulation rate (more than 800 mm water equivalent) reported for coastal regions in the south-eastern Bellingshausen Sea is much higher than in the rest of the continent (Zwally and Giovinetto 1995). Air masses moving southwards from the subtropics may loose some of their water and contaminant burden through precipitation in the Southern Ocean. However, it seems likely that the western sector of Antarctica and the Antarctic Peninsula will receive increasing amounts of airborne contaminants from South America. Although recent data on Coats Land snow (Planchon et al. 2002a) indicate a decreasing input of Pb, Cr, Ag and U, concentrations of Cu, Zn and other metals are still increasing in snow. Human activity in the Southern Hemisphere, particularly the mining and smelting of non-ferrous metals in Chile, Peru and other countries, will probably be the main sources of Cu, Zn and other trace metals to the Antarctic environment during the next decade.

This book highlights how volcanic emissions in Antarctica and elsewhere in the Southern Hemisphere are possible sources of metals to Antarctic snow. More reliable estimates of the amount of trace elements emitted by volcanoes are needed. In general, most available estimates relate to total sulphur emission data and disregard the fact that proportions of emitted metals and metal-to-sulphur ratios may vary by orders of magnitude in different types of volcanoes or even during the different phases of activity of a single volcano (Hinkley et al. 1999). Recent research on natural and anthropogenic sources of metals to Antarctic snow (e.g. Vallelonga et al. 2003; Planchon et al. 2003) highlights the inadequacy of $nssSO_4^{2-}$ concentrations in snow as proxies for volcanic emissions, and reveals metal contributions from Mt. Erebus and volcanoes in the Antarctic Peninsula in samples of snow and ice from Law Dome and Coats Land.

There is only one report of tropospheric Hg depletion during and after spring sunrise in Antarctica (Ebinghaus et al. 2002). Although this process has only recently been recognised, it seems likely that Antarctic ecosystems, like coastal Arctic ecosystems, are receiving a disproportionate amount of the atmospheric Hg burden in the Southern Hemisphere. While anthropogenic emissions of Hg are decreasing in the Northern Hemisphere, the growing demand for energy, the burning of coal and biomass, the extraction of gold and the lack of emission control technologies in Asia, Africa and South America are probably increasing the atmospheric burden of Hg. As suggested by Lindberg et al. (2002), the role of polar regions as "cold traps" for Hg may also be enhanced by global warming through an increase in the natural outgassing of Hg from continents and oceans, and by changes in sea-ice cover and in the pattern and amount of atmospheric precipitation. The chemical forms of Hg deposited on Antarctic snow and in terrestrial ecosystems are easily bioavailable, and it cannot be excluded that the enhanced deposition of

Hg during and after the polar sunrise may pose a threat to organisms in Antarctic freshwater and terrestrial ecosystems.

Most airborne contaminants from lower latitudes will probably be deposited in the continental fringes of Antarctica. However, increasing snow deposition on the polar plateau may also increase the input of black carbon, trace metals and POPs in inner Antarctica. Model studies indicate that precipitation on the plateau can be due to large-amplitude long waves of warm, moist air from mid latitudes (20–40° S; Ciais et al. 1995). Moreover, cyclones in the Southern Hemisphere may raise air masses and aerosols to the upper troposphere, allowing their penetration in the polar anticyclone and the subsequent deposition of contaminants on the Antarctic plateau. Basile et al. (1997), for instance, found that most atmospheric dust trapped in East Antarctica ice has a Patagonian origin (i.e. from the only sizeable landmass in the Southern Hemisphere belt of westerlies; Iriondo 2000).

8.2.2 The Unpredictable Pathway and Temporal Trend of POPs

Predicting the possible consequences of human activity and climate change on the pathway and environmental fate of POPs in the Southern Hemisphere is much more challenging than for other persistent contaminants. POPs include a large number of chemical compounds with a wide range of physico-chemical properties, and they are released by many different human activities from all continents. The extent to which POPs are associated with aerosols generally plays an essential role in their atmospheric transport to polar regions. The association with particles may reduce or slow the transport of POPs through their temporary or permanent deposition on the ground. However, the association with small particles may also protect an organochlorine compound from oxidation during its transit to the south. Temperature changes may alter partitioning between gaseous and particulate phases, especially of compounds with higher log values of the octanol–air partition coefficient (K_{OA}). In the Arctic atmosphere, for instance, over 70 % of DDT is bound to aerosols in winter (air temperature=–30 °C), while DDT occurs almost exclusively in the gaseous phase in summer (temperature=0 °C; Bidleman et al. 2003). An increase in air temperature of only a few degrees can increase the volatility of many POPs and, consequently, their atmospheric transport potential. One expected effect of global warming is therefore increased atmospheric cycling of POPs. A fraction of the POP atmospheric burden is lost during transport through photolytic oxidation by OH, O_3 or NO_3. Although photolytic reactions are scarcely affected by temperature, global warming is predicted to increase cloud cover (IPCC 2001), which is already higher over the Southern Ocean than elsewhere in the Southern Hemisphere (more than 85 % cloud cover throughout the year along the 60° S parallel; King and Turner 1997). A further increase in cloud cover promoted by global warming would probably significantly reduce hydroxyl radical concentrations and POP removal from the atmosphere.

It is even more difficult to forecast the environmental fate of POPs in the Southern Hemisphere because some of their chemical properties, such as volatility, phase partitioning and degradation kinetics, are affected not only by changes in temperature or other climatic and meteorological factors but also by environmental characteristics of the ocean. Recent findings (e.g. Li et al. 2002) emphasise the importance of atmosphere–ocean coupling in the transfer of POPs from release points in tropical and temperate regions to polar regions. Models which combine the transport of semi-volatile POPs in air and seawater and consider the continuous exchange between the two compartments show an overall accelerated transport of POPs to remote regions with respect to those which treat air and water separately (Beyer and Matthies 2001). Other models (e.g. Dachs et al. 2002) suggest that levels of primary and secondary productivity in surface ocean waters may contribute to the deposition of POPs at mid–high latitudes.

Available data show that the main sources of POPs in the Southern Hemisphere are urbanised areas, those with intensive agriculture, and tropical or subtropical regions where spraying is used for disease vector control. In the Northern Hemisphere the emission of most POPs of environmental and toxicological concern peaked in the 1970s and 1980s and thereafter generally declined or ceased. In contrast, in the Southern Hemisphere the demand and use of many POPs were still increasing in the 1990s. In several developing countries, considerable quantities of PCBs are used in older electrical devices and deposited as landfill. There is evidence (e.g. Kallenborn et al. 1998) that air masses with high PCB concentrations from South America may reach the sub-Antarctic islands, the Antarctic Peninsula and the northern Weddell Sea. South America has historically been among the heaviest users of DDT, toxaphene and lindane. In general, more volatile POPs such as HCHs and HCB are dominant in the polar atmosphere; they are also deposited in the Bellingshausen, Scotia and Weddell Seas through the global distillation process. Much higher concentrations of HCB, for instance, were measured in preen oil from seabirds nesting in the sub-Antarctic islands than in samples from related species of seabirds from the Northern Hemisphere (van den Brink 1997). Concentrations of HCB in fish species from the Antarctic Peninsula were as high as those in *Limanda limanda* from the North Sea (Weber and Goerke 1996). These data suggest that during the next decade marine and terrestrial ecosystems in the Antarctic Peninsula will probably still be affected by enhanced warming and deposition of persistent atmospheric contaminants.

Biomass burning in South America and southern Africa produces large amounts of hydrocarbons, and there is evidence that emitted trace gases may affect tropospheric concentrations of O_3 in southern areas of the Pacific and Atlantic oceans (Thompson et al. 1996; Schultz et al. 1999). Natural and human-related combustion processes produce PAHs which can be transported over long distances and, as in the case of many other POPs, an increase in air temperature may shift the equilibrium from particulate to vapour phases for compounds such as pyrene, fluoranthene, phenanthrene and anthracene. Moreover, it seems likely that warmer temperatures will increase the occurrence of forest fires, especially in the interior of Southern Hemisphere continents.

The above scenario suggests that deposition of some POPs in Antarctica and the Southern Ocean will increase in the near future. As discussed in the preceding chapter, the melting of sea ice and snow cover during the austral summer enhances the transfer of POPs from surface seawater to lipids of planktonic organisms. The chemicals are transferred through krill to organisms at higher trophic levels, such as seabirds and marine mammals. The potential toxicological effects for these animals are further exacerbated by starvation cycles during which their fat store is reduced.

8.3 The Development of Large-Scale Monitoring Networks

According to IPCC (2001), climate change in polar regions is expected to be greater and more rapid than in other regions and will cause major physical and ecological impacts, especially in the Antarctic Peninsula and Southern Ocean. The break-up of ice shelves further south in the Antarctic Peninsula and changes in marine and terrestrial ecosystems (e.g. the introduction of exotic plants and animal species) are among the most probable effects of warming and increased precipitation. The physical oceanography and ecology of the Southern Ocean will change, and the projected reduction in sea-ice extent will alter phytoplankton spring blooms in the marginal sea-ice zone, with profound impacts on krill and all levels of the pelagic food chain. Marine mammals and birds linked to specific breeding sites will be affected by environmental changes or changes in the availability of prey species. Warming (by 4–5 °C) of the western Antarctic Peninsula over the past 50 years changed spatio-temporal patterns of winter sea-ice formation and probably contributed to a significant increase in chinstrap penguin populations and a reduction in Adélie penguin populations (Fraser et al. 1992; R.C. Smith et al. 1999). The two species of seabirds have similar diets and breeding ranges in the peninsula, but the increasing availability of open water is unfavourable to an obligate inhabitant of pack ice such as the Adélie penguin. This is just one example of the importance of long-term monitoring programmes on penguins and other key Antarctic species.

Within the framework of the CCAMLR Ecosystem Monitoring Program (CEMP) and in collaboration with Italian researchers, since 1989 the Australian Antarctic Division has undertaken long-term monitoring studies on

Adélie penguins in East Antarctica (e.g. Gardner et al. 1997; Clarke et al. 1998). Although the main aim of the CCAMLR programme was to evaluate whether krill harvesting adversely affects elements of the Antarctic marine food chain, studies on penguin populations (especially those on food consumption and breeding success) may also help detect annual and long-term effects of climate change on sea-ice extent, krill distribution and abundance, and on inputs of persistent contaminants from other continents of the Southern Hemisphere. Fraser and Hofmann (2003) examined the long-term foraging response of Adélie penguins to ice-induced changes in krill recruitment and availability near Palmer Station (western Antarctic Peninsula). They found a causal relationship between change in ice cover, krill availability and penguin foraging strategies. During the last two decades, krill populations were sustained by strong age classes which emerged episodically every 4 or 5 years. Fraser and Hofmann (2003) hypothesise that cohort senescence has become an additional ecosystem stressor in an environment where enhanced warming is deteriorating sea-ice conditions conducive to good krill recruitment. Their results suggest that at least one "senescence event" has already occurred in the western Antarctic Peninsula region, with a decrease in krill abundance, penguin foraging, and breeding performance and populations. It was therefore suggested that krill longevity should be incorporated into models on causal links between climate, physical forcing and ecosystem response.

Monitoring is an essential element of environmental management, and a number of national Antarctic programmes perform environmental monitoring near scientific stations. The most common monitoring activities include: the determination of atmospheric pollutants associated with station activities; the quality of sewage and wastewater; levels of hydrocarbons, trace metals and other pollutants in snow, water, soils or sediments; animal population counts and breeding success of penguins and other Antarctic birds; studies on benthic marine communities and the use of some species of benthic organisms as bioaccumulators or biomarkers of persistent pollutants in more impacted marine coastal ecosystems. A summary of monitoring activities performed by national programmes in Antarctica has been prepared by COMNAP-AEON (2001). Standardised approaches are essential to determine spatio-temporal variations in environmental contaminants or organism populations. During the last decade, efforts were made to develop international coordination of monitoring procedures, quality assurance and data management (e.g. the handbooks published by SCAR/COMNAP 1996, 2000). Human activities and their impact must be monitored to improve Antarctic environmental management, meet the legal requirements of the Protocol on Environmental Protection to the Antarctic Treaty, and evaluate the effectiveness of existing conservation measures. These surveys are generally performed in relatively small areas around scientific stations or field camps, but in Antarctica there is also a need for large-scale monitoring of global phenomena.

Regional- or continental-scale surveys are necessary to establish baselines, to improve our understanding of environmental processes and functioning of Antarctic ecosystems, and to verify predictions concerning atmospheric processes and the deposition of long-range transported contaminants. Monitoring of global changes and observed changes in the Antarctic environment should ensure that the environmental impact during sampling and fieldwork is minimal and sustainable (i.e. that it does not compromise the significance of the study area for future observations on global-change processes). Therefore, whenever possible, it is advisable to adopt sampling strategies based on large time intervals (possibly some years) and not requiring the use of electric power generators or other sources of environmental contaminants.

8.3.1 Regional Baseline Concentrations of Persistent Contaminants

Although dry and occult (fog and ice crystals) deposition contributes significantly to the input of contaminants to Antarctica and the Southern Ocean, snow undoubtedly represents the main cleansing mechanism for the Antarctic atmosphere. Snow, firn and ice are the most widespread environmental matrices used to indirectly assess spatio-temporal variations in contaminant deposition throughout the continent. A number of techniques have been introduced for the collection of clean samples and for reliable analytical determinations. However, as discussed in Chapter 5, concentrations of persistent contaminants in Antarctic snow and ice are very low and can be determined in a relatively small number of laboratories. The post-deposition fate of organic chemicals in snow is largely unknown, and variations in snow accumulation over time can determine significant change in its chemical composition, even in the absence of any change in the composition of aerosol over the study area. Snow and ice are thus the only available environmental matrices to study spatio-temporal trends of atmospheric contaminants on the Antarctic plateau and most of the continent. Without better knowledge of their transfer and post-depositional processes, however, the analysis of snow and ice can provide only a qualitative picture of the chemical composition of the overlying atmosphere. Moreover, data on climate change and on contaminant input from other continents in the Southern Hemisphere show that there are considerable differences among various Antarctic regions, and it would not be wise to assign a unique baseline concentration to each persistent contaminant in Antarctic snow or ice. Based on data reported in the previous chapters, it would be opportune to at least differentiate among snow samples collected in the Antarctic Peninsula, West Antarctica, East Antarctica and Victoria Land.

As ice-free areas in Antarctica mainly occur in coastal regions, soils, lake sediments and cryptogams cannot be used to complete environmental surveys throughout the continent. However, most atmospheric deposition occurs on continental fringes, and environmental matrices in ice-free areas receive a large proportion of contaminants deposited on the continent after long-range transport. Although very few data exist on concentrations of chemicals in abiotic and biotic matrices of Antarctic freshwater and terrestrial ecosystems, they could play an important role in regional monitoring programmes. Large-scale surveys in Greenland and other polar regions (e.g. AMAP 1997, 2002) show that freshwater sediments and various species of organisms can be used to detect the deposition of atmospheric contaminants in polar ecosystems. As a rule, the sampling of lake sediments or biota is performed over rather large intervals of time (often 5 years), and contaminant concentrations in these materials are usually some orders of magnitude higher than in snow. These approaches therefore have many advantages and determine much less environmental impact than monitoring surveys based on sampling and analysis of snow or aerosols. As the composition of the latter matrices is extremely variable in space and time, monitoring requires repeated sampling (every few days or weeks) and sophisticated analytical techniques. The determination of very low concentrations of elements or compounds implies errors and large variability of results, and narrow spatio-temporal variations in atmospheric deposition of contaminants often cannot be detected. Furthermore, several countries involved in Antarctic monitoring programmes do not have adequate technical skills, laboratories and instruments for a reliable determination of persistent contaminants in snow and ice.

As discussed in Chapter 5, in ice-free areas of continental Antarctica the chemical weathering of exposed substrata occurs only to a very limited extent, and the environmental biogeochemistry of terrestrial and freshwater ecosystems is largely dominated by elements and compounds in snow and atmospheric dry and occult deposition. As most Antarctic lakes and ponds are located in poorly developed rocky catchments and usually lack outlets, they are the main sink for melting water, solutes and contaminants deposited in the surrounding environment. Preliminary comparative studies in northern Victoria Land (Triulzi et al. 1990; Fuoco et al. 1996), for instance, indicate that lake sediments accumulate higher concentrations of PCBs and ^{137}Cs than soils or marine sediments from the same region. Sediments and algal mats behave as natural integrators of soluble and particulate chemicals deposited in the watershed, and can be important indicators of the impact of local human activity and/or contribute to regional surveys on long-term changes in climatic conditions, biogeochemical cycles or inputs of long-range transported contaminants.

Lichens and mosses are perennial organisms which constitute most of the biomass in Antarctic terrestrial ecosystems (Fig. 49). Due to their slow growth rate, cation exchange capacity and lack of roots, these organisms depend on atmospheric deposition for their metabolism. Together with essential ions, they accumulate persistent atmospheric pollutants from melting snow to levels well above those in snow or aerosols. Several species of cryptogams are

Fig. 49. Lichens and mosses constitute most of the biomass in Antarctic ice-free areas, and these organisms can be used as biomonitors of persistent atmospheric contaminants

used worldwide as biomonitors of airborne metals, pesticides and radionuclides, especially for large-scale surveys in remote regions of the Northern Hemisphere. Although Antarctic lichens were found to behave as reliable biomonitors of long-range transported chlorinated hydrocarbons (Bacci et al. 1986), during the last two decades these organisms have mainly been used to biomonitor trace metals, PCBs and PAHs around scientific stations. Some species of Antarctic cryptogams have a circumpolar distribution and could be very useful in establishing long-term biomonitoring networks of persistent atmospheric pollutants in different regions of Antarctica. Predicted changes in temperature and atmospheric precipitation will probably affect the growth rate, colonisation pattern and chemical composition of Antarctic lichens and mosses, and these organisms could constitute an effective early-warning system to detect environmental changes in Antarctic terrestrial ecosystems.

8.3.2 Circumpolar Biomonitoring of Coastal Marine Ecosystems

Like Antarctica, the Southern Ocean is a mosaic of distinct subsystems with specific physico-chemical and biologic characteristics. Marine sediments are probably one of the most useful environmental matrices for gaining better knowledge of the deposition and cycling of major and trace elements and POPs in different regions of the Southern Ocean. Although preliminary studies by Angino (1966) on the geochemistry of pelagic sediments from the Ross, Amundsen and Bellingshausen seas revealed marked geographical differences in their chemical composition, very few large-scale studies have been performed during the last decades to assess spatio-temporal variations in the deposition of biogenic and terrigenous materials and associated contaminants. Most research on Southern Ocean sediments has been performed on surface samples collected from polluted marine areas near scientific stations and/or disused whaling stations. As in the case of snow and terrestrial ecosystems, long-term, large-scale studies throughout the Southern Ocean are necessary to better understand the role of this ocean in global processes and its physico-chemical links to other oceans.

Chapters 7 and 8 showed how the environment of the Southern Ocean is scarcely affected by anthropogenic contaminants, except in very localised areas adjacent to human settlements or in areas affected by incidental oil spills. However, in several species of marine organisms, in waters and in sediments, concentrations of potentially toxic metals such as Cd and Hg are not necessarily low, and values measured in the Antarctic marine environment cannot always be taken as global reference values. While in sediment samples from different seas one can reliably compare measured concentrations of trace metals and POPs by normalising values to grain size (or specific surface area) and to organic matter content, it is much more difficult to compare data on marine organisms collected in different seas and assess global reference values.

Antarctic marine organisms underwent long evolutionary processes in isolation and in an ocean with unique physico-chemical features. The biogeochemical cycles of major and trace elements in different regions of the Southern Ocean, together with the unique ecophysiological features of Antarctic marine organisms and the structure of food chains, enhance the natural accumulation of Cd, Hg, As and other elements in some species of invertebrates, seabirds and marine mammals. Most Antarctic organisms are slow-growing and long-lived, and some may show large differences in age, feeding behaviour, growth, reproductive cycle, and species-specific detoxification and excretion mechanisms. Contaminant concentrations in organs or tissues of organisms from remote and uncontaminated areas of the Southern Ocean are therefore not easily compared and/or used as baseline values for concentrations measured in related species from other seas. However, the long-term monitoring of trace metal and POP concentrations in key species of Antarctic pelagic food chains could be used to detect the possible effects of changes in sea-ice cover, food availability and input of persistent contaminants from lower latitudes. Large-scale surveys on the chemical composition of krill, some species of seabirds and marine mammals representative of different regions of the Southern Ocean, performed using standardised procedures at regular intervals of time (e.g. 5–6 years), are probably among the few possible means of detecting spatio-temporal changes in the nature and amount of atmospheric persistent pollutant deposition in the Antarctic environment. Given their extreme mobility, albatrosses, petrels and marine mammals behave as spatial and temporal integrators of persistent contaminants over large areas of the Southern Ocean. Moreover, in recent years, the standardisation of non-destructive procedures for sampling feathers, blood, uropygial oil and excreta, and skin biopsies (in marine mammals) has allowed the assessment of environmental contamination and the study of feeding, migratory, reproductive and genetic characteristics of different animal populations. Data on trace metal and POP concentrations in wandering albatrosses, southern giant petrels, southern minke whales, crabeater seals and on species restricted to the Antarctic environment such as Adélie and Emperor penguins or Weddell seals (Fig. 50) will allow the detection of spatio-temporal trends in the deposition of persistent contaminants, especially those of POPs such as lindane which are still largely used in the Southern Hemisphere.

Several species of invertebrates and fish have been proposed as reliable biomonitors of environmental contamination in Antarctic coastal areas. Seaweeds such as *Phyllophora antarctica* and *Iridaea cordata*, the amphipod *Paramoera walkeri*, filter-feeding molluscs such as *Adamussium colbecki* and *Laternula elliptica*, and benthic and pelagic fish (e.g. *Trematomus bernacchii* and *Pleuragramma antarcticum*) have a circum-Antarctic distribution and most of the features required of sensitive accumulative biomonitors of persistent contaminants. In the Antarctic Peninsula and sub-Antarctic islands, these species can be substituted with algae belonging to the genus *Desmaretia*, the limpet *Nacella concinna* or the fish *Notothenia coriiceps*. Besides the assessment of environmental pollution near scientific stations, these species could be very useful in establishing regional and circumpolar biomonitoring networks.

Recent research on different species of mussels such as *Perna viridis*, *Mytilus galloprovincialis*, *M. edulis*, *M. grayamus* and *M. trossulus* from different seas in the Northern and the Southern Hemispheres (e.g. Blackmore and Wang 2003; Monirith et al. 2003) indicates that these molluscs can be used as

Fig. 50. Adélie and Emperor penguins and the Weddell seal are restricted to the Antarctic environment and allow to detect spatio-temporal changes of climate and contaminant deposition

reliable biomonitors of persistent contaminants at different local and global scales. The uptake rate, clearance rate and assimilation efficiency for Cd and Zn were similar in mussels from sub-Arctic, temperate and tropical regions when measured under the same laboratory conditions and after body-size corrections. Blackmore and Wang (2003) found that when other factors such as salinity are also corrected for, different mussel species from different marine areas may yield directly comparable biomonitoring data. These results strongly support Mussel Watch programmes and stimulate further research on Antarctic molluscs for the development of a circum-Antarctic biomonitoring network. Trace metal and POP concentrations in *A. colbecki* from remote coastal sites are useful baselines to evaluate environmental pollution in coastal areas impacted by human activity, and to detect spatial and temporal variations in atmospheric deposition of contaminants. These data may also be used to detect the possible effects of climate change, such as the increased melting of snow and ice and changes in leaching and drainage processes in coastal terrestrial ecosystems. Nigro et al. (1997), for instance, found variations in metal concentrations in the shells, gills and digestive glands of Antarctic scallops collected at increasing distances from the outlet of a meltwater stream. Comparisons between the elemental composition of organs and tissues of *A. colbecki* from different Victoria Land coastal environments showed that samples from an oligotrophic habitat affected by meltwater runoff (Explorers Cove) had significantly higher concentrations of Cd, Cu, Mn, Ni and Zn in the shells, and of Cu, Mn, Fe and Zn in soft parts. On the contrary, Cd concentrations were higher in soft tissues of scallops from coastal habitats with enhanced upwelling of nutrient- and Cd-rich marine waters. These preliminary results indicate that long-term circumpolar biomonitoring networks based on an adequate knowledge of mollusc ecophysiology could also be useful in detecting changes in the hydrology and geochemistry of nearshore terrestrial ecosystems.

8.4 Global Environmental Challenges and the Reduction of Adverse Impacts in Antarctica

The recurring formation of the ozone hole over Antarctica is the best exemplification of the fact that the main environmental problems in the continent are determined by human activity in the rest of the world. The new century will increasingly be characterised by global environmental challenges due to the increasing global population (at present, about a billion people every 11–12 years), the unprecedented exploitation of natural resources and the production of enormous quantities of solid, liquid and gaseous wastes. Besides the destruction of stratospheric O_3 by chlorofluorocarbons, other global environmental problems in the next decades will be the increase in greenhouse gas concentrations, climate change, loss of biological diversity, land and water degradation, and the worldwide diffusion of hazardous chemicals. In spite of its remoteness, Antarctica will be increasingly affected by global environmental threats due to the enhanced effects of climate change in polar regions and the large population increase in poor countries of the Southern Hemisphere. Recent history has shown that the major causes of environmental degradation are linked to aspects of economic growth and development in industrialised countries (e.g. fossil fuel energy, increased mining and smelting activities, intensive agriculture and clearing of forests). If the reduction of poverty and the improvement of the standard of living in poor countries of the Southern Hemisphere are pursued through the same industrial, agricultural and energy policies which produced economical growth in a limited number of countries (mostly in the Northern Hemisphere), the Southern Ocean and Antarctica will probably be affected by many of the environmental problems which currently afflict Arctic ecosystems. Having burdened the global environment, rich countries now have the moral responsibility to assist poorer southern nations in addressing environmental threats, through financial aid and the transfer of new technologies and know-how.

The term globalisation is one of the most widely used worldwide, its common definition being "economic interchange and/or interdependence". Present and future environmental threats to Antarctica, the remotest region on Earth, indicate that global environmental challenges and ecological interdependencies are even more a reality than the acknowledged economical interdependence. The global environment, even more than the economy, has an urgent need for coordinated global-scale political, scientific and technological action. In the last decade there have been many multilateral negotiations to address global environmental issues, and several international agreements launched an ongoing process of reporting and reviewing new scientific evidence and the policies of different nations. The development of a system of governance of the global environment thus largely depends on scientific research, technology, and the involvement and interest of society and mass media.

News of the recurring formation of the "ozone hole" and the enhanced disintegration of some ice shelves in Antarctica, spread worldwide by mass media, helped increase public awareness of the possible environmental impact of human activity. Antarctica is also the region where nongovernmental organisations, especially Greenpeace, have shown that they can play a very important role in international negotiations by mobilising public opinion and providing governments with additional or alternative perspectives. In recent years several nongovernmental organisations have gained worldwide prominence and greater weight in negotiations on climate change, natural resources and environmental threats. Their interest in Antarctic affairs began in the 1980s (e.g. Herr 1996; Hansom and Gordon 1998). Since then, they have greatly helped increase international public awareness of Antarctic issues, ban mining activity and Antarctic resource exploitation, adopt better waste-management practices, and interfere with the building of airfields or works of notable environmental impact. For several years nongovernmental organisations have monitored the compliance of national Antarctic expeditions with environmental guidelines; in the past they also strongly advocated the establishment of an independent Antarctic Environmental Protection Agency, but the request was not accepted by Antarctic Treaty member states. However, their suggestions for long-term conservation of the Antarctic environment were incorporated in the Protocol on Environmental Protection to the Antarctic Treaty, and the World Conservation Union and SCAR produced recommendations for the development of the system of protected areas in Antarctica. During the last decade, Greenpeace has checked the compliance of several scientific stations with the Protocol on Environmental Protection, and ASOC (Antarctic and Southern Ocean Coalition, an alliance of environmental organisations from over 40 countries) have lobbied hard for more effective regulation of Southern Ocean fisheries under CCAMLR (Hansom and Gordon 1998). The Protocol on Environmental Protection progressively eliminated the large differences in environmental awareness and attitudes of Antarctic nations. Although there is still concern that some of the relatively poorer nations lack resources to undertake the required baseline studies and monitoring of human impacts, most nations are adopting much more responsible environmental management policies and have introduced measures to minimise impacts. Thus, although the presence of humans in Antarctica is necessarily accompanied by environmental impact, in the future persistent pollutants in the Antarctic environment will increasingly arise from human activity in the Southern Hemisphere and the rest of the world, rather than from local sources. Greenpeace and other nongovernmental organisations are aware of this risk, and are therefore engaged in industrialised and developing countries to raise wider awareness of the possible effects of human activity on climate change and on other global processes. Without further efforts to counterbalance powerful and well-financed economic interests and to adopt new models of development, especially in poor countries of the Southern Hemisphere, threats to the Antarctic environment cannot be significantly reduced. Technology can play an important role in reconciling economic development with the reduction of environmental impacts. Nongovernmental organisations must influence society and politicians involved in environmental negotiations through scientific arguments, proposals for technological innovation, and different perspectives at all levels of the decision-making process.

8.5 Science and the Protection of the Antarctic Environment

The Antarctic environment is now a prime focus of interest in terms of its value for scientific research, nature conservation and tourism. While for scientists Antarctica is unique for its role in global processes and for its ecological and environmental value, many other people perceive it as the last unspoiled corner of the world and a symbol of global conservation. This book shows that "pure" air and snow do not exist, even in Antarctica. Scientific work on this continent should increase social awareness of the fact that, if global environmental threats are not addressed, near-pristine Antarctic ecosystems will also be at risk. Although scientific research in Antarctica is of paramount importance in addressing climatic and environmental challenges, there is no doubt that the value of Antarctica for science should be weighed against the environmental impact of scientific work and its logistic support. The unique environmental characteristics and global significance of Antarctica and the Southern Ocean are such that research should be carried out in these regions only when it cannot be done as effectively elsewhere. Another essential pre-requisite of research in Antarctica is that the expected output should justify the inevitable environmental impact. As for environmental research, scientific programmes should deal with processes of regional or global significance.

In spite of intensive international research on climate change and its pos-sible effects on the Antarctic environment, there are still many challenges and opportunities for research on global processes. This book emphasises the study of long-range transport of persistent contaminants and possible modi-fications due to variations in climate or anthropogenic activity in the South-ern Hemisphere. Despite the trend towards the internationalisation of scien-tific research which addresses multidisciplinary questions of wider global relevance, routine measurements and reliable spatio-temporal trends of basic parameters such as air temperature, atmospheric deposition, sea-ice cover and primary productivity in the Southern Ocean are lacking. The shortage of temporal and spatial time-series data hampers the development of reliable models of climate change and of atmospheric contaminant pathways. Possible connections between the Antarctic climate and meteorological phenomena with the ENSO phenomenon and the rest of the global atmosphere are largely unknown.

Although the atmosphere is the most important pathway for the transport of persistent contaminants in Antarctica, recent studies indicate that the jour-ney of many POPs from their zone of emission to polar ecosystems is affected by complex interactions among air, ice, seawater and plankton organisms. These interactions are poorly understood and can be easily modified by changes in climatic and meteorological factors such as temperature, precipi-tation, wind, ice cover, surface currents and the productivity of ocean surface waters. Reliable regional monitoring networks of contaminant deposition in Antarctica should therefore be established through the collaboration of scien-tists working on contaminant issues with those working on climate change, glaciology and oceanography. Antarctic regions are likely to be the ultimate sink of most volatile POPs and some trace metals; the specific composition of these persistent contaminants may yield information on either the probable source or the history of these compounds (through metabolites and their par-ent compounds), and on mineral deposits and volcanic sources. The results of long-term monitoring and the assessment of geographically extensive pat-terns will be very useful in pointing out possible sources and pathways in the

Southern Hemisphere, and will complement research on climatic and meteorological processes.

The creation of a relevant Antarctic monitoring network for POPs and trace metals requires the development of a specifically designed international programme with standardised procedures for sampling and analysis of environmental matrices such as lake sediments and algae, cryptogamic organisms, preen oil or feathers from albatrosses, petrels, Adélie and/or Emperor penguins or skin biopsies from Weddell seals and other widespread species of marine mammals. Sampling strategies should take into account the extreme seasonality of the Antarctic environment and the ecophysiology of organisms, and must entail the collection of types and amounts of samples which do not exacerbate damage to terrestrial, freshwater and marine ecosystems. Quality control of sampling procedures is necessary to obtain meaningful, comparable results. Although in the last decade increasing attention has been paid to modern, extremely efficient analytical systems, quality control of analytical results has not always been coupled with the development of reliable, standardised procedures for representative sampling and careful preparation of Antarctic environmental matrices. Data collected during monitoring surveys should be managed and archived so that they are accessible and can be effectively used in global assessments. Models of atmospheric transport should be developed to include region-specific dynamics, such as deposition in snow, freshwater sediments and cryptogamic organisms.

This book highlights the study of environmental processes of concern such as gaseous Hg depletion at polar sunrise, which accompanies surface-level ozone depletion. Mercury depletion poses a special challenge to the modelling of the transport and deposition of metals in Antarctica and of their environmental fate in terrestrial and freshwater ecosystems. Environmental research in the next few years should aim to improve the sources inventory for the Southern Hemisphere, refine pathways, gain better knowledge of air–surface exchange, regional- or continental-scale budgets and models, and assess the occurrence in the Antarctic environment of "newer" classes of POPs (e.g. chlorinated paraffins or toxaphene) and elements (e.g. platinum-group elements from cars equipped with catalytic converters). Further efforts are also needed to identify and quantify new or hitherto unknown POPs, notably metabolites, stereoisomers and other polar POPs.

8.6 Summary

One of the main values of Antarctica for science is the nearly pristine environment, with biotic and abiotic matrices representing ideal archives of data on past and current trends in global processes. The Protocol on Environmental Protection to the Antarctic Treaty sets a high standard of protection for the

Antarctic environment and will probably help further reduce the impact of human activity in Antarctica. However, this book shows that most persistent airborne contaminants in Antarctica and the Southern Ocean originate from anthropogenic sources in other continents, mainly in the Southern Hemisphere. These regions are characterised by an extraordinary population explosion, and it is very likely that their contribution to the global burden of greenhouse gases and persistent contaminants will increase significantly in the near future. Richer countries must adequately assist developing nations in addressing global environmental threats through financial aid and transfer of technologies, because the development of past Northern Hemisphere production and consumption practices would pose real risks to the Antarctic climate and environment.

Although Antarctica and the Southern Ocean have been set aside by the international community for protection and are seen as symbols of global conservation, environmental pollution in many Arctic ecosystems teaches that global challenges must be addressed and economic development must be reconciled with the environment in order to effectively protect the last unspoiled corner of the world. Nongovernmental organisations can play an important role in the pursuit of these objectives, while the role of science is to achieve adequate knowledge of global processes and technological innovation in order to assist governance of the global environment.

In the last decade, striking phenomena such the "ozone hole" in Antarctica raised public awareness of the seriousness of the global-scale impact of human activity. Nevertheless, the task of Antarctic researchers in educating the public and increasing their awareness of global processes has only just begun. The ability of the international scientific community to respond to emerging issues is still limited by the scarce knowledge of the basic properties and functioning of Antarctic ecosystems and of how they are affected by climate change and contaminant inputs. This chapter highlights the need to establish high-quality, long-term baseline datasets of environmental conditions in different Antarctic regions, and promote collaboration between scientists working on climate change with those studying contamination from long-range transported atmospheric pollutants. The development of international programmes which address global issues with minimum environmental impact is critical for Antarctic research. Knowledge of the Antarctic environment must be transferred to other sectors of society in order to favour coordinated global-scale political, scientific and technological action.

Aid to developing countries in the Southern Hemisphere in adopting alternative sources of energy and food, thus avoiding past mistakes in the Northern Hemisphere, is vital to the preservation of the Antarctic environment. If the environment becomes a priority of global concern, environmental problems facing all segments of international society will be reduced and future generations will live in a better world. This book indicates that at present the deposition of long-range transported pollutants does not seem to pose a threat to Antarctic ecosystems. Nevertheless, the recent development of the precautionary principle requires that action be taken when there is potential for serious or irreversible harm, even in the face of scientific uncertainty. On the grounds of a precautionary approach, present conditions in the Antarctic environment suggest the need for science to monitor ecosystems and gain better knowledge of their functioning and possible responses to global processes. Although scientific work alone cannot change the perspective of environmental contamination in Antarctica and the rest of the world, this book was written in the hope that it can facilitate the development of a cleaner, sustainable environment.

References

Abyzov SS (1993) Microorganisms in the Antarctic ice. In: Friedmann EI (ed) Antarctic microbiology. Wiley-Liss, New York, pp 265–295

Ackley SF, Sullivan CW (1994) Physical controls on the development and characteristics of Antarctic sea ice biological communities. Deep-Sea Res 41:1583–1604

Adams J, Maslin M, Thomas E (1999) Sudden climate transitions during the Quaternary. Prog Phys Geogr 23:1–36

Adams PJ, Seinfeld JH, Koch D, Mickley L, Jacob D (2001) General circulation model assessment of direct radiative forcing by the sulphate–nitrat-ammonium-water inorganic aerosol system. J Geophys Res 106:1087–1111

Adamson E, Adamson H (1992) Possible effects of global climate change on Antarctic terrestrial vegetation. Impact of climate change on Antarctica. Australian Department of Arts, Sport, Environment and Territories, Australian Government Publ Ser, Canberra, pp 52–62

Adriano DC (2001) Trace elements in terrestrial environments. Biogeochemistry, bioavailability, and risks of metals, 2nd edn. Springer, Berlin Heidelberg New York

AECOM (1992) Final fortress rocks landfill remediation report for McMurdo Station, Antarctica. AECOM Technol Corp Antarctic Support Associates, Englewood, CO

Aguilar A, Ingemansson T, Magnien E (1998) Extremophile microorganisms as cell factories: support from the European Union. Extremophiles 2:367–373

Ahn I-Y, Lee SH, Kim KT, Shim JH, Kim D-Y (1996) Baseline heavy metal concentrations in the Antarctic clam, *Laternula elliptica* in Maxwell Bay, King George Island, Antarctica. Mar Pollut Bull 32:592–598

Ainley DG, Ribic CA, Fraser WR (1994) Ecological structure among migrant and resident seabirds of the Scotia-Weddell confluence region. J Anim Ecol 63:347–364

Ainley DG, Wilson PR, Barton KK, Ballard G, Nar N, Karl B (1998) Diet and foraging effort of Adélie penguins in relation to pack-ice conditions in the southern Ross Sea. Polar Biol 20:311–319

Aislabie J, Balks M, Astori N, Stevenson G, Symons R (1999) Polycyclic aromatic hydrocarbons in fuel-oil contaminated soils, Antarctica. Chemosphere 39:2201–2207

Alam IA, Sadiq M (1993) Metal concentrations in Antarctic sediment samples collected during the trans-Antarctica 1990 Expedition. Mar Pollut Bull 26:523–527

Albert MR, Grannas A, Shepson P, Bottenheim J (2002) Processes and properties of snow-air transfer with application to interstitial ozone at Alert. Atmos Environ 36:2779–2787

Aleksandrova VD (1988) Vegetation of the Soviet polar deserts. Cambridge University Press, Cambridge

Allan RJ, Lindesay JA, Reason CJC (1995) Multidecadal variability in the climate system over the Indian Ocean region during the austral summer. J Climate 8:1853–1873

Alleman LY, Curch TM, Ganguli P, Véron AJ, Hamelin B, Flegal AR (2001) Role of oceanic circulation on contaminant lead distribution in the South Atlantic. Deep-Sea Res 48:2855–2876

Allen MR (2002) Natural climate fluctuations. In: Goudie AS, Cuff DJ (eds) Encyclopedia of global change and human society, vol 1. Oxford University Press, Oxford, pp 123–132

Alley RB, Mayewski PA, Sowers T, Suiver M, Taylor KC, Clark PV (1997) Holocene climate instability. A prominent widespread event 8,200 years ago. Geology 25:483–486

Allison I, Kerry K, Wright S (1985) Observations of water mass modification in the vicinity of an iceberg. Iceberg Res 1:3–9

AMAP (1997) AMAP Greenland 1994–1996. Ministry of Environment and Energy, Danish Environmental Protection Agency, Copenhagen

AMAP (1998) AMAP assessment report: Arctic pollution issues. Arctic Monitoring and Assessment Programme, Oslo

AMAP (2000) AMAP report on issues of concern: updated information on human health, persistent organic pollutants, radioactivity, and mercury in the Arctic. Arctic Monitoring and Assessment Programme, Oslo

AMAP (2002) Arctic pollution 2002: persistent organic pollutants, heavy metals, radioactivity, human health, changing pathways. Arctic Monitoring and Assessment Programme, Oslo

AMAP (2003) AMAP assessment 2002: the influence of global change on contaminant pathways to, within, and from the Arctic. Arctic Monitoring and Assessment Programme, Oslo

Amyot M, Gill GA, Morel FMM (1997) Production and loss of dissolved gaseous mercury in coastal seawater. Environ Sci Technol 31:3606–3611

Ancora S, Volpi V, Olmastroni S, Focardi S, Leonzio C (2002) Assumption and elimination of trace elements in Adélie penguins from Antarctica: a preliminary study. Mar Environ Res 54:341–344

Anderlini VC, Connors PG, Risebrough RW, Martin JH (1972) Concentrations of heavy metals in some Antarctic and North American sea birds. In: Parker BC (ed) Proc Coll Conservation Problems in Antarctica. Virginia Polytechnic State University, Blacksburg, pp 49–62

Anderson C (1991) In troubled waters. Nature 350:293

Anderson JB (1991) The Antarctic continental shelf: results from marine geological and geophysical investigations. In: Tingey RJ (ed) The geology of Antarctica. Clarendon Press, Oxford, pp 285–334

Anderson JB (1999) Antarctic marine geology. Cambridge University Press, Cambridge

Anderson DE, Binney HA, Smith MA (1998) Evidence for abrupt climate change in northern Scotland between 3900 and 3500 calendar years BP. Holocene 7:97–103

André JM, Ribeyre F, Boudou A (1990) Mercury contamination levels and distribution in tissues and organs of delphinids (*Stenella attenuata*) from the eastern tropical Pacific, in relation to biological and ecological factors. Mar Environ Res 30:43–72

Andreae MO (1990) Ocean-atmosphere interactions in the global biogeochemical sulfur cycle. Mar Chem 30:1–29

Andreae MO, Crutzen PJ (1997) Atmospheric aerosols: biogeochemical sources and role in atmospheric chemistry. Science 276:1052–1058

Andres RJ, Kasgnoc AD (1998) A time-averaged inventory of sub-aerial volcanic sulphur emissions. J Geophys Res Atmos 103:25251–25261

Andriashev AP (1970) Cryopelagic fishes of the Arctic and Antarctic and their signifi-
    cance in polar ecosystems. In: Holdgate MW (ed) Antarctic ecology, vol 1. Academic
    Press, London, pp 297–304
Angino EE (1966) Geochemistry of Antarctic pelagic sediments. Geochim Cosmochim
    Acta 30:939–961
Angino EE, Andrews RS (1968) Trace element chemistry, heavy minerals, and sediment
    statistics of Weddell Sea sediments. J Sediment Petrol 38:634–642
ANL (1992) Preliminary site investigations at McMurdo Station, Ross Island, Antarctica.
    Final Report Argonne Natl Lab National Science Foundation, Washington, DC
Antarctic Division (1993) Davis Station management plan. Department of Environment,
    Sport and Territories Australia, Canberra
Aono S, Tanabe S, Fujise Y, Kato H, Tatsukawa R (1997) Persistent organochlorines in
    minke whale (Balaenoptera acutorostrata) and their prey species from the Antarctic
    and the North Pacific. Environ Pollut 98:81–89
Archer DE, Kheshgi H, Maier-Reimer E (1997) Multiple timescales for neutralization of
    fossil fuel CO_2. Geophys Res Lett 24:405–408
Archer D, Winguth A, Lea D, Mahowald N (2000) What caused the glacial/interglacial
    CO_2 cycles? Rev Geophys 38:159–189
Arimoto R, Duce RA, Ray BJ, Hewitt AD, Williams J (1987) Trace elements in the atmos-
    phere of American Samoa: concentrations and deposition to the tropical South
    Pacific. J Geophys Res 92:8465–8479
Arking A (1991) The radiative effects of clouds and their impact on climate. Bull Am
    Meteorol Soc 72:795–813
Armstrong AJ, Siegfried WR (1991) Consumption of Antarctic krill by minke whales.
    Antarct Sci 3:13–18
Armstrong TE, Roberts B, Swithinbank CWM (1973) Illustrated glossary of snow and
    ice. Scott Polar Institute, Cambridge
Arnaud PM (1977) Adaptations within the Antarctic marine benthic ecosystem. In:
    Llano G (ed) Adaptations within Antarctic ecosystems. Smithsonian Institution,
    Washington, DC, pp 135–157
Arntz WE, Brey T, Gallardo VA (1994) Antarctic zoobenthos. Oceanogr Mar Biol
    32:241–304
Arrigo KR (1994) Impact of ozone depletion on phytoplankton growth in the Southern
    Ocean: large-scale spatial and temporal variability. Mar Ecol Prog Ser 114:1–12
Arrigo KR, Robinson DH, Worthen DL, Schieber B, Lizotte MP (1998a) Primary produc-
    tion in Southern Ocean waters. J Geophys Res 103:15587–15600
Arrigo KR, Worthen DL, Dixon P, Lizotte MP (1998b) Primary productivity of near sur-
    face communities within Antarctic pack ice. In: Lizotte MP, Arrigo KR (eds) Antarc-
    tic sea ice: biological processes, interactions and variability. American Geophysical
    Union, Washington, DC, vol 73, pp 23–43
Arrigo KR, Robinson DH, Worthen DL, Dunbar RB, di Tullio GR, van Woert M, Lizotte
    MP (1999) Phytoplankton community structure and the drawdown of nutrients and
    CO_2 in the Southern Ocean. Science 283:365–367
Artaxo P, Rabello ML, Maenhaut W, van Grieken R (1992) Trace elements and individual
    particles analysis of atmospheric aerosols from the Antarctic Peninsula. Tellus
    44:318–334
Asmund G, Johansen P, Fallis BW (1991) Disposal of mine wasters containing Pb and Zn
    near the ocean: an assessment of associated environmental implications in the Arc-
    tic. Chem Ecol 5:1–15
ATCP (1988) Convention on the regulation of Antarctic mineral resource activities. Final
    Rep 4th Special Antarctic Treaty Consultative Meet Antarctic Mineral Resources,
    Wellington

ATCP (1992) Protocol on environmental protection to the Antarctic treaty. Final Rep
    11th Antarctic Treaty Special Consultative Meet, 22–30 April 1991, 17–22 June 1991,
    3–4 October 1991, Madrid. Ministerio de Asuntos Exteriores, Madrid

Atlas RM (1991) Microbial hydrocarbon degradation – bioremediation of oil spills. J
    Chem Technol Biotechnol 52:149–156

Auburn FM (1982) Antarctic law and politics. Hurst, London

Bacci E (1994) Ecotoxicology of organic contaminants. CRC Press, Boca Raton

Bacci E, Calamari D, Gaggi C, Fanelli R, Focardi S, Morosini M (1986) Chlorinated hydro-
    carbons in lichen and moss samples from the Antarctic Peninsula. Chemosphere
    15:747–754

Bacon CE, Jarman WM, Costa DP (1992) Organochlorine and polychlorinated biphenyl
    levels in pinniped milk from the Arctic, the Antarctic, California and Australia.
    Chemosphere 24:779–791

Bailey RE (2001) Global hexachlorobenzene emissions. Chemosphere 43:167–182

Baker JE (ed) (1997) Atmospheric deposition of contaminants to the Great Lakes and
    coastal waters. SETAC Press, New York

Bakker DCE, de Baar HJW, Bathmann UV (1997) Changes of carbon dioxide in surface
    waters during spring in the Southern Ocean. Deep-Sea Res 44:91–127

Ballschmiter K, Hackenberg R, Jarman WM, Looser R (2002) Man-made chemicals
    found in remote areas of the world: the experimental definition for POPs. Environ Sci
    Pollut Res 9:274–288

Banse K (1996) Low seasonality of low concentrations of surface chlorophyll in the sub-
    antarctic water ring: underwater irradiance, iron, or grazing? Prog Oceanogr
    37:241–291

Barbante C, Turetta C, Bellomi T, Gambaro A, Piazza R, Moret I, Scarponi G (1997) Possi-
    ble sources and origins of lead in present-day East Antarctic snow. Geogr Fis Dinam
    Quat 20:199–202

Barbante C, Turetta C, Gambaro A, Capodaglio G, Scarponi G (1998) Sources and origins
    of aerosols reaching Antarctica as revealed by lead concentrations profiles in shallow
    snow. Ann Glaciol 27:674–678

Barbante C, Cozzi G, Capodaglio G, van de Velde K, Ferrari C, Veysseyre A, Boutron CF,
    Scarponi G, Cescon P (1999) Determination of Rh, Pd, and Pt in polar and Alpine
    snow and ice by double-focusing ICPMS with microconcentric nebulization. Anal
    Chem 71:4125–4133

Barbante C, Cozzi G, Capodoglio G, Cescon P (2000) Trace element determination in a
    candidate reference material (Antarctic krill) by ICP-sector field MS. J Anal Atom
    Spectrom 15:377–382

Barbante C, Turetta C, Capodaglio G, Cescon P, Hong S, Candelone J-P, van de Velde K,
    Boutron CF (2001a) Trace element determination in polar snow and ice. An overview
    of the analytical processes and application in environmental and paleoclimatic stud-
    ies. In: Caroli S, Cescon P, Walton DWH (eds) Environmental contamination in
    Antarctica. A challenge to analytical chemistry. Elsevier, Amsterdam, pp 55–85

Barbante C, Veysseyre A, Ferrari C, van de Velde K, Morel C, Capodoglio G, Cescon P,
    Scarponi G, Boutron CF (2001b) Greenland snow evidence of large scale atmospheric
    contamination for platinum, palladium, and rhodium. Environ Sci Technol
    35:835–839

Barber DG, Nghiem SV (1999) The role of snow on the thermal dependence of backscat-
    ter over sea ice. J Geophys Res 104:25789–25803

Bargagli R (1989) Determination of metal deposition patterns by epiphytic lichens. Tox-
    icol Environ Chem 18:249–256

Bargagli R (1990) Mercury emission in an abandoned mining area: assessment by epi-
    phytic lichens. In: Cheremisinoff PN (ed) Encyclopedia of environmental control

technology, vol 4. Hazardous waste containment. Gulf Publ Co, Houston, Texas, pp 613–640

Bargagli R (1993) Cadmium in marine organisms from the Tyrrhenian Sea: no evidence of pollution or biomagnification. Oebalia 19:13–25

Bargagli R (1995) The elemental composition of vegetation and the possible incidence of soil contamination of samples. Sci Total Environ 176:121–128

Bargagli R (1998) Trace elements in terrestrial plants: an ecophysiological approach to biomonitoring and biorecovery. Springer, Berlin Heidelberg New York

Bargagli R (1999) Mercury in the environment. In: Alexander DE, Fairbridge RW (eds) Encyclopedia of environmental sciences. Kluwer, Dordrecht, pp 314–316

Bargagli R (2000) Trace metals in Antarctica related to climate change and increasing human impact. Rev Environ Contam Toxicol 166:129–173

Bargagli R (2001) Trace metals in Antarctic organisms and the development of circum-polar biomonitoring networks. Rev Environ Contam Toxicol 171:53–110

Bargagli R, Barghigiani C (1991) Lichen biomonitoring of mercury emission and deposition in mining, geothermal and volcanic areas of Italy. Environ Monit Assess 16:265–275

Bargagli R, Mikhailova I (2002) Accumulation of inorganic contaminants. In: Nimis PL, Scheidegger C, Wolseley PA (eds) Monitoring with lichens – Monitoring lichens. Kluwer, Amsterdam, pp 65–84

Bargagli R, Ferrara R, Maserti BE (1988) Assessment of mercury distribution and partitioning in recent sediments of the western Mediterranean basin. Sci Total Environ 72:123–130

Bargagli R, Battisti E, Focardi S, Formichi P (1993) Preliminary data on environmental distribution of mercury in northern Victoria Land, Antarctica. Antarct Sci 5:3–8

Bargagli R, Brown DH, Nelli L (1995) Metal biomonitoring with mosses: procedures for correcting for soil contamination. Environ Pollut 89:169–175

Bargagli R, Broady PA, Walton DWH (1996a) Preliminary investigation of the thermal biosystem of Mt. Rittmann fumaroles (northern Victoria Land, Antarctica). Antarct Sci 8:121–126

Bargagli R, Nelli L, Ancora S, Focardi S (1996b) Elevated cadmium accumulation in marine organisms from Terra Nova Bay (Antarctica). Polar Biol 16:513–520

Bargagli R, Monaci F, Sanchez-Hernandez JC, Cateni D (1998a) Biomagnification of mercury in an Antarctic marine coastal food web. Mar Ecol Prog Ser 169:65–76

Bargagli R, Sanchez-Hernandez JC, Monaci F (1998b) Baseline concentrations of elements in the Antarctic macrolichen Umbilicaria decussata. Chemosphere 38:475–487

Bargagli R, Sanchez-Hernandez JC, Martella L, Monaci F (1998 c) Mercury, cadmium and lead accumulation in Antarctic mosses growing along nutrient and moisture gradients. Polar Biol 19:316–322

Bargagli R, Corsolini S, Fossi MC, Sanchez-Henrnandez JC, Focardi S (1998d) Antarctic fish Trematomus bernacchii as biomonitor of environmental contaminants at Terra Nova Bay Station (Ross Sea). Mem Natl Inst Polar Res 52:220–229

Bargagli R, Smith RIL, Martella L, Monaci F, Sanchez-Hernandez JC, Ugolini FC (1999) Solution geochemistry and behaviour of major and trace elements during summer in a moss community at Edmonson Point, Victoria Land, Antarctica. Antarct Sci 11:3–12

Bargagli R, Sanchez-Hernandez JC, Monaci F, Focardi S (2000) Environmental factors promoting bioaccumulation of Hg and Cd in Antarctic marine and terrestrial organisms. In: Davison W, Howard-Williams C, Broady P (eds) Antarctic ecosystems: models for wider ecological understanding. Caxton Press, Christchurch, pp 308–314

Bargagli R, Borghini F, Monaci F (2001) Environmental biogeochemistry of major and trace elements in terrestrial ecosystems of Victoria Land (continental Antarctica). In:

Weber J, Jamroz E, Drozd J, Karczewska (eds) Biogeochemical processes and cycling
    of elements in the environment. Polish Soc Humic Substances, Wroclaw, pp 229–230
Bargagli R, Borghini F, Monaci F (2003) The sea as major source of ions to lichens in ter-
    restrial ecosystems of Victoria Land. In: Huiskes AHL, Gieskes WWC, Rozema J,
    Schorno RML, van der Vies SM, Wolff WJ (eds) Antarctic biology in a global context.
    Backhuys, Leiden, pp 157–160
Barinaga M (1990) Eco-quandary: what killed the skuas? Science 249:243
Barnola JM, Raynaud D, Korotkevich YS, Lorius C (1987) Vostok ice core provides
    160,000-year record of atmospheric CO_2. Nature 329:408–414
Baroni C (1991) Glaciers and glacial history. In: Baroni C (ed) Antarctica – the Earth's
    white heart. Enrico Rainero, Florence, pp 92–99
Baroni C (2001) Lineamenti geografici. In: Baroni C (ed) Antartide. Terra di Scienza e
    Riserva Naturale. Terra Antartica, Siena, pp 9–22
Baroni C, Frezzotti M, Orombelli G (2001) I ghiacciai e la storia glaciale recente. In:
    Baroni C (ed) Antartide. Terra di Scienza e Riserva Naturale. Terra Antartica, Siena,
    pp 64–85
Barrett JP, Stoffers P, Glasby GP, Plueger WL (1984) Texture, mineralogy and composition
    of four sediment cores from Granite Harbour and New Harbour, southern Victoria
    Land, Antarctica. NZ J Geol Geophys 27:477–485
Barrett PJ, Hambrey MJ, Robinson PR (1991) Cenozoic glacial and tectonic history from
    CIROS-1, McMurdo Sound. In: Thomson MRA, Crame JA, Thomson JW (eds) Geo-
    logical evolution of Antarctica. Cambridge University Press, Cambridge, pp 651–656
Barrett PJ, Adams CJ, McIntosh WC, Swisher CC III, Wilson GS (1992) Geochronological
    evidence supporting Antarctic deglaciation three million years ago. Nature
    359:816–818
Barron J, Leg 119 Shipboard Scientific Party (1988) Early glaciation in Antarctica. Nature
    333:303–304
Barry RG, Chorley RJ (1992) Atmosphere, weather and climate. Methuen, London
BAS (1987) British Antarctic survey annual report 1986/87. British Antarctic Survey,
    Cambridge
BAS (1989) Proposed construction of a crushed rock airstrip at Rothera Point, Adelaide
    Island, Bristish Antarctic Territory. British Antarctic Survey, Natl Environ Res Coun-
    cil, Cambridge
Basile I, Grousset F, Revel M, Petit J, Biscaye J, Barkov N (1997) Patagonian origin of
    glacial dust deposited in East Antarctica (Vostok and Dome C) during glacial states 2,
    4, and 6. Earth Planet Sci Lett 146:573–589
Bates TS, Kapustin VN, Quinn PK, Covert DS, Coffman DJ, Mari C., Durkee PA, De Bruyn
    WJ, Saltman ES (1998) Processes controlling the distribution of aerosols particles in
    the lower marine boundary layer during the First Aerosol Characterization Experi-
    ment (ACE-1). J Geophys Res 103:16369–16383
Becker PR (2000) Concentrations of chlorinated hydrocarbons and heavy metals in
    Alaska Arctic marine mammals. Mar Pollut Bull 40:819–829
Beek B (ed) (2000) The handbook of environmental chemistry, vol 2, part J. Bioaccumu-
    lation: new aspects and developments. Springer, Berlin Heidelberg New York
Bell RE, Studinger M, Tikku AA, Clarke GKC, Gutner MM, Meertens C (2002) Origin and
    fate of Lake Vostok water frozen to the base of the East Antarctic ice sheet. Nature
    416:307–310
Beltramino JCM (1993) The structure and dynamics of Antarctic populations. Vantage
    Press, New York
Benn DI, Evans DJA (1998) Glaciers and glaciation. Arnold, London
Benninghoff WS, Benninghoff AS (1985) Wind transport of electrostatically charged
    particles and minute organisms in Antarctica. In: Siegfried WR, Condy PR, Laws RM

(eds) Antarctic nutrient cycles and food webs. Springer, Berlin Heidelberg New York, pp 592–596

Benson AA, Summons RE (1981) Arsenic accumulation in Great Barrier Reef invertebrates. Science 6:497–506

Bentley CR (1987) Antarctic ice streams. A review. J Geophys Res 92:8843–8858

Bentley CR, Giovinetto MB (1991) Mass balance of Antarctica and sea level change. In: Weller G, Wilson J, Severin B (eds) Role of polar regions in global change. In: Proc Int Conf University of Alaska. Geophys Inst Center Global Change Arctic System Res, University of Alaska, Fairbanks, pp 489–494

Berge JA (1990) Macrofauna recolonization of subtidal sediments. Experimental studies on defaunated sediment contaminated with crude oil in two Norwegian fjords with unequal eutrophication status. I. Community responses. Mar Ecol Prog Ser 66:103–115

Berkman PA (1992) The Antarctic marine ecosystem and humankind. Rev Aquat Sci 6:295–333

Berkman PA, Nigro M (1992) Trace metal concentrations in scallops around Antarctica: extending the mussel watch programme to the Southern Ocean. Mar Pollut Bull 24:322–323

Beyer A, Matthies M (2001) Long-range transport potential of semivolatile organic chemicals in coupled air-water systems. Environ Sci Pollut Res 8:173–179

Beyer L, Bölter M (2000) Chemical and biological properties, formation, occurrence and classification of Spodic Cryosols in a terrestrial ecosystem of East Antarctica (Wilkes Land). Catena 39:95–119

Beyer L, Bölter M (eds) (2002) Geoecology of Antarctic ice-free coastal landscapes. Springer, Berlin Heidelberg New York

Beyer L, Pingpank K, Wriedt G, Bölter M (2000) Soil formation in coastal continental Antarctica (Wilkes Land). Geoderma 95:283–304

Bibby DM, Patterson JE, O'Neill S (1998) Levels of atmospheric mercury at two sites in the Wellington area, New Zealand. Environ Technol Lett 9:71–74

Bicego MC, Weber RR, Gonçalves IR (1996) Aromatic hydrocarbons on surface waters of Admiralty Bay, King George Island, Antarctica. Mar Pollut Bull 32:549–553

Bidigare RR (1989) Potential effects of UV-B radiation on marine organisms of the Southern Ocean, distributions of phytoplankton and krill during austral spring. Photochem Photobiol 50:469–477

Bidigare RR, Ondrusek ME, Kennicutt MC II, Iturriaga R, Harvey HR, Hoham RW, Macko SA (1993) Evidence for a photoprotective function for secondary carotenoids of snow algae. J Phycol 29:427–434

Bidleman TF, Walla MD, Roura R, Carr E, Schmidt S (1993) Organochlorine pesticides in the atmosphere of the Southern Ocean and Antarctica, January-March 1990. Mar Pollut Bull 26:258–262

Bidleman TF, Mcdonald RW, Stow JP (2003) Canadian Arctic contaminants assessment report II: sources, occurrence, trends and pathways in the physical environment. Indian and Northern Affairs, Ottawa

Bidoglio G, Stumm W (eds) (1994) Chemistry of aquatic systems: local and global perspectives. Kluwer, Dordrecht

Bindoff NL, McDougall TJ (2000) Decadal changes along an Indian Ocean section at 32° S and their interpretation. J Phys Oceanogr 30:1207–1222

Birkenmajer K (1988) Geochronology of Tertiary glaciations on King George Island, West Antarctica. Bull Polish Acad Sci Earth Sci 37:27–48

Blackmore G, Wang W-X (2003) Comparison of metal accumulation in mussels at different local and global scales. Environ Toxicol Chem 22:388–395

Blay SKN (1992) New trends in the protection of the Antarctic environment: the 1991 Madrid Protocol. Am J Int Law 86:377–399

Block W (1990) Cold tolerance of insects and other arthropods. Philos Trans R Soc Ser B 326:613–633

Block W (1994) Terrestrial ecosystems: Antarctica. Polar Biol 14:293–300

Blume H-P, Bölter M (1993) Soils of Casey Station (Wilkes Land, Antarctica). In: Gilichinsky DA (ed) Cryosols: the effects of cryogenesis on the processes and peculiarities of soil formation. In: Proc Int Conf Cryopedol, Russian Academy of Sciences, Pushchino, pp 96–103

Blume H-P, Beyer L, Bölter M, Erlenheuser H, Kalk E, Kneesch S, Pfisterer U, Schneider D (1997) Pedogenic zonation in soils of Southern circumpolar region. Adv Geol Ecol 30:69–90

Bockheim JG (1982) Properties of a chronosequence of ultraxerous soils in the Trans-Antarctic Mountains. Geoderma 47:59–77

Bockheim JG (1995) Permafrost distribution in the Southern Circumpolar Region and its relation to the environment: a review and recommendations for future research. Permafrost Periglac 6:27–45

Bockheim JG (1997) Properties and classification of cold desert soils from Antarctica. Soil Sci Soc Am J 61:224–231

Bockheim JG, Ugolini FC (1990) A review of pedogenic zonation in well-drained soils of the southern circumpolar region. Quat Res 34:47–66

Bockheim JG, Wilson SC (1992) Soil-forming rates and processes in cold desert soils of Antarctica. In: Gilichinsky DA (ed) Cryosols: the effects of cryogenesis on the processes and peculiarities of soil formation. Proc Int Conf Cryopedol, Russian Academy of Sciences, Pushchino, pp 42–56

Bodeker G (1997) UV radiation in polar regions. In: Lyons WB, Howard-Williams C, Hawes I (eds) Ecosystem processes in Antarctic ice-free landscapes. Balkema, Rotterdam, pp 23–42

Bodhaine BA (1996) Central Antarctica: atmospheric chemical composition and atmospheric transport. In: Wolff EW, Bales RC (eds) Chemical exchange between the atmosphere and polar snow. Springer, Berlin Heidelberg New York, pp 145–172

Boer GJ, Flato G, Ramsden D (2000) A transient climate change simulation with greenhouse gas and aerosol forcing: projected climate for the 21st century. Climate Dynam 16:427–450

Bolshov MA, Rudniev SN, Candelone J-P, Boutron CF (1994) Ultratrace determination of Bi in Greenland snow by laser atomic fluorescence spectrometry. Spectrochim Acta 49B:1445–1452

Booth CR, Lucas TB, Morrow JH, Weiler CS, Penhale PA (1994) The United States National Science Foundation's polar network for monitoring ultraviolet radiation. In: Weiler CS, Penhale PA (eds) Ultraviolet radiation in Antarctica: measurements and biological effects. American Geophysical Union, Washington, DC, vol 62, pp 17–37

Bordin G, Appriou P, Treguer P (1987) Répartitions horizontale et verticale de cuivre, du manganese et du cadmium dans le secteur indien de l'Océan Antarctique. Oceanologr Acta 10:411–420

Borys RD, del Vecchio D, Jaffrezo J-L, Davidson CI, Mitchell Dl (1993) Assessment of ice particle growth processes at Dye-3, Greenland. Atmos Environ 27A:2815–2822

Bottenheim JW, Dibb JE, Honrath Re, Shepson PB (2002) An introduction to the Alert 2000 and Summit 2000 Arctic research studies. Atmos Environ 36:2467–2469

Boucher N, Prézelin BB (1996) An *in situ* biological weighting function for UV inhibition of phytoplankton carbon fixation in the Southern Ocean. Mar Ecol Prog Scr 144:223–236

Boucher O, Tanré D (2000) Estimation of the aerosol perturbation to the Earth's radiative budget over oceans using POLDER satellite aerosol retrievals. Geophys Res Lett 27:1103–1106

Boutron CF, Patterson CC (1987) Relative levels of natural and anthropogenic lead in recent Antarctic snow. J Geophys Res 92:8454–8464

Boutron CF, Wolff (1989) Heavy metal and sulphur emissions to the atmosphere from human activities in Antarctica. Atmos Environ 23:1669–1675

Boutron CF, Patterson CC, Petrov VN, Barkov NI (1987) Preliminary data on changes of lead concentrations in Antarctic ice from 155000 to 26000 years BP. Atmos Environ 21:1197–1202

Boutron CF, Görlach U, Candelone J-P, Bolshov MA, Delmas RJ (1991) Decrease in anthropogenic lead, cadmium and zinc in Greenland snows since the late 1960s. Nature 353:153–156

Boutron CF, Rudniev SN, Bolshov MA, Koloshnikov VG, Patterson CC, Barkov NI (1993) Changes in cadmium concentrations in Antarctic ice and snow during the past 155,000 years. Earth Planet Sci Lett 117:431–441

Bowen HJM (1979) Environmental chemistry of the elements. Academic Press, London

Boyd PW (2002) The role of iron in the biogeochemistry of the Southern Ocean and equatorial Pacific: a comparison on in situ enrichments. Deep-Sea Res II 49:1803–1821

Boyd PW, La Roche J, Gall M, Frew R, McKay RM (1999) Role of iron, light, and silicate in controlling algal biomass in subantarctic waters SE of New Zealand. J Geophys Res 104:13395–13408

Boyd PW, Watson A, Law CS, Abraham E, Trull T, Murdoch R, Bakker DCE, Bowie AR, Buesseler K, Chang H, Charette M, Croot P, Downing K, Frew R, Gall M, Hadfield M, Hall J, Harvey M, Jameson G, La Roche J, Liddicoat M, Ling R, Maldonado M, McKay RM, Nodder S, Pickmere S, Pridmore R, Rintoul S, Safi K, Sutton P, Strzepek R, Tanneberger K, Turber S, Waite A, Zeldis J (2000) A mesoscale phytoplankton bloom in the polar Southern Ocean stimulated by iron fertilization. Nature 407:695–702

Boyle EA (1988) Cadmium: chemical traces of deepwater paleoceanography. Paleoceanography 3:47–489

Boyle EA (1992) Cadmium and $\delta^{13}C$ paleochemical ocean distribution during the stage 2 glacial maximum. Annu Rev Earth Planet Sci 20:245–287

Boyle EA (1998) Pumping iron makes thinner diatoms. Nature 393:733–734

Boyle EA, Edmond JM (1975) Copper in surface waters south of New Zealand. Nature 253:107–110

Boyle EA, Sclater FR, Edmond JM (1976) On the marine geochemistry of cadmium. Nature 263:42–45

Boyle EA, Chapnik SD, Shen GT (1986) Temporal variability of lead in the western north Atlantic. J Geophys Res 91:8573–8593

Boyle EA, Sherrell RM, Bacon MP (1994) Lead variability in the western North Atlantic ocean and central Greenland ice: implications for the search for decadal trends in anthropogenic emissions. Geochim Cosmochim Acta 58:3227–3238

Bracher AU, Wiencke C (2000) Simulation of the effects of naturally enhanced UV radiation on photosynthesis of Antarctic phytoplankton. Mar Ecol Prog Ser 196:127–141

Braddock JF, Lindstrom JE, Brown EJ (1995) Distribution of hydrocarbon-degrading microorganisms in sediments from Prince William Sound, following *Exxon Valdes* oil spill. Mar Pollut Bull 30:125–132

Braithwaite RJ, Olesen OB (1989) Calculation of glacier ablation from air temperature, west Greenland. In: Oerlemans J (ed) Glacier fluctuations and climatic change. Kluwer, Dordrecht, pp 219–233

Breivik K, Sweetman A, Pacyna JM, Jones KC (2002) Towards a global historical emission inventory for selected PCB congeners – a mass balance approach. 2. Emissions. Sci Total Environ 290:199–224

Bréon F-M, Tanré D, Generoso S (2002) Aerosol effect on cloud droplet size monitored from satellite: Science 295:834–838

Brey T, Mackensen A (1997) Stable isotopes prove shell growth bands in the Antarctic bivalve *Laternula elliptica* to be formed annually. Polar Biol 17:465–468

Broady PA (1993) Soils heated by volcanism. In: Friedmann EI (ed) Antarctic microbiology. Wiley-Liss, New York, pp 413–432

Broady PA, Given D, Greenfield L, Thompson K (1987) The biota and environment of fumaroles on Mt. Melbourne, northern Victoria Land. Polar Biol 7:97–113

Bromwich DH (1988) Snowfall in high southern latitudes. Rev Geophys 26:149–168

Bromwich DH (1989) An extraordinary katabatic wind regime at Terra Nova Bay, Antarctica. Month Weather Rev 117:688–695

Bromwich DH (1990) Estimates of Antarctic precipitation. Nature 343:627–629

Bromwich DH, Parish TR (1998) Meteorology of the Antarctic. In: Karoly DJ, Vincent DG (eds) Meteorology of the Southern Hemisphere. American Meteorological Society, Boston, Meteorological Monographs, vol 27, pp 175–200

Bromwich DH, Weaver CJ (1983) Latitudinal displacement from main moisture sources controls $\delta^{18}O$ of snow in coastal Antarctica. Nature 301:145–147

Bromwich DH, Parish TR, Pellegrini A, Stearns CR, Weidner GA (1993) Spatial and temporal characteristics of the intense katabatic winds at Terra Nova Bay, Antarctica. In: Bromwich DH, Stearns CR (eds) Antarctic meteorology and climatology: studies based on automatic weather stations. American Geophysical Union, Washington, DC, vol 61, pp 47–68

Bromwich DH, Robasky FM, Cullather RI, van Woert ML (1995) The atmospheric hydrologic cycle over the Southern Ocean and Antarctica from operational numerical analyses. Ann Glaciol 21:149–156

Bromwich DH, Zhong L, Rogers AN (1998) Winter atmospheric forcing of the Ross Sea polynya. In: Jacobs SS, Weiss RF (eds) Ocean, ice and atmosphere interactions at the Antarctic continental margin. American Geophysical Union, Washington, DC, vol 75, pp 101–133

Brown DH (1984) Uptake of mineral elements and their use in pollution monitoring. In: Dyer AF, Duckett JG (eds) The experimental biology of bryophytes. Academic Press, London, pp 229–255

Bruland KW (1980) Oceanographic distributions of cadmium, nickel, and copper in the North Pacific. Earth Planet Sci Lett 47:176–198

Bruland KW, Franks RP (1983) Manganese, nickel, copper, zinc and cadmium in the western North Atlantic. In: Wong CS, Boyle E, Bruland KW, Burton JD, Goldberg ED (eds) Trace metals in sea water. Plenum Press, New York, pp 395–414

Bruland KW, Donat JR, Hutchins DA (1991). Interactive influences of bioactive trace metals on biological production in oceanic waters. Limnol Oceanogr 36:1557–1577

Budd WF, Wu X (1998) Modeling long term global and Antarctic changes resulting from increased greenhouse gases. In: Meighen PJ (ed) Coupled climate modelling. Bureau of Meteorology, Canberra, vol 69, pp 71–74

Buffon GLL (1749–1767) Histoire naturelle: générale et particulière; avec la description du Cabinet de Roi, 15 vols. Imprimerie Royale, Paris

Bunch JN (1987) Effects of petroleum releases on bacterial numbers and microheterotrophic activity in the water and sediment of an Arctic marine ecosystem. Arctic 40:172–183

Burgess JS, Kaup E (1997) Some aspects of human impact on lakes in the Larsemann Hills, Princess Elizabeth Land, Eastern Antarctica. In: Lyons WB, Howard-Williams C,

Hawes I (eds) Ecosystem processes in Antarctic ice-free landscapes. Balkema, Rotterdam, pp 259–264

Burns JM, Trumble SJ, Castellini MA, Testa JW (1998) The diet of Weddell seals in McMurdo Sound, Antarctica as determined from scat collections and stable isotopes analysis. Polar Biol 19:272–282

Burton HR (1981) Chemistry, physics, and evolution of Antarctic saline lakes. Hydrobiologia 82:339–362

Bustamante P, Cherel Y, Caurant F, Miramand P (1998) Cadmium, copper and zinc in octopuses from Kerguelen Islands, Southern Indian Ocean. Polar Biol 19:264–271

Bustamante P, Bocher P, Chérel Y, Miramand P, Caurant F (2003) Distribution of trace elements in the tissues of benthic and pelagic fish from the Kerguelen Islands. Sci Total Environ 313:25–39

Butler JH, Battle M, Bender ML, Montzka SA, Clark AD, Saltzman ES, Sucher CM, Severinghaus JP, Elkins JW (1999) A record of atmospheric halocarbons during the twentieth century from polar firn air. Nature 399:749–755

Cabanes A, Legagneux L, Dominé F (2002) Evolution of the specific surface area and of crystal morphology of Arctic fresh snow during the ALERT 2000 campaign. Atmos Environ 36:2767–2777

Cai W, Gordon HB (1998) Transient responses of the CSIRO climate model to two different rates of CO_2 increase. Climate Dynam 14:503–506

Caldeira K, Duffy PB (2000) The role of the Southern Ocean in uptake and storage of anthropogenic carbon dioxide. Science 287:620–622

Calkin PE (1964) Geomorphology and glacial geology of the Victoria Valley system, southern Victoria Land, Antarctica. Ohio State Univ Inst Polar Stud Rep 10

Cameron RE (1972) Pollution and conservation of the Antarctic terrestrial ecosystems. In: Parker BC (ed) Proc Coll Conservation Problems in Antarctica. Allen Press, Lawrence, Kansas, pp 267–305

Cameron RE, Honour RC, Morelli FA (1977) Environmental impact studies of Antarctic sites. In: Llano (ed) Adaptations within Antarctic ecosystems. Smithsonian Institution, Washington, DC, pp 1157–1176

Campbell IB, Claridge GGC (1968) Soils in the vicinity of Edisto Inlet, Victoria land, Antarctica. NZ J Sci 11:498–520

Campbell IB, Claridge GGC (1987) Antarctica: soils, weathering processes and environment. Elsevier, Amsterdam

Campbell IB, Claridge GGC, Campbell DI, Balks MR (1998) The soil environment of the McMurdo Dry Valleys, Antarctica. In: Priscu JC (ed) Ecosystem dynamics in a polar desert. The McMurdo Dry Valleys, Antarctica. American Geophysical Union, Washington, DC, vol 72, pp 297–322

Candelone J-P, Hong S, Pellone C, Boutron CF (1995) Post-Industrial Revolution changes in large-scale atmospheric pollution of the Northern Hemisphere by heavy metals as documented in central Greenland snow and ice. J Geophys Res 100:16605–16616

Candelone J-P, Jaffrezo J-L, Hong S, Davidson CI, Boutron CF (1996) Seasonal variations in heavy metals concentrations in present day Greenland snow. Sci Total Environ 193:101–110

Canfield DE, Green WJ, Nixon P (1995) ^{210}Pb and stable lead through the redox transition zone of an Antarctic lake. Geochim Cosmochim Acta 59:2459–2468

Capodaglio G, Scarponi G, Cescon P (1991) Lead speciation in the Antarctic Ocean. Anal Proced 28:76–77

Capodaglio G, Toscano G, Scarponi G, Cescon P (1994) Copper complexation in surface seawater of Terra Nova Bay (Antarctica). Int J Environ Anal Chem 55:129–148

Capon RJ, Elsbury K, Butler MS, Lu CC, Hooper JNA, Rostas JAP, O'Brien HJ, Mudge L-M, Sim ATR (1993) Extraordinary levels of cadmium and zinc in a marine sponge, *Tedania charcoti* Topsent: inorganic chemical defense agents. Experientia 49:263–264

Carginale V, Capasso A, Capasso C, Kille P, Parisi E, Passaretti GL, di Prisco G, Raggio M, Scudiero R (1998) Metallothionein in Antarctic fish. In: di Prisco G, Pisano E, Clarke A(eds) Fishes of Antarctica. A biological overview. Springer, Berlin Heidelberg New York, pp 151–161

Caricchia AM, Chiavarini S, Cremisini S, Morabito R, Perini A, Pezza M (1995) Determination of PAH in atmospheric particulates in the area of the Italian base in Antarctica: report on monitoring activities during the last three scientific expeditions. Environ Pollut 87:345–356

Carman KR, Means JC, Pomarico SC (1996) Response of sedimentary bacteria in a Louisiana salt marsh to contamination by diesel fuel. Aquat Microb Ecol 10:231–241

Caroli S, Senofonte O, Caimi S, Pauwels J, Kramer GN (1996) Planning and certification of new multielemental reference materials for research in Antarctica. Mikrochim Acta 123:119–128

Caron DA, Dennet MR, Lonsdale DJ, Moran DM, Shalapyonok L (2000) Microzooplankton herbivory in the Ross Sea, Antarctica. Deep-Sea Res II 47:3249–3272

Carrera G, Fernández P, Vilanova RM, Grimalt JO (2001) Persistent organic pollutants in snow from European high mountain areas. Atmos Environ 35:245–254

Carroll JJ (1982) Long-term means and short-time variability of the surface energy balance components at the South Pole. J Geophys Res 87:4277–4286

Carroll JJ (1994) Observations and model studies of episodic events over the south polar plateau. Antarct J US 29:322–323

Carsey FD (1980) Microwave observations of the Weddell polynya. Month Weather Rev 108:2032–2044

Carson RL (1962) Silent spring. Houghton Mifflin, Boston

Carter MW, Moghissi MAA (1977). Three decades of nuclear testing. Health Phys 33:55–71

Carter TR, Hulme M, Crossley JF, Malyshev S, New MG, Schlesinger ME, Tuomenvirta H (2000) Climate change in the 21st century – interim characterizations based on the new IPCC emission scenarios. Finnish Environmental Institute, Helsinki, Finnish Environment no 433

Casaux R, Baroni A, Carlini A (1997) The diet of the Weddell seal *Leptonychotes weddelli* at Harmony Point, South Shetland Islands. Polar Biol 18:371–375

Cassidy W, Harvey R, Schutt J, Delisle G, Yanai K (1992) The meteorite collection sites of Antarctica. Meteorites 27:490–525

Castello M, Nimis PL (1997) Diversity of lichens in Antarctica. In: Battaglia B, Valencia J, Walton DWH (eds) Antarctic communities: species, structure and survival. Cambridge University Press, Cambridge, pp 15–21

Catteneo-Vietti R, Chiantore M, Schiaparelli S, Albertelli G (2000) Shallow and deep-water mollusc distribution at Terra Nova Bay (Ross Sea, Antarctica). Polar Biol 23:173–182

Cazenave AK, Dominh MC, Gennero MC, Ferret B (1998) Global mean sea level from Topex-Poseidon and ERS-1. Phys Chem Earth 23:1069–75

CCAMLR (2003) Statistical bulletin, vol 15. Commission for the Conservation of Antarctic Marine Living Resources, Hobart

Cess RD, Zhang MH, Potter GL, Alekseev V, Barker HW, Bony S, Colman RA, Dazlich DA, del Genio AD, Deque M, Dix MR, Dymnikov V, Esch M, Fowler LD, Fraser JR, Galin V, Gates WL, Hack JJ, Ingram WJ, Kiehl JT, Kim Y, Le Treut H, Lo KK-W, McAvancy BJ, Mcleshko VP, Morcrette J-J, Randall DA, Roeckner E, Royer J-F, Schlesinger ME, Sporyshev PV, Timbal B, Volodin EM, Taylor KE, Wang W, Wang WC, Wetherald RT

(1997) Comparison of seasonal change in cloud-radiative forcing from atmospheric general circulation models and satellite observations. J Geophys Res 102:16593–16603

Charlson RJ, Heintzenberg J (eds) (1995) Aerosol forcing of climate. Wiley, Chichester

Charlson RJ, Lovelock JE, Andreae MO, Warren SG (1987) Oceanic phytoplankton, atmospheric sulfur, cloud albedo and climate. Nature 326:655–661

Chaturvedi S (1996) The polar regions. A political geography. Wiley, Chichester

Cheam V, Lawson G, Lechner J, Desrosiers R (1998) Recent metal pollution in Agassiz Ice Cap. Environ Sci Technol 32:3974–3979

Chen B, Smith SR, Bromwich DH (1996) Evolution of the tropospheric split jet over the South Pacific Ocean during the 1986–1989 ENSO cycle. Month Weather Rev 124:1711–1731

Cherel Y, Kooyman GL (1998) Food of emperor penguins (Aptenodytes forsteri) in the western Ross Sea, Antarctica. Mar Biol 20:335–344

Chiarenzelli JR, Aspler LB, Ozarco DL, Hall GEM, Powis KB, Donaldson JA (1997) Heavy metals in lichens, southern District of Keewatin, Northwest Territories, Canada. Chemosphere 35:1329–1341

Chillrud SN, Bopp RF, Simpson HJ, Ross JM, Shuster EL, Chaky DA, Walsh DC, Choy CC, Tolley L-R, Yarme A (1999) Twentieth century atmospheric metal fluxes into Central Park Lake, New York City. Environ Sci Technol 33:657–662

Chinn TJ (1993) Physical hydrology of the Dry Valley lakes. In: Green WJ, Friedmann EI (eds) Physical and biogeochemical processes in Antarctic lakes. American Geophysical Union, Washington, DC, vol 59, pp 1–51

Chown SL, Smith VR (1993) Climate change and the short-term impact of feral house mice at the sub-Antarctic Prince Edward Islands. Oecologia 96:508–516

Chuan RL (1994) Dispersal of volcano-derived particles from Mount Erebus in the Antarctic atmosphere. In: Kyle P (ed) Volcanological and environmental studies of Mount Erebus, Antarctica. American Geophysical Union, Washington, DC, vol 66, pp 97–102

Ciais P, White JCW, Jouzel J, Petit JR (1995) The origin of present-day Antarctic precipitation from surface snow deuterium excess data. J Geophys Res 100:18917–18927

Claridge GGC (1965) The clay mineralogy and chemistry of some soils from the Ross Dependency, Antarctica. NZ J Geol Geophys 8:186–220

Claridge GGC, Campbell IB (1977) The salts in Antarctic soils, their distribution and relationship to soil processes. Soil Sci 123:377–384

Claridge GGC, Campbell IB, Powell HKJ, Amin ZH, Balks MR (1995) Heavy metal contamination in some soils of the McMurdo region, Antarctica. Antarct Sci 7:9–14

Clark PU (1995) Fast glacier flow over soft beds. Science 267:43–44

Clark JS, Cachier H, Goldammer JG, Stocks BJ (eds) (1997) Sediment records of biomass burning and global change. Springer, Berlin Heidelberg New York

Clarke A, Leakey RJG (1996) The seasonal cycle of phytoplankton, macronutrients and the microbial community in a nearshore Antarctic marine ecosystem. Limnol Oceanogr 41:1281–1299

Clarke JC, Manly B, Kerry K, Gardner H, Franchi E, Corsolini S, Focardi S (1998) Sex differences in Adélie penguin foraging strategies. Polar Biol 20:248–258

Claussen M, Kubatzki C, Brovkin V, Ganopolski A (1999) Simulation of an abrupt change in Saharan vegetation in the mid-Holocene. Geophys Res Lett 26:2037–2040

Cleveland L, Little EE, Petty JD, Johnson BT, Lebo JA, Orazio CE, Dionne J, Crockett A (1997) Toxicological and chemical screening of Antarctica sediments: use of whole sediment toxicity tests, Microtox, Mutatox and semipermeable membrane devices (SPMDs). Mar Pollut Bull 34:194–202

Clow GD, Saltus RW, Waddington ED (1996) A new, high-precision borehole temperature logging system used at GISP2 Greenland and Taylor Dome, Antarctica. J Glaciol 142:576–584

Coale KH, Johnson KS, Fitzwater SE, Gordon RM, Tanner S, Chavez FP, Ferioli L, Sakamoto C, Rogers P, Millero F, Steinberg P, Nightingale P, Cooper D, Cochlan WP, Landry MR, Constantinou G, Rollwagen G, Trasvina A, Kudela R (1996) A massive phytoplankton bloom induced by an ecosystem-scale iron fertilization experiment in the equatorial Pacific Ocean. Nature 383:495–501

Coates DA, Stricker GD, Landis ER (1990) Coal geology, coal quality and coal resources in Permian rocks of the Beacon Supergroup, Transantarctic Mountains, Antarctica. In: Splettstoesser JF, Dreschhoff GAM (eds) Mineral resources potential of Antarctica. American Geophysical Union, Washington, DC, vol 51, pp 133–162

Colhoun EA (1991) Geological evidence for changes in East Antarctic ice sheet (60° -120° E) during the last glaciation. Polar Record 27:345–355

Collier R, Dymond J, Honjo S, Manganini S, Francois R, Dunbar R (2000) The vertical flux of biogenic and lithogenic material in the Ross Sea: moored sediment trap observations 1996–1998. Deep-Sea Res II 47:3491–3520

Comiso JC (1994) Surface temperatures in the polar regions from Nimbus 7 temperature humidity infrared radiometer. J Geophys Res 99:5181–5200

Comiso JC (1999) Variability and trends in Antarctic surface temperatures from in situ and satellite infrared measurements. J Climate 13:1674–1696

Comiso JC, Gordon AL (1987) Recurring polynyas over the Cosmonaut Sea and the Maud Rise. J Geophys Res 92:2819–2833

COMNAP-AEON (2001) Summary of environmental monitoring activities in Antarctica. Council of Managers of National Antarctic Programs, Hobart

Conlan KE, Rau GH, McFeters GA, Kvitek RG (2000) Influence of McMurdo Station sewage on Antarctic marine benthos: evidence from stable isotopes, bacteria, and biotic indices. In: Davison W, Howard-Williams C, Broady P (eds) Antarctic ecosystems: model for wider ecological understanding. Caxton Press, Christchurch, pp 315–318

Connolley VM, King JC (1993) Atmospheric water vapour transport to Antarctica inferred from radiosonde data. Q J R Meteorol Soc 119:325–342

Conroy JWH, French MC (1974) Organochlorine levels in two species of Antarctic birds. Bull Br Antarct Surv 38:43–48

Convey P (1997) Environmental change: possible consequences for the life histories of Antarctic terrestrial biota. Korean J Polar Res 8:127–144

Convey P (2000) Environmental change and Antarctic life histories: fact and prediction. In: Davison W, Howard-Williams C, Broady P (eds) Antarctic ecosystems: model for wider ecological understanding. Caxton Press, Christchurch, pp 243–251

Convey P, Smith RIL (1993) Investment in sexual reproduction by Antarctic mosses. Oikos 68:293–302

Correia A, Freydier R, Delmas RJ, Simões JC, Taupin J-D, Dupré B, Artaxo P (2003) Trace elements in South America during 20th century inferred from a Nevado Illimani ice core, Eastern Bolivian Andes (6350 m a.s.l.). Atmos Chem Phys Discuss 3:2143–2177

Corsolini S, Focardi S (2000) Bioconcentration of polychlorinated biphenyls in the pelagic food chain of the Ross Sea. In: Faranda FM, Guglielmo L, Ianora A (eds) Ross Sea ecology. Springer, Berlin Heidelberg New York, pp 575–584

Corsolini S, Romeo T, Ademollo N, Greco S, Focardi S (2002) POPs in key species of marine Antarctic ecosystem. Microchem J 73:187–193

Corsolini S, Olmastroni S, Ademollo N, Minucci G, Focardi S (2003) Persistent organic pollutants in stomach contents of Adélie penguins from Edmonson Point (Victoria Land, Antarctica). In: Huiskes AHL, Gieskes WWC, Rozema J, Schorno RML, van der

Vies SM, Wolff WJ (2003) Antarctic biology in a global context. Backhuys, Leiden, pp 296–300

Cota GF, Kottmeier ST, Robinson DH, Smith WO, Sullivan CW (1990) Bacterioplankton in the marginal ice zone of the Weddell Sea: biomass, production and metabolic activities during austral autumn. Deep-Sea Res 37:1145–1167

Couch TL, Sumner AL, Dassau TM, Shepson PB, Honrath RE (2000) An investigation of the interaction of carbonyl compounds with the snowpack. Geophys Res Lett 27:2241–2244

Court GS, Davis LS, Focardi S, Bargagli R, Fossi MC, Leonzio C, Marsili L (1997) Chlorinated hydrocarbons in the tissue of South Polar skua (*Catharacta maccormicki*) and Adélie penguins (*Pygoscelis adeliae*) from Ross Sea, Antarctica. Environ Pollut 97:285–301

Crame JA (ed) (1989) Origins and evolution of the Antarctic biota. Geol Soc Lond Spec Publ 47

Crame JA (1992) Evolutionary history of the polar regions. Histor Biol 6:37–60

Cremisini C, Orlandi C, Torcini S (1990) Major, minor and trace elements in the surface waters at Terra Nova Bay. In: Cescon P (ed) Environmental impact in Antarctica. Consiglio Nazionale delle Ricerche, Rome, pp 7–16

Crête M, Lefebvre MA, Zikovsky L, Walsh P (1992) Cadmium, lead, mercury and ^{137}Cs in fruticose lichens of northern Québec. Sci Total Environ 121:217–230

Cripps GC (1989) Problems in the identification of anthropogenic hydrocarbons against natural background levels in the Antarctic. Antarct Sci 1:307–312

Cripps GC (1992a) The extent of hydrocarbon contamination in the marine environment from a research station in the Antarctic. Mar Pollut Bull 25:288–292

Cripps GC (1992b) Natural and anthropogenic hydrocarbons in the Antarctic marine environment. Mar Pollut Bull 25:266–273

Cripps GC, Shears JR (1997) The fate in the marine environment of a minor diesel fuel spill from an Antarctic research station. Environ Monit Assess 46:221–232

Crockett AB (1998) Background levels of metals in soils, McMurdo Station, Antarctica. Environ Monit Assess 50:289–296

Croxall JP (1984) Seabirds. In: Laws RM (ed) Antarctic ecology, vol 2. Academic Press, London, pp 533–616

Croxall JP (1987) The status and conservation of Antarctic seas and seabirds: a review. Environ Int 13:55–70

Croxall JP (1992) Southern Ocean environmental changes: effects on seabird, seal and whale populations. Philos Trans R Soc Lond 338:319–328

Croxall JP, Hill HJ, Lidstone-Scott R, O'Connell MJ, Prince PA (1988) Food and feeding ecology of Wilson's storm petrel Oceanites oceanicus at South Georgia. J Zool Lond 216:83–102

Cuff DJ (2002) Global warming. In: Goudie AS, Couff DJ (eds) Encyclopedia of global change. Environmental change and human society, vol 1. Oxford University Press, Oxford, pp 543–545

Cullen JJ, Neale PJ, Lesser MP (1992) Biological weighting function for the inhibition of phytoplankton photosynthesis by ultraviolet radiation. Science 258:646–650

Cullen JT, Lane TW, Morel FMM, Sherrell RM (1999) Modulation of cadmium uptake in phytoplankton by seawater CO_2 concentration. Nature 402:165–167

Cutter GA, Bruland KW, Risebrough RW (1979) Deposition and accumulation of plutonium isotopes in Antarctica. Nature 279:628–629

Dachs J, Lohmann R, Ockenden WA, Méjanelle L, Eisenreich SJ, Jones KC (2002) Oceanic biogeochemical controls on global dynamics of persistent organic pollutants. Environ Sci Technol 36:4229–4237

D'Almeida GA (1991) Atmospheric aerosols: global climatology and radiative characteristics. Deepak, Hampton, VA

Daly KL (1998) Physioecology of juvenile Antarctic krill (*Euphausia superba*) during
spring in ice-covered seas. In: Lizotte MP, Arrigo KR (eds) Antarctic sea ice: biological processes, interactions and variability. American Geophysical Union, Washington,
DC, vol 73, pp 183–198

Dalziel JA (1995) Reactive mercury in the eastern North Atlantic and southeast Atlantic.
Mar Chem 49:307–314

Danielsson L-G, Magnusson B, Westerlund S (1985) Cadmium, copper, iron, nickel and
zinc in the North Atlantic Ocean. Mar Chem 17:23–41

Davidson CI, Bergin MH, Kuhns HD (1996) The deposition of particles and gases to ice
sheets. In: Wolff EW, Bales RC (eds) Chemical exchange between the atmosphere and
polar snow. Springer, Berlin Heidelberg New York, pp 275–306

Davison W, Howard-Willimas C, Broady P (eds) (2000) Antarctic ecosystems: models for
wider ecological understanding. Caxton Press, Christchurch

Dayton PK (1990) Polar benthos. In: Smith WO (ed) Polar oceanography, part B. Chemistry, biology and geology. Academic Press, San Diego, pp 631–685

Dayton PK, Robilliard GA (1971) Implications of pollution to the McMurdo Sound benthos. US Antarct J 8:53–56

Dayton PK, Watson D, Palmisano A, Barry JP, Oliver JS, Rivera D (1986) Distribution patterns of benthic microalgal standing stock at McMurdo Sound, Antarctica. Polar Biol
6:207–213

Deacon G (1983) Antarctic Ocean. Interdisc Sci Rev 2:109–123

Deacon G (1984) The Antarctic Circumpolar Ocean. Cambridge University Press, Cambridge

Dearborn JH (1965) Food of Weddell seals at McMurdo Sound, Antarctica. J Mammal
46:37–43

De Baar HJW, Saager PM, Nolting RF, van der Meer J (1994) Cadmium versus phosphate
in the world ocean. Mar Chem 46:261–281

De Baar HJW, de Jong JTM, Löscher BM, Veth C, Bathmann U, Smetaceck V (1995)
Importance of iron for plankton blooms and carbon dioxide drawdown in the Southern Ocean. Mar Ecol Prog Ser 65:105–122

De Baar HJW, de Jong JTM, Nolting RF, Timmermans KR, van Leeuwe MA, Bathmann U,
van der Loeff MMR, Sildam J (1999) Low dissolved Fe and the absence of diatom
blooms in remote Pacific waters of the Southern Ocean. Mar Chem 66:1–34

De la Mare WK (1997) Abrupt mid-twentieth century decline in Antarctic sea ice extent
from whaling records. Nature 389:57–61

Delille D, Delille B (2000) Field observations on the variability of crude oil impact on
indigenous hydrocarbon-degrading bacteria from sub-Antarctic intertidal sediments. Mar Environ Res 49:403–417

Dell RK (1972) Antarctic benthos. Adv Mar Biol 10:1–216

Delmas RJ (1995) Ice core studies of global biogeochemical cycles. Springer, Berlin Heidelberg New York

Delmas RJ, Kirchner S, Palais JM, Petit J-R (1992) 1000 years of explosive volcanism
recorded et the South Pole. Tellus 44B:335–350

De Long EF, Wu KY, Prézelin BB, Jovine RVM (1994) High abundance of Archaea in
Antarctic marine picoplankton. Nature 371:695–697

DeMaster DJ, Ragueneau O, Nittrouer CA (1996) Preservation efficiencies and accumulation rates for biogenic silica and organic C, N, and P in high-latitude sediments: the
Ross Sea. J Geophys Res 101:18501–18518

De Mora SJ, Bibby DM, Patterson JE (1993) Baseline atmospheric mercury studies at
Ross Island, Antarctica. Antarct Sci 5:323–326

Denton GH, Prentice ML, Burckle LH (1991) Cenozoic history of the Antarctic ice sheet. In: Tingey RJ (ed) The geology of Antarctica. Clarendon Press, Oxford, pp 365–433

Deprez PP, Arens M, Locher E (1999) Identification and assessment of contaminated sites at Casey Station, Wilkes Land, Antarctica. Polar Record 35:299–316

Desideri P, Lepri L, Checchini L (1989) Identification and determination of organic compounds in seawater in Terra Nova Bay (Antarctica). Ann Chim (Rome) 79:589–605

Desideri P, Lepri L, Checchini L (1990) Organic compounds in Antarctic matrices: sea water, particulate, pack and sediments. In: Proc Conf Environmental Impact in Antarctica, 8–9 June 1990, Consiglio Nazionale delle Richerche, Rome, pp 63–70

DeVries AL (1971) Glycoproteins as biological antifreeze agents in Antarctic fishes. Science 172:1152–1155

De Wit MJ (1985) Minerals and mining in Antarctica. Clarendon Press, Oxford

Dibb JE (1996) Overview of field data on the deposition of aerosol-associated species to the surface snow of polar glaciers (emphasizing recent work in Greenland). In: Wolff EW, Bales RC (eds) Processes of chemical exchange between the atmosphere and polar snow. Springer, Berlin Heidelberg New York, pp 249–274

Dibb JE, Jaffrezo JL (1997) Air-snow exchange investigations at Summit, Greenland: an overview. J Geophys Res 102:26795–26807

Diekmann B, Kuhn G, Rachold V, Abelmann A, Brathauer U, Fütterer KD, Gersonde R, Grobe H (2000) Terrigenous sediment supply in the Scotia Sea (Southern Ocean): response to Late Quaternary ice dynamics in Patagonia and on the Antarctic Peninsula. Palaeogeogr Palaeoclimatol Palaeoecol 162:357–387

Dietz R, Riget F, Johansen P (1996) Lead, cadmium, mercury and selenium in Greenland marine animals. Sci Total Environ 186:67–93

Dietz R, Nørgaard J, Hansen JC (1998) Have Arctic marine mammals adapted to high cadmium levels? Mar Pollut Bull 36:490–492

Dietz R, Riget F, Born EW (2000) An assessment of selenium to mercury ration in Greenland marine animals. Sci Total Environ 245:15–24

Di Giulio RT, Washburn PC, Wenning RJ, Winston GW, Jewell CS (1989) Biochemical responses in aquatic animals: a review of determinants of oxidative stress. Environ Toxicol Chem 8:1103–1123

Dignon J, Penner JE (1991) Biomass burning: a source of nitrogen oxides in the atmosphere. In: Levine JS (ed) Global biomass burning: atmospheric, climatic, and biospheric implications. MIT Press, Cambridge, MA

Di Prisco G, Maresca B, Tota B (eds) (1991) Biology of Antarctic fish. Springer, Berlin Heidelberg New York

Di Prisco G, Pisano E, Clarke A (eds) (1998) Fishes of Antarctica: a biological overview. Springer, Berlin Heidelberg New York

Di Tullio GR, Smith WO (1995) Relationship between dimethylsulfide and phytoplankton pigment concentrations in the Ross Sea, Antarctica. Deep-Sea Res 42:873–892

Di Tullio GR, Garrison DL, Mathot S (1998) Dimethylsulfoniopropionate in sea ice algae from the Ross Sea polynya. In: Lizotte MP, Arrigo KR (eds) Antarctic sea ice. Biological processes, interactions and variability. American Geophysical Union, Washington, DC, vol 73, pp 139–145

Di Tullio GR, Grebmeier JM, Arrigo KR, Lizotte MP, Robinson DH, Leventer A, Barry JP, VanWoert ML, Dunbar RB (2000) Rapid and early export of *Phaeocystis antarctica* blooms in the Ross Sea, Antarctica. Nature 404:595–598

Dixon RW, Mosimann L, Oberholzer B, Staehelin J, Waldvogel A, Collet JL (1995) The effect of riming on the ion concentrations of winter precipitation. A quantitative analysis of field measurements. J Geophys Res 100:11517–11527

Doake CSM, Vaughan DG (1991) Rapid disintegration of the Wordie Ice Shelf in response to atmospheric warming. Nature 350:328–30

Dodge CW (1973) Lichen flora of the Antarctic continent and adjacent islands. Phoenix, Canaan, New Hampshire

Domack EW, McClennen CE (1996) Accumulation of glacial marine sediments in fjords of the Antarctic Peninsula and their use as late Holocene paleoenvironmental indicators. In: Ross RM, Hofmann EE, Quetin LB (eds) Foundations for ecological research west of the Antarctic Peninsula. American Geophysical Union, Washington, DC, vol 70, pp 135–154

Domack EW, Jull AJ, Nakao S (1991) Advance of East Antarctic outlet glaciers during the Hypsithermal: implications for the volume state of the Antarctic ice sheet under global warming. Geology 19:1059–1062

Dominé F, Shepson PB (2002) Air-snow interactions and atmospheric chemistry. Science 297:1506–1510

Doney SC (1999) Major challenges confronting marine biogeochemical modeling. Global Biogeochem Cycles 13:705–714

Döring T, Schwikowski M, Gäggeler HW (1997) Determination of lead concentrations and isotope ratios in recent snow samples from high alpine sites with a double focusing ICP-MS. Fresenius J Anal Chem 359:382–384

Drake F (1995) Stratospheric ozone depletion – an overview of the scientific debate. Prog Phys Geogr 19:1–17

Drewry DJ (ed) (1983) Antarctica: glaciological and geophysical folio. Scott Polar Research Institute, Cambridge

Drewry DJ (1991) The response of the Antarctic ice sheet to climatic change. In: Harris CM, Stonehouse B (eds) Antarctica and global climatic change. Belhaven Press, London, pp 90–106

Dubovik O, Holben BN, Eck TF, Smirnov A, Kaufman YJ, King MD, Tanré D, Slutsker I (2002) Variability of absorption and optical properties of key aerosol types observed in worldwide locations. J Atmos Sci 59:590–608

Duce RA, Tindale NW (1991) Atmospheric transport of iron and its deposition in the ocean. Limnol Oceanogr 36:1715–1726

Duce RA, Hoffman GL, Zoller WH (1975) Atmospheric trace metals at remote northern and southern hemispheric sites: pollution or natural? Science 187:59–61

Dudgale RC, Wilkerson FP, Minas HJ (1995) The role of a silicate pump in driving new production. Deap-Sea Res 42:697–719

Duhamel G, Hureau J-C (1990) Changes in fish populations and fisheries around the Kerguelen Islands during the last decade. In: Kerry KR, Hempel G (eds) Antarctic ecosystems: ecological change and conservation. Springer, Berlin Heidelberg New York, pp 323–333

Dutton EG, Stone RS, Nelson DW, Mendonca BG (1991) Recent interannual variations in solar radiation, cloudiness and surface temperature at the South Pole. J Climate 4:848–858

Eastman JT (1993) Antarctic fish biology. Evolution in a unique environment. Academic Press, San Diego

Eastman JT, Clarke A (1998) Radiations of Antarctic and non-Antarctic fish. In: Di Prisco G, Pisano E, Clarke A (eds) Fishes of Antarctica: a biological overview. Springer, Berlin Heidelberg New York, pp 3–26

Eastman JT, Hubold G (1999) The fish fauna of the Ross Sea, Antarctica. Antarct Sci 11:293–304

Ebinghaus R, Kock HH, Temme C, Einax JW, Lowe AG, Richter A, Burrows JP, Schroeder WH (2002) Antarctic springtime depletion of atmospheric mercury. Environ Sci Technol 36:1238–1244

Edwards CA (1973) Persistent pesticides in the environment, 2nd edn. CRC Press, Cleveland

Ehrenberg CG (1844) Einige vorläufige Resultate seiner Untersuchungen der ihm von der Südpolreise des Kapitän Ross, so wie von der Herren Schayer und Darwin zugekommenen Materialien. Abh K Preuss Akad Wissen, Berlin, pp 182–207

Ehrhardt M, Klugsøyr J, Law RJ (1991) Hydrocarbons: review of methods for analysis in sea water, biota and sediments. International Council for the Exploration of the Sea, Copenhagen, Tech Mar Environ Res no 12

Eisma D (1973) Sediment cores from Breid Bay and Brekilen, Antarctica. Neth J Sea Res 6:327–338

Ekau W (1991) Morphological adaptations and mode of life in high Antarctic fish. In: Di Prisco G, Maresca B, Tota B (eds) Biology of Antarctic fish. Springer, Berlin Heidelberg New York, pp 23–39

Ekman S (1953) Zoogeography of the sea. Sidgwick and Jackson, London

Elinder C-G, Järup L (1996) Cadmium exposure and health risks: recent findings. Ambio 25:370–373

Ellis J (1991) Antarctica and global climatic change: review of prominent issues. In: Harris CM, Stonehouse B (eds) Antarctica and global climatic change. Belhaven Press, London, pp 11–20

Ellis-Evans JC, Wynn-Williams D (1996) A great lake under the ice. Nature 381:644–646

Ellis-Evans JC, Laybourn-Parry J, Bayliss PR, Perriss ST (1997) Human impact on an oligotrophic lake in the Lansermann Hills. In: Battaglia B, Valencia J, Walton DWH (eds) Antarctic communities: species, structure, and survival. Cambridge University Press, Cambridge, pp 396–404

Ellwood MJ, Hunter KA (1999) Determination of the Zn/Si ratio in diatom opal: a method for the separation, cleaning and dissolution of diatoms. Mar Chem 66:149–160

El-Sayed SZ (1968) Primary productivity of the Antarctic and sub-Antarctic. In: Bushnell V (ed) Antarctic map folio series, Folio 10. American Geography Society, New York, pp 1–6

El-Sayed SZ (ed) (1994) Southern Ocean ecology: the BIOMASS perspective. Cambridge University Press, Cambridge

El-Sayed SZ, Fryxell GA (1994) Phytoplankton. In: Friedmann EI (ed) Antarctic microbiology. Wiley-Liss, New York, pp 65–122

El-Sayed SZ, Stockwell PA, Reheim HA, Taguchi S, Mayer MA (1979) On the productivity of the southwestern Indian Ocean. CNFRA 44:1–43

Embleton C, King CAM (1975) Glacial geomorphology. Arnold, London

Emslie SD, Fraser W, Smith RC, Walker W (1998) Abandoned penguin colonies and environmental change in the Palmer Station area, Anvers Island, Antarctic Peninsula. Antarct Sci 10:257–268

Enomoto H, Ohmura A (1990) The influence of atmospheric half-yearly cycle on the sea ice extent in the Antarctic. J Geophys Res 95:9497–9511

Eppley ZA, Rubega MA (1989) Indirect effects of an oil spill. Nature 345:513

Estep KW, Nejstgaard JC, Skjoldal HR, Rey F (1990) Grazing of copepods upon *Phaeocystis* colonies as a function of the physiological state of the prey. Mar Ecol Prog Ser 67:235–249

Everson I, Goss C (1991) Krill fishing activity in the southwest Atlantic. Antarct Sci 3:351–358

Falkowski PG, Barber RT, Smetacek V (1998) Biogeochemical controls and feedbacks on ocean primary production. Science 281:200–206

Fallis BW (1982) Trace metals in sediments and biota from Strathcona Sound, NWT; Nanisivik Mar Monit Prog 1974–1979. Can Tech Rep Fish Aquat Sci 39:1082–1116

Fant ML, Nyman M, Helle E, Rudbäck E (2001) Mercury, cadmium, lead and selenium in ringed seals (*Phoca hispida*) from the Baltic Sea and from Svalbard. Environ Pollut 111:493–501

Faranda FM, Guglielmo L, Ianora A (eds) (2000) Ross Sea ecology. Springer, Berlin Heidelberg New York

Farman JC, Gardiner BG, Shanklin JD (1985) Large losses of total ozone in Antarctica reveal seasonal ClO_x/NO_x interaction. Nature 315:207–210

Ferguson SH, Franzmann PD, Snape I, Revill AT, Trefry MG, Zappia LR (2003) Effects of temperature on mineralisation of petroleum in contaminated Antarctic terrestrial sediments. Chemosphere 52:975–987

Fergusson JE (1990) The heavy elements: chemistry, environmental impact and health effects. Pergamon Press, Oxford

Ferrara R, Maserti BE, Bargagli R (1988) Mercury in the atmosphere and in lichens in a region affected by a geochemical anomaly. Environ Technol Lett 9:689–694

Ferrari CP, Clotteau T, Thompson LG, Barbante C, Cozzi G, Cescon P, Hong S, Bourgoin L-M, Francou B, Boutron CF (2001) Heavy metals in ancient tropical ice: initial results. Atmos Environ 35:5809–5815

Ferraro SP, Cole FA, DeBen WA, Swartz RC (1989) Power-cost efficiency of height macrobenthic sampling schemes in Puget Sound, Washington, USA. Can J Fish Aquat Sci 46:2157–2165

Fimreite N (1979) Accumulation and effects of mercury on birds. In: Nriagu JO (ed) The biogeochemistry of mercury in the environment. Elsevier/North-Holland, Amsterdam, pp 601–627

Fischer W, Hureau JC (eds) (1985) FAO species identification sheets for fishery purposes. Southern Ocean (fishing area 48,58 and 88), vol 1. FAO, Rome

Fitzgerald WF, Mason RP (1997) Mercury and its environment and biology. In: Siegel A, Siegel KH (eds) Metals ions in biological systems, vol 34. Dekker, New York, pp 53–111

Fitzgerald WF, Engstrom DR, Mason RP, Nater EA (1998) The case of atmospheric mercury contamination in remote areas. Environ Sci Technol 32:1–7

Fitzwater SE, Johnson KS, Gordon RM, Coale KH, Smith WO (2000) Trace metal concentrations in the Ross Sea and their relationship with nutrients and phytoplankton growth. Deep-Sea Res II 47:3159–3179

Flegal AR, Maring H, Niemeyer S (1993) Anthropogenic lead in Antarctic sea water. Nature 365:242–244

Focardi S, Gaggi, C, Chemello G, Bacci E (1991) Organochlorine residues in moss and lichen samples from two Antarctic areas. Polar Record 27:241–244

Focardi S, Lari L, Marsili L (1992a) PCB congeners, DDTs and hexacholorobenzene in Antarctic fish from Terra Nova Bay (Ross Sea). Antarct Sci 4:151–154

Focardi S, Fossi MC, Lari L, Marsili L, Leonzio C, Casini S (1992b) Induction of mixed function oxidase (MFO) system in two species of Antarctic fish from Terra Nova Bay (Ross Sea). Polar Biol 12:721–725

Focardi S, Bargagli R, Fossi MC, Leonzio C, Marsili L, Court G, Davis L (1992c) Mixed function oxidase activity and chlorinated hydrocarbon residues in Antarctic seabirds: south polar skua and Adélie penguin. Mar Environ Res 34:201–205

Focardi S, Bargagli R, Corsolini S (1993) Organochlorines in Antarctic marine food chain at Terra Nova Bay (Ross Sea). Korean J Polar Res 4:73–77

Focardi S, Bargagli R, Corsolini S (1995a) Isomer-specific analysis and toxic potential evaluation of polychlorinated biphenyls in Antarctic fish, seabirds and Weddell seals from Terra Nova Bay (Ross Sea). Antarct Sci 7:31–35

Focardi S, Fossi MC, Lari L, Casini S, Leonzio C, Meidel SK, Nigro M (1995b) Induction of MFO activity in the Antarctic fish *Pagothenia bernacchii*: preliminary results. Mar Environ Res 39:97–100

Focardi S, Fossi MC, Lari L, Casini S, Bargagli R (1997) Investigation of mixed function oxidase activity in Antarctic organisms. In: Battaglia B, Valencia J, Walton DWH (eds) Antarctic communities: species, structure and survival. Cambridge University Press, Cambridge, pp 405–408

Fogg GE (1967) Observations on the snow algae of the South Orkney Islands. Philos Trans R Soc B 252:279–287

Fogg GE (1998) The biology of polar habitats. Oxford University Press, Oxford

Folco L, Capra A, Chiappini M, Frezzotti M, Mellini M, Tabacco IE (2002) The Frontier Mountain meteorite trap (Antarctica). Meteorit Planet Sci 37:209–228

Foldvik A, Gammelsrød T (1988) Notes on Southern Ocean hydrography, sea-ice and bottom water formation. Palaeogeogr Palaeoclimatol Palaeoecol 67:3–17

Ford AB (1990) The Dufek intrusion of Antarctica. In: Splettstoesser JF, Dreschhoff GAM (eds) Mineral resources potential of Antarctica. American Geophysical Union, Washington, DC, vol 51, pp 15–32

Forget G (1991) Pesticides and the third world. J Toxicol Environ Health 32:11–31

Förstner U, Wittmann GTW (1983) Metal pollution in the aquatic environment. Springer, Berlin Heidelberg New York

Fortuin JPF, Oerlemans J (1990) Parameterisation of the annual surface temperature and mass balance of Antarctica. Ann Glaciol 14:78–84

Fossi MC, Casini S, Savelli C, Lan L, Corsi I, Sanchez-Hernadez JC, Mattei N, Franchi E, Depledge M, Bamber S (1996) Multi-trial biomarkers approach using *Carcinus aestuarii* to evaluate toxicological risk due to Mediterranean contaminants: field and experimental studies. Fresenius Environ Bull 5:706–711

Fowbert JA, Smith RIL (1994) Rapid population increases in native vascular plants in the Argentine Island, Antarctic Peninsula. Arctic Alpine Res 26:290–296

Fowler SW (1986) Trace metal monitoring of pelagic organisms from the open Mediterranean Sea. Environ Monit Assess 7:59–78

Fowler SW (1990) Critical review of selected heavy metal and chlorinated hydrocarbon concentrations in the marine environment. Mar Environ Res 29:1–64

Frache R, Abelmoschi ML, Baffi F, Ianni C, Magi E, Soggia F (2001) Trace metals in particulate and sediments. In: Caroli S, Cescon P, Walton DWH (eds) Environmental contamination in Antarctica. Elsevier, Amsterdam, pp 219–236

Francioni F (1993) The Madrid Protocol on the Protection of the Antarctic environment. Texas Int Law J 28:47–72

Francis JE (1991) Paleoclimatic significance of Cretaceous-early Tertiary fossil forests of the Antarctic Peninsula. In: Thomson MRA, Crame JA, Thomson JW (eds) Geological evolution of Antarctica. Cambridge University Press, Cambridge, pp 623–628

Francis P (1993) Volcanoes. A planetary perspective. Clarendon Press, Oxford

Franz TP, Eisenreich SJ (1998) Snow scavenging of polychlorinated biphenyls and polycyclic aromatic hydrocarbons in Minnesota. Environ Sci Technol 32:1771–1778

Fraser FC (1936) On the development and distribution of young stages of krill *Euphausia superba*. Discovery Rep 14:3–190

Fraser WR, Ainley DG (1986) Ice edges and seabird occurrence in Antarctica. Bioscience 36:258–263

Fraser WR, Hofmann EE (2003) A predator's perspective on causal links between climate change, physical forcing and ecosystem response. Mar Ecol Prog Ser 265:1–15

Fraser WR, Trivelpiece WZ, Ainley DG, Trivelpiece SG (1992) Increases in Antarctic penguin populations: reduced competition with whales or a loss of sea ice due to environmental warming? Polar Biol 11:525–531

Frenot Y, van Vliet-Lanoe B, Gloaguen J-C (1995) Particle translocation and initial soil development on a glacier foreland, Kerguelen Islands, Subantarctic. Arctic Alpine Res 27:107–115

Frew RD (1995) Antarctic bottom water formation and the global cadmium to phosphorus relationship. Geophys Res Lett 22:2349–2352

Frew RD, Bowie A, Croot P, Pickmere S (2001) Macronutrient and trace metal geochemistry of an in situ iron-induced Southern Ocean bloom. Deep-Sea Res II 48:2467–2481

Friedmann EI (ed) (1993) Antarctic microbiology. Wiley-Liss, New York

Fritsch FE (1912) Freshwater algae collected in the South Orkney Islands by RNR Brown, 1902–1904. J Linn Soc Bot 40:293–338

Fulford-Smith SP, Sikes EL (1996) The evolution of Ace Lake, Antarctica, determined from sedimentary diatom assemblages. Paleo 124:73–86

Fullager PJ (1976) McCormick's skua, Catharacta maccormicki, in the North Atlantic. Australasian Seabird Group Newslett 7:18–19

Fumanti B, Cavacini P, Alfinito S (1997) Benthic algal mats of some lakes of Inexpressible Island (northern Victoria Land, Antarctica). Antarct Sci 17:25–30

Fung IY, Meyn SK, Tegen I, Doney SC, John J, Bishop JKB (2000) Iron supply and demand in the upper ocean. Global Biogeochem Cycles 14:281–295

Fuoco R, Ceccarini A (2001) Polychlorobiphenyls in Antarctic matrices. In: Caroli S, Cescon P, Walton DWH (eds) Environmental contamination in Antarctica. A challenge to analytical chemistry. Elsevier, Amsterdam, pp 237–273

Fuoco R, Colombini MP, Abete C (1994) Determination of polychlorobiphenils in environmental samples from Antarctica. Int J Environ Anal Chem 55:15–25

Fuoco R, Colombini MP, Ceccarini A, Abete C (1996) Polychlorobiphenils in Antarctica. Microchem J 54:384–390

Furness RW (1987) The skuas. Poyser, Carlton

Furness RW (1993) Birds as monitor of pollutants. In: Furness RW, Greenwood JJD (eds) Birds as monitor of environmental change. Chapman and Hall, London, pp 86–143

Furness R, Hutton M (1979) Pollutant levels in the great skua Catharacta skua. Environ Pollut 13:261–268

Gaino E, Bavestrello G, Cattaneo-Vietti R, Sarà M (1994) Scanning electron microscope evidence for diatom uptake by two Antarctic sponges. Polar Biol 14:55–58

Gambi MC, Bussotti S (1999) Composition, abundance and stratification of soft-bottom macrobenthos from selected areas of the Ross Sea shelf (Antarctica). Polar Biol 21:347–354

Gambi MC, Lorenti M, Russo GF, Scipione MB (1994) Benthic associations of the shallow hard bottoms of Terra Nova Bay, Ross Sea: zonation, biomass and population structure. Antarct Sci 6:449–462

Gardner BD (1983) Chlorinated pesticides in cats on Marion Island. S Afr J Sci 80:43–44

Gardner BD, Siegfried WR, Connell AD (1985) Chlorinated hydrocarbons in seabird eggs from the Southern Atlantic and Indian oceans. In: Siegfried WR, Condy PR, Laws RM (eds) Antarctic nutrient cycles and food webs. Springer, Berlin Heidelberg New York, pp 647–651

Gardner H, Kerry K, Riddle M (1997) Poultry virus infection in Antarctic penguins. Nature 387:245

Garrison DL (1991) Antarctic sea ice biota. Am Zool 4:17–33

Garrison DL, Buck KR (1991) Surface-layer assemblages in Antarctic pack ice during the austral spring: environmental conditions, primary production and community structure. Mar Ecol Prog Ser 75:161–172

Garty J, Galun M, Kessel M (1979) Localization of heavy metals and other elements accumulated in the lichen thallus. New Phytol 82:159–168

Gauthier J, Sears R (1999) Behavioural response of four species of balaenopterid whales to biopsy sampling. Mar Mammal Sci 15:85–101

Gee CT (1989) Permian Glossoppteris and Elatocladus megafossil floras from the English Coast, eastern Ellsworth Land, Antarctica. Antarct Sci 1:35–44

George AL (2002) Seasonal factors affecting surfactants biodegradation in Antarctic coastal waters: comparison of polluted and pristine site. Mar Environ Res 53:403–415

George JL, Frear DEH (1966) Pesticides in the Antarctic. J Appl Ecol 3:155–167

Germani MS (1980) Selected studies of four high temperature air-pollution sources. PhD Thesis, University of Maryland, College Park

Gibson JA, Garrick RC, Burton HR, McTaggart AR (1990) Dimethylsulfide and the alga *Phaeocystis pouchetii* in Antarctic coastal water. Mar Biol 104:339–346

Gilichinsky DA, Rivkina EM, Vorobyova EA, Hoover RB, Spirina E (1999) Ancient viable microbial communities from Antarctic permafrost and subsurface ice. In: Proc SPIE Technical Conf 3755 Insruments, Methods, and Missions for Astrobiology II, 20–22 July, Denver, Abstr 16

Gill AE (1982) Atmosphere-ocean dynamics. Academic Press, New York

Gill GA, Fitzgerald WF (1988) Vertical mercury distributions in the oceans. Geochim Cosmochim Acta 52:1719–1728

Gillespie A, Burns WCG (eds) (2000) Climate change in the South Pacific: impacts and responses in Australia, New Zealand and small island states. Kluwer, Dordrecht

Giordano R, Lombardi G, Ciaralli L, Beccaloni E, Sepe A, Ciprotti M, Costantini S (1999) Major and trace elements in sediments from Terra Nova Bay, Antarctica. Sci Total Environ 227:29–40

Giovinetto MB, Bentley CR (1985) Surface balance in ice drainage systems of Antarctica. Antarct J US 20:6–13

Giovinetto MB, Waters NM, Bentley CR (1990) Dependence of Antarctic surface mass balance on temperature, elevation and distance to open water. J Geophys Res 95:3517–3531

Glasby GP, Barrett PJ, McDougall JC, McKnight DC (1975) Localized variations in sedimentation characteristics in the Ross Sea and McMurdo Sound regions, Antarctica. NZ J Geol Geophys 18:605–621

Glasby GP, Hunt JL, Renner RM (1985) Trace element analyses of marine sediments from the Southwest Pacific. NZ Soil Bureau Rep no 53, pp 1–62

Gloersen P (1995) Modulation of hemispheric sea-ice cover by ENSO events. Nature 373:503–506

Gloersen P, Campbell WJ (1988) Variations in the Arctic, Antarctic and sea ice covers during 1978–1987 as observed with the Nimbus 7 scanning multichannel microwave radiometer. J Geophys Res 93:3564–3572

Gloersen P, Campbell WJ, Cavalieri DJ, Comiso JC, Parkinson CL, Zwally HJ (1992) Arctic and Antarctic sea ice 1978–1987. NASA, Washington, DC

Godfrey JS, Rintoul SR (1998) The role of the oceans in Southern Hemisphere climate. In: Karoly DJ, Vincent DG (eds) Meteorology of the Southern Hemisphere. American Meteorological Society, Boston, vol 27, pp 283–306

Godoy JM, Schuch LA, Nordemann DJR, Reis VRG, Ramalho M, Recio JC, Brito RRA, Olech MA (1998) ^{137}Cs, [226, 228]Ra, ^{210}Pb, [40]K concentrations in Antarctic soil, sediment and selected moss and lichen samples. J Environ Radioactivity 41:33–45

Goldman CR (1970) Antarctic freshwater ecosystems. In: Holdgate MW (ed) Antarctic ecology. Academic Press, London, pp 609–620

Gon O, Heemstra PC (eds) (1990) Fishes of the Southern Ocean. JLB Smithsonian Institute of Ichthyology, Grahamstown, South Africa

Gonzáles-Solís J, Sanpera C, Ruiz X (2002) Metals and selenium as bioindicators of geographic and trophic segregation in giant petrels *Macronectes* spp. Mar Ecol Prog Ser 244:257–264

Gordon AL (1981) Seasonality of Southern Ocean sea ice. J Geophys Res 86:4193–4197

Gordon AL (1988) Spatial and temporal variability within the Southern Ocean. In: Sahrhage D (ed) Antarctic Ocean and resources variability. Springer, Berlin Heidelberg New York, pp 41–56

Gordon AL, Molinelli EM (1982) The Southern Ocean atlas: thermohaline and chemical distributions and the atlas data set. Columbia University Press, New York

Gordon HB, O'Farrell SP (1997) Transient climate change in the CSIRO coupled model with dynamic sea ice. Month Weather Rev 25:875–907

Gore DB, Creagh DC, Burgess JS, Colhoun EA, Spate AP, Baird AS (1996) Composition, distribution and origin of surficial salts in the Vestfold Hills, East Antarctica. Antarct Sci 8:73–84

Görlach U, Boutron CF (1992) Variations in heavy metals concentrations in Antarctic snows from 1940 to 1980. J Atmos Chem 14:205–222

Goudie AS (2002) Abrupt climate change. In: Goudie AS, Cuff DJ (eds) Encyclopedia of global change. Environmental change and human society, vol 1. Oxford University Press, Oxford, pp 172–175

Graf H-F, Feichter J, Langmann B (1997) Volcanic sulfur emissions: estimates of source strength and its contribution to the global sulfate distribution. J Geophys Res 102:10727–10738

Graf H-F, Langmann B, Feichter J (1998) The contribution of Earth degassing to the atmospheric sulphur budget. Chem Geol 147:131–145

Gran HH (1931) On the conditions for the production of plankton in the sea. Rapp P-V Réunion Conseil Int Explor Mer 75:37–46

Green K, Burton H (1987) Seasonal and geographical variation in the food of Weddell seals *Leptonychotes weddelli* in Antarctica. Aust Wildlife Res 14:475–489

Green WJ, Friedmann EI (1993) Physical and biogeochemical processes in Antarctic lakes. American Geophysical Union, Washington, DC, Antarct Res Ser 59

Green G, Nichols PD (1995) Hydrocarbons and sterols in marine sediments and soils at Davis Station, Antarctica: a survey for human-derived contaminants. Antarct Sci 7:137–144

Green WJ, Ferdelman TG, Gardner TJ, Varber LC, Angle MP (1986) The residence times of eight trace metals in a closed-basin Antarctic lake: Lake Hoare. Hydrobiologia 134:249–255

Green G, Skerratt JH, Leeming R, Nichols PD (1992) Hydrocarbons and coprostanol levels in seawater, sea-ice algae and sediments near Davis Station in eastern Antarctica: a regional survey and preliminary results for a field fuel spill experiment. Mar Pollut Bull 25:293–302

Green WJ, Canfield DE, Nixon P (1998) Cobalt cycling and fate in Lake Vanda. In: Priscu JC (ed) Ecosystem dynamics in a polar desert. The McMurdo Dry Valleys, Antarctica. American Geophysical Union, Washington, DC, pp 205–215

Gregor D, Peters AJ, Teixeira C, Jones N, Spencer C (1995) The historical residue trend of PCBS in the Agassiz Ice Cap, Ellesmere Island, Canada. Sci Total Environ 160/161:117–126

Gregory JM, Oerlemans J (1998) Simulated future sea level rise due to glacial melt based on regionally and seasonally resolved temperature changes. Nature 391:474–476

Griffiths RP, Caldwell BA, Broich WA, Morita RY (1981) Long-term effects of crude oil on uptake and respiration of glucose and glutamate in Arctic and subarctic marine sediments. Appl Environ Microbiol 42:792–801

Grimm AM, Barros VR, Doyle ME (2000) Climate variability in South America associated with El Niño and La Niña events. J Climate 13:35–58

Grobe CW, Ruhland CT, Day TA (1997) A new population of *Colobanthus quitensis* near Arthur Harbour, Antarctica: correlating recruitment with warmer summer temperatures. Arctic Alpine Res 29:217–221

Grossi SM, Kottmeier ST, Moe RL, Taylor GT, Sullivan CW (1987) Sea ice microbial communities. VI. Growth and primary productivity in bottom ice under graded snow cover. Mar Ecol Prog Ser 35:153–164

Grotti M, Soggia F, Abelmoschi ML, Rivaro P, Magi E, Frache R (2001) Temporal distribution of trace metals in Antarctic coastal water. Mar Chem 76:189–209

Grotti M, Soggia F, Dalla Riva S, Magi E, Frache R (2003) An in situ filtration system for trace element determination in suspended particulate matter. Anal Chim Acta 498:165–173

Gruzov YN (1977) Seasonal alterations in coastal communities in the Davis Sea. In: Llano GA (ed) Adaptations within Antarctic ecosystems. Smithsonian Institution, Washington, DC, pp 263–278

Guinet C, Jouventin P, Georges J-Y (1994) Long term changes of fur seals *Arctocephalus gazella* and *Arctocephalus tropicalis* on subantarctic (Crozet) and subtropical (St Paul and Amsterdam) islands and their possible relationship to El Niño Southern oscillation. Antarct Sci 6:473–478

Gutt J (2001) On the direct impact of ice on marine benthic communities, a review. Polar Biol 24:553–564

Hagen W, Vleet ESV, Kattner G (1996) Seasonal lipid storage as overwintering strategy of Antarctic krill. Mar Ecol Prog Ser 134:85–89

Hall MM, Bryden HL (1982) Direct estimates and mechanisms of ocean heat transport. Deep-Sea Res 29:339–360

Hall A, Manabe S (1999) The role of water vapour feedback in unperturbed climate variability and global warming. J Climate 12:2327–2346

Hall K, Walton DWH (1992) Rock weathering, soil development and colonisation under a changing climate. Philos Trans R Soc B 33:269

Hamelin B, Ferrand JL, Alleman L, Nicolas E, Véron AJ (1997) Isotopic evidence of pollutant lead transport from North-America to the Subtropical North Atlantic Gyre. Geochim Cosmochim Acta 61:4423–4428

Hamilton W (1964) Diabase sheets differentiated by liquid fractionation. Taylor Glacier Region, southern Victoria Land. In: Adje RJ (ed) Antarctic geology. North Holland, Amsterdam, pp 442–454

Hammer CU (1977) Past volcanism revealed by Greenland ice sheet impurities. Nature 270:482–486

Hammer CU (1982) The history of atmospheric composition as recorded in ice sheets. In: Goldberg ED (ed) Atmospheric chemistry. Springer, Berlin Heidelberg New York, pp 119–134

Hammer CU, Clausen HB, Dansgaard W (1980) Greenland ice sheet evidence of postglacial volcanism and its climatic impact. Nature 288:230–235

Hanna E (1996) The role of the Antarctic sea ice in global climate change. Prog Phys Geogr 20:371–401

Hansell DA, Feely RA (2000) Atmospheric intertropical convergence impacts on surface ocean carbon and nitrogen biogeochemistry in the western tropical Pacific. Geophys Res Lett 27:1013–1016

Hansen CT, Nielsen CO, Dietz R, Hansen MM (1990) Zinc, cadmium, mercury and selenium in minke whales, belugas and narwhals from west Greenland. Polar Biol 10:529–539

Hansom JD, Gordon JE (1998) Antarctic environments and resources. A geographical perspective. Addison Wesley Longman, New York

Hanson RB, Lowery HK (1983) Nucleic acid synthesis in oceanic microplankton from the Drake Passage, Antarctica: evaluation of steady-state growth. Mar Biol 73:79–89

Hara K, Kikuchi T, Furuya K, Hayashi M, Fuji Y (1996) Characterization of Antarctic aerosol particles using laser microprobe mass spectrometry. Environ Sci Technol 30:385–391

Harrington HJ (1958) Nomenclature of rock units in the Ross Sea region. Nature 182:290

Harris CM (1991) Environmental effects of human activities on King George Island, South Shetland Islands, Antarctica. Polar Record 27:313–324

Harris JE, Fabris GJ (1979) Concentrations of suspended matter and particulate cadmium, copper, lead and zinc in the Indian sector of the Antarctic Ocean. Mar Chem 8:163–179

Harris AJL, Wright R, Flynn LP (1999) Remote monitoring of Mount Erebus volcano, using polar orbiters: progress and prospects. Int J Remote Sensing 20:3051–3071

Hart TJ (1934) On the phytoplankton of the Southwest Atlantic and the Bellingshausen Sea 1929–1931. Discovery Rep 8:1–268

Harvey MJ, Fisher GW, Lechner IS, Isaac P, Flower NE, Dick AL (1991) Summertime aerosol measurements in the Ross Sea region of Antarctica. Atmos Environ 25:569–580

Hawes I (1963) Turbulent mixing and its consequences on phytoplankton development in two ice covered lakes. Bull Antarct Surv 60:69–82

Hawes I (2001) Aquatic habitats. In: Waterhouse E (ed) Ross Sea Region 2001. A state of the environment report for the Ross Sea region of Antarctica. New Zealand Antarctic Institute, Christchurch, pp 53–68

Hays JD (1978) A review of the late Quaternary climatic history of Antarctic seas. In: Van Zinderen Bakker EM (ed) Antarctis glacial history and world palaeoenvironment. Balkema, Rotterdam, pp 57–71

Hedgecock IM, Pirrone N (2001) Mercury photochemistry in the marine boundary layer. Modeling studies for in-situ production of reactive gas phase mercury. Atmos Environ 35:3055–3062

Hedgecock IM, Pirrone N, Sprovieri F, Pesenti E (2003) Reactive gaseous mercury in the marine boundary layer: modelling and experimental evidence of its formation in the Mediterranean region. Atmos Environ 37:S41–S49

Hedgpeth JW (1970) Marine biogeography of the Antarctic regions. In: Holdgate MW (ed) Antarctic ecology, vol 1. Academic Press, London, pp 97–104

Hefu Y, Kirst GO (1997) Effect of UV-radiation on DMSP content and DMS formation of *Phaeocystis antarctica*. Polar Biol 18:402–409

Heilmayer O, Brey T, Chiantore M, Cattaneo-Vietti R, Arntz WE (2003) Age and productivity of the Antarctic scallop, *Adamussium colbecki*, in Terra Nova Bay (Ross Sea, Antarctica). J Exp Mar Biol Ecol 288:239–256

Heintzenberg J, Charlson RJ, Clarke AD, Liousse C, Ramaswamy V, Shine KP, Wendisch M, Helas G (1997) Measurements and modelling of aerosol single-scattering albedo: progress, problems and prospects. Contrib Atmos Phys 70:249–263

Helbling EW, Villafañe V, Holm-Hansen O (1994) Effects of ultraviolet radiation on Antarctic marine phytoplankton photosynthesis with particular attention to the influence of mixing. In: Weiler CS, Penhale PA (eds) Ultraviolet radiation in Antarctica: measurements and biological effects. American Geophysical Union, Washington, DC, vol 62, pp 207–227

Held IM, Soden BJ (2000) Water vapour feedback and global warming. Annu Rev Energy Environ 25:441–475

Hempel G (1985) Antarctic marine food webs. In: Siegfried WR, Condy PR, Laws RM (eds) Antarctic nutrient cycles and food webs. Springer, Berlin Heidelberg New York, pp 266–270

Hennig HF-KO, Eagle GA, McQuaid CD, Rickett LH (1985) Metal concentrations in Antarctic zooplankton species. In: Siegfried WR, Condy PR, Laws RM (eds) Antarctic nutrient cycles and food webs. Springer, Berlin Heidelberg New York, pp 656–661

Hermann JR, Newman PA, Larko D (1995) Meteor-3/TOMS observations of the 1994 ozone hole. Geophys Res Lett 22:3227–3229

Herr R (1996) The changing roles of non-governmental organisations in the Antarctic Treaty System. In: Stokke OS, Vidas D (eds) Governing the Antarctic the effectiveness and legitimacy of the Antarctic Treaty System. Cambridge University Press, Cambridge, pp 91–110

Heumann KG (1993) Determination of inorganic and organic traces in the clean room compartment of Antarctica. Anal Chim Acta 283:230–245

Heumann KG (2001) Biomethylation in the Southern Ocean and its contribution to the geochemical cycle of trace elements in Antarctica. In: Caroli S, Cescon P, Walton DWH (eds) Environmental contamination in Antarctica. A challenge to analytical chemistry. Elsevier, Amsterdam, pp 181–217

Hewitt RP, Low EHL (2000) The fishery on Antarctic krill: defining an ecosystem approach to management. Rev Fish Sci 8:235–298

Heywood RB (1984) Antarctic inland waters. In: Laws RM (ed) Antarctic ecology, vol 1. Academic Press, London, pp 275–344

Hieke Merlin O, Longo Salvator G, Menegazzo Vitturi L, Pistolato M, Ramazzo G (1989) Preliminary results on trace element geochemistry of sediments from the Ross Sea, Antarctica. Boll Oceanol Teor Appl 7:97–108

Hindell MA, Brothers N, Gales R (1999) Mercury and cadmium concentrations in the tissues of three species of southern albatrosses. Polar Biol 22:102–108

Hinkley TK, Le Cloarec M-F, Lambert G (1994) Fractionation of families of major, minor and trace metals across the melt-vapor interface in volcanic exhalations. Geochim Cosmochim Acta 58:3255–3263

Hinkley TK, Pertsiger F, Zavjalova L (1997) The modern atmospheric background dust load: recognition in central Asian snowpack, and compositional constraints. Geophys Res Lett 170:315–325

Hinkley TK, Lamothe PJ, Wilson SA, Finnegan DL, Gerlach TM (1999) Metal emissions from Kilauea, and a suggested revision of the estimated worldwide metal output by quiescent degassing volcanoes. Earth Planet Sci Lett 170:315–325

Hinkley TK, Lamothe PJ, Meeker GP, Jiang X, Miller ME, Fulton R (2002) Trace elements deposited with dusts in Southwestern US – enrichment, fluxes, comparison with records from elsewhere. In: Lee JA, Zobeck TM (eds) Proc ICAR5/GCTE-SEN Joint Conf, Texas Technical University, Lubbock, Texas, pp 34–39

Hirst AC (1999) The Southern Ocean response to global warming in the CSIRO coupled ocean atmosphere model: special issue on global change. Environ Model Softw 14:227–242

Hiscock MR, Marra J, Smith WO, Goericke R, Measures C, Vink S, Olson RJ, Sosik HM, Barber RT (2003) Primary productivity and its regulation in the Pacific Sector of the Southern Ocean. Deep-Sea Res II 50:533–558

Hobbie JE (1984) Polar limnology. In: Taub FB (ed) Lakes and reservoirs. Elsevier, New York, pp 63–105

Hobbs PV (ed) (1993) Aerosol-cloud-climate interactions. Academic Press, San Diego, CA

Hobbs PV, Deepak A (eds) (1981) Clouds, their formation, optical properties, and effects. Academic Press, New York

Hobbs JE, Lindesay JA, Bridgemann HA (eds) (1998) Climates of the southern continents. Wiley, New York

Hoff JT, Wania F, Mackay D, Gillham R (1995) Sorption of nonpolar organic vapors by ice and snow. Environ Sci Technol 29:1982–1989

Hofmann DJ, Pyle JA (1999) Predicting future ozone changes and detection of recovery. WMO, Geneva, Global Ozone Research and Monitoring Project, vol 44, pp 1–57

Holben BN, Eck TF, Slutsker I, Tanré D, Buis JP, Stezer A, Vermote E, Reagan JA, Kaufman UJ, Nakajima T, Lavenu F, Jankowiak I, Smirnov A (1998) AERONET – a federated instrument network and data archive for aerosol characterization. Remote Sensing Environ 66:1–16

Holm-Hansen O (1985) Nutrient cycles in Antarctic marine ecosystems. In: Siegfried WR, Condy PR, Laws RM (eds) Antarctic nutrient cycles and food webs. Springer, Berlin Heidelberg New York, pp 6–10

Holm-Hansen O, Huntley M (1984) Feeding requirements of krill in relation to food sources. J Crustacean Biol 4:156–173

Honda K, Yamamoto Y, Hidaka H, Tatsukawa R (1986) Heavy metal accumulations in Adélie penguin, *Pygoscelis adeliae*, and their variations with the reproductive process. Mem Natl Inst Polar Res 40:443–453

Honda K, Yamamoto Y, Kato H, Tatsukawa R (1987) Heavy metal accumulations and their recent changes in southern minke whales *Balaenoptera acutorostrata*. Arch Environ Contam Toxicol 16:209–216

Hong S, Candelone J-P, Boutron CF (1996a) Deposition of atmospheric heavy metals to the Greenland ice sheet from the 1783–1784 volcanic eruption of Laki, Iceland. Earth Planet Sci Lett 144:605–610

Hong S, Candelone J-P, Patterson CC, Boutron CF (1996b) History of copper smelting pollution during Roman and medieval times recorded in Greenland ice. Science 272:246–248

Hong S, Candelone J-P, Soutif M, Boutron CF (1996c) A reconstruction of changes in copper production and copper emissions to the atmosphere during the past 7000 years. Sci Total Environ 188:183–193

Hong S, Boutron CF, Edwards R, Morgan VI (1998) Heavy metals in Antarctic ice from Law Dome: initial results. Environ Res A 78:94–103

Hong S, Kang CY, Kang J (1999) Lichen biomonitoring for the detection of local heavy metal pollution around King Sejong Station, King George Island, Antarctica. Korean J Polar Res 10:17–24

Honrath R, Guo S, Peterson MC, Dziobak MP, Dibb JE (2000) Photochemical production of gas phase NO_x from ice crystal NO_3^-. J Geophys Res 105:24183–24190

Hooker JD (1847) Daitomaceae. The botany of the Antarctic voyage of HM discovery ships "Erebus" and "Terror", years 1839–1843, vol 106. Reeve, London

Horner RA (1985) Sea ice biota. CRC Press, Boca Raton, Florida

Houdier S, Perrier S, Dominé F, Cabanes A, Lagagneux L, Grannas AM, Guimbaud C, Shepson PB, Boudries H, Bottenheim JW (2002) Acetaldehyde and acetone in Arctic snowpack during the ALERT 2000 campaign. Snowpack composition, incorporation processes and atmospheric impact. Atmos Environ 36:2609–2618

Howard PS (1991) Handbook of environmental fate and exposure data for organic chemicals. Lewis, Chelsea, MI

Howard-Williams C, Hawes I, Schwarz A-M (1997) Sources and sinks of nutrients in a polar desert stream, the Onyx River, Antarctica. In: Lyons WB, Howard-Williams C, Hawes I (eds) Ecosystem processes in Antarctic ice-free landscapes. Balkema, Rotterdam, pp 155–170

Hsu NC, Herman JR, Torres O, Holben BN, Tanré D, Eck TF, Smirnov A, Chatenet B, Lavenu F (1999) Comparisons of the TOMS aerosol index with Sun-photometer aerosol optical thickness: results and applications. J Geophys Res 104:6269–6279

Huiskes AHL, Gieskes WWC, Rozema J, Schorno RML, van der Vies SM, Wolff WJ (eds) (2003) Antarctic biology in a global context. Backhuys, Leiden

Hung T-C, Meng P-J, Wu S-J (1993) Species of copper and zinc in sediments collected from the Antarctic Ocean and the Taiwan Erhjin Chi coastal area. Environ Pollut 80:223–230

Hureau J-C (1994) The significance of fish in the marine Antarctic ecosystems. Polar Biol 14:307–313

Husar RB, Prospero JM, Stowe LL (1997) Characterization of tropospheric aerosols over the oceans with the NOAAA advanced very high resolution radiometer optical thickness operational product. J Geophys Res 102:16889–16909

Hutchins DA, Bruland KW (1998) Iron-limited growth and Si:N ratios in a coastal upwelling regime. Nature 393:561–564

Hutchins DA, Witter AE, Butler A, Luther GW (1999) Competition among marine phytoplankton for different chelated iron species. Nature 400:858–861

Hutter K (1983) Theoretical glaciology, material science of ice and the mechanics of glaciers and ice sheets. Reidel, Tokyo

Hutterli MA, McConnell JR, Bales RC, Stewart RW (2003) Sensitivity of hydrogen peroxide (H_2O_2) and formaldehyde (HCHO) preservation in snow to changing environmental conditions: implications for ice core records. J Geophys Res 108:4023–4029

Hutzinger O, Safe S, Zitko V (1974) The chemistry of PCBs. CRC Press, Cleveland

Huybrecths P, de Wolde J (1999) The dynamic response of the Greenland and Antarctic ice sheets to multiple-century climatic warming. J Climate 12:2169–2188

Iacozza J, Barber DG (1999) Modelling the distribution of snow on sea ice using variograms. Atmos Oceans 37:21–51

Ikegawa M, Kimura M, Honda K, Makita K, Fujii Y, Itokawa Y (1997) Springtime peaks of trace metals in Antarctic snow. Environ Health Perspec 105:654–659

Ikegawa M, Kimura M, Honda K, Akabane I, Makita K, Motoyama H, Fuji Y, Itokawa Y (1999) Geographical variations of major trace elements in East Antarctica. Atmos Environ 33:1457–1467

Imura S, Bando T, Saito S, Seto K, Kanda H (1999) Benthic moss pillars in Antarctic lakes. Polar Biol 22:137–140

Inomata ONK, Montone RC, Lara WH, Wever RR, Toledo HHB (1996) Tissue distribution of organochlorine residues – PCBs and pesticides – in Antarctic penguins. Antarct Sci 8:253–255

IPCC (1996) The science of climate change. Cambridge University Press, Cambridge

IPCC (2000) Emissions scenarios. A special report of working group III of the IPCC. Cambridge University Press, Cambridge

IPCC (2001) Climate change 2001. The scientific basis. Cambridge University Press, Cambridge

Iriondo M (2000) Patagonian dust in Antarctica. Quat Int 68–71:83–86

Isaksson E, Karlén W (1994) High resolution climatic information from short firn cores, western Dronning Maud Land. Climatic Change 26:421–434

Iwata H, Tanabe S, Sakai N, Nishimura A, Tatsukawa R (1994) Geographical distribution of persistent organochlorines in air, water and sediments from Asia and Oceania, and their implications for global redistribution from lower latitudes. Environ Pollut 85:15–33

Jacka TH, Budd WF (1998) Detection of temperature and sea ice extent changes in the Antarctic and the Southern Ocean, 1949–96. Ann Glaciol 27:553–559

Jacobi H-W, Frey MM, Hutterli MA, Bales RC, Schrems O, Cullen NJ, Steffen K, Koehler C (2002) Measurements of hydrogen peroxide and formaldehyde exchange between the atmosphere and surface snow at Summit, Greenland. Atmos Environ 36:2619–2628

Jacobs SS, Weiss RF (eds) (1998) Ocean, ice, and atmosphere. Interactions at the Antarctic continental margin. American Geophysical Union, Washington, DC, Antarct Res Ser 75

Jacobs SS, McAyeal DR, Ardal JD (1986) The recent advance of the Ross Ice Shelf, Antarctica. J Glaciol 32:464–474

Jacobs SS, Helmer HH, Doake CSM, Jenkins A, Frolich RM (1992) Melting of ice shelves and the mass balance of Antarctica. J Glaciol 38:375–387

Jacques G, Fukuchi M (1994) Phytoplankton of the Indian Antarctic Ocean. In: El-Sayed SZ (ed) Southern Ocean ecology. The BIOMASS perspective. Cambridge University Press, Cambridge, pp 63–78

Jakeman AJ, Beck MB, McAleer MJ (eds) (1993) Modelling change in environmental systems. Wiley, New York

James IN (1989) The Antarctic drainage flow: implications for hemispheric flow on the southern hemisphere. Antarct Sci 1:279–290

Jarre-Teichmann A, Brey T, Bathmann UV, Dahm C, Dieckmann GS, Gorny M, Klages M, Pagés F, Plötz J, Schnack-Schiel SB, Stille M, Arntz WE (1997) Trophic flows in the benthic shelf community of eastern Weddell Sea, Antarctica. In: Battaglia B, Valencia J, Walton DWH (eds) Antarctic communities: species, structure and survival. Cambridge University Press, Cambridge, pp 118–134

Jeffries MO (ed) (1998) Antarctic sea ice: physical processes interactions and variability. American Geophysical Union, Washington, DC, Antarct Res Ser 74

Jenkins A, Vaughan DG, Jacobs SS, Hellmer HH, Keys JR (1997) Glaciological and oceanographic evidence of high melt rates beneath Pine Island Glacier, West Antarctica. J Glaciol 43:114–121

Jennings JC, Gordon LI, Nelson DM (1984) Nutrient depletion indicates high primary productivity in the Weddell Sea. Nature 309:51–54

JGOFS (1996) Protocols for the Joint Global Ocean Flux Study (JGOFS) core measurements. Intergovernmental Oceanographic Commission, Bergen, Norway, Rep no 1

Jia G, Triulzi C, Nonnis-Marzano F, Belli M, Sansone U, Maghi M (1999) Plutonium, [241]Am, ^{90}Sr, and ^{137}Cs concentrations in some Antarctic matrices. Biol Trace Element Res 71–72:349–357

Johannessen OM, Muench RD, Overland JE (eds) (1994) The polar oceans and their role in shaping the global environment. Nansen Centennial Symp. American Geophysical Union, Washington, DC

Johansen P, Hansen MM, Asmud G, Nielsen PB (1991) Marine organisms as indicators of heavy metal pollution: experience from 16 years of monitoring at a lead-zinc mine in Greenland. Chem Ecol 5:35–55

Johnson DW, Osborne S, Wood R, Suhre K, Johnson R, Businger S, Quinn PK, Weidensohler A, Durkee PA, Russell LM, Andreae MO, O'Dowd C, Noone KJ, Bandy B, Rudolph J, Rapsomanikis S (2000) An overview of the Lagrangian experiments undertaken during the North Atlantic regional Aerosol Characterization Experiment (ACE-2). Tellus 52B:290–320

Joiris C, Overloop W (1991) PCBs and organochlorine pesticides in phytoplankton and zooplankton in the Indian sector of the Southern ocean. Antarct Sci 3:371–377

Jones PD (1990) Antarctic temperature over the present century. A study of the early expedition record. J Climate 3:1193–1203

Jones PD (1994) Hemispheric surface air temperature variations. A reanalysis and an update to 1993. J Climate 7:1794–1802

Jones PD (1995) Recent variations in mean temperature range in the Antarctic. Geophys Res Lett 22:1345–1348

Jones PD, Allan RJ (1998) Climatic change and long-term climate variability. In: Karoly DJ, Vincent DG (eds) Meteorology of the Southern Hemisphere. American Meteorological Society, Boston, Meteorological Monographs, vol 27, no 49, pp 337–363

Jones KC, de Voogt P (1999) Persistent organic pollutants (POPs): state of the science. Environ Pollut 100:209–221

Jones AE, Shanklin JD (1995) Continued decline of total ozone over Halley, Antarctica, since 1985. Nature 376:409–411

Jones PD, Raper SC, Wigley TML (1986) Southern Hemisphere surface air temperature variations 1851–1984. J Climatol Appl Meteorol 25:1213–1230

Jones PD, Marsh R, Wigley TML, Peel DA (1993) Decadal timescale links between Antarctic Peninsula ice-core oxygen-18, deuterium and temperature. Holocene 3:14–26

Jones HG, Pomeroy JW, Walker DA, Hoham RW (eds) (2001) Snow ecology. Cambridge University Press, Cambridge

Jouventin P, Stahl JC, Weimerskirch H, Mougin JL (1984) The seabirds of the French Sub-Antarctic islands and Adélie Land, their status and conservation. In: Croxall JP, Evans PGH, Schreiber RW (eds) Status and conservation of the world's seabirds. International Council for Bird Preservation, Cambridge, Tech Publ no 2, pp 609–625

Jouzel J, Merlivat L, Pourchet M, Lorius C (1979) A continuous record of artificial tritium fallout at the South Pole (1954–1978). Earth Planet Sci Lett 45:188–200

Jouzel J, Lorius C, Petit C, Genthon C, Barkov NI, Kotlyakov VM, Petrov VM (1987) Vostok ice core: a continuous isotope temperature record over the last climatic cycle (160,000 years). Nature 329:403–408

Joyce TM, Patterson SL (1977) Cyclonic ring formation at the Antarctic Polar Front in the Drake Passage. Nature 256:131–133

Kahle J, Zauke G-P (2002) Bioaccumulation of trace metals in the calanoid copepod *Metridia gerlachei* from the Weddell Sea (Antarctica). Sci Total Environ 295:1–16

Kahle J, Zauke G-P (2003a) Trace metals in Antarctic copepods from the Weddell Sea (Antarctica). Chemosphere 51:409–417

Kahle J, Zauke G-P (2003b) Bioaccumulation of trace metals in the Antarctic amphipod *Orchomene plebs*: evaluation of toxicokinetic models. Mar Environ Res 55:359–384

Kaimal JC, Finnigan JJ (1994) Atmospheric boundary layer flows: their structure and measurement. Cambridge University Press, Cambridge

Kalina MF, Puxbaum H (1994) A study of the influence of riming of ice crystals on snow chemistry during different seasons in precipitating continental clouds. Atmos Environ 28:3311–3328

Kallenborn R, Oehme M, Wynn-Williams DD, Schiabach M, Harris J (1998) Ambient air levels and atmospheric long-range transport of persistent organochlorines to Signy Island, Antarctica. Sci Total Environ 220:167–180

Kanakidou M, Tsigaridis K, Dentener FJ, Crutzen PJ (2000) Human activity enhances the formation of organic aerosols by biogenic hydrocarbon oxidation. J Geophys Res 105:9243–9254

Kapitsa AP, Ridley JK, Robin G de Q, Siegert MJ, Zotuk IA (1996) A large deep fresh-water lake beneath the ice of central east Antarctica. Nature 381:684–686

Karentz D (1991) Cell survival characteristics and molecular responses of Antarctic phytoplankton to ultraviolet-B radiation exposure. J Phycol 27:326–341

Karentz D (1994) Ultraviolet tolerance mechanisms in Antarctic marine organisms. In: Weiler CS, Penhale PA (eds) Ultraviolet radiation in Antarctica: measurements and biological effects. American Geophysical Union, Washington, DC, vol 62, pp 93–110

Karl DM, Bird DF, Bjorkman K, Houlihan T, Shackelford R, Tupas L (1999) Microorganisms in the accreted ice of Lake Vostok, Antarctica. Science 286:2144–2147

Karolewski MA, Lukowski AB, Halba R (1987) Residues of chlorinated hydrocarbons in the adipose tissue of the Antarctic pinnipeds. Polish Polar Res 8:189–197

Karoly DJ, Vincent DG (eds) (1998) Meteorology of the Southern Hemisphere. American Meteorological Society, Boston, Meteorological Monographs, vol 27

Kaufman YJ, Tanré D, Boucher O (2002) A satellite view of aerosols in the climate system. Nature 419:215–223

Kawano M, Inoue T, Hidaka H, Tatsukawa R (1984) Chlordane compounds residues in Weddell seals (*Leptonychotes weddelli*) from the Antarctic. Chemosphere 13:95–100

Kawano M, Tanabe S, Inoue T, Tatsukawa R (1985) Chlordane compound found in the marine atmosphere from the Southern Hemisphere. Trans Tokyo Univ Fish 6:59–66

Kear AJ (1992) The diet of Antarctic squid: comparison of conventional and serological gut contents analyses. J Exp Mar Biol Ecol 156:161–178

Kejna M, Láska K, Caputa Z (1998) Recession of Ecology Glacier (King George Island) in the period 1961–1996. Polish Polar Studies 25th Int Symp, Warszawa, pp 121–128

Kellogg DE, Kellogg TB (1984) Diatoms from the McMurdo Ice Shelf, Antarctica. Antarct J US 19:76–77

Kelly BC, Gobas FAPC (2001) Bioaccumulation of persistent organic pollutants in lichen-caribou-wolf food chains of Canada's central and western Arctic. Environ Sci Technol 35:325–334

Kemler K, Jones PD, Giesy JP (2000) Risk assessment of 2,3,7,8-tetrachlorodibenzo-*p*-dioxin equivalents in tissue samples from three species in the Denver metropolitan area. Human Ecol Risk Assess 6:1087–1099

Kemp DD (2002) Global warming: an overview. In: Goudie S, Cuff DJ (eds) Encyclopedia of global change and human society, vol 1. Oxford University Press, Oxford, pp 540–555

Kennedy AD (1993) Water as limiting factor in the Antarctic terrestrial environment: a biogeographical synthesis. Arctic Alpine Res 25:308–315

Kennedy AD (1995) Antarctic terrestrial ecosystems responses to global environmental change. Annu Rev Ecol Syst 26:683–704

Kennicutt MC, Champ MA (eds) (1992) Environmental awareness in Antarctica: history, problems, and future solutions. Mar Pollut Bull 25:219–233

Kennicutt MC, McDonald SJ (1996) Marine disturbance – contaminants. In: Ross RM, Hofmann EE, Quetin LB (eds) Foundations for ecological research west of the Antarctic Peninsula. American Geophysical Union, Washington, DC, vol 70, pp 401–415

Kennicutt MC, Sweet ST (1992) Hydrocarbon contamination on the Antarctic Peninsula. III. The *Bahia Paraiso* – two years after the spill. Mar Pollut Bull 25:303–306

Kennicutt MC, Fraser W, Culver ME et al. (1990) Oil spillage in Antarctica. Initial report of the National Science Foundation-sponsored quick response team on the grounding of the *Bahia Paraiso*. Environ Sci Technol 24:620–624

Kennicutt MC, Sweet ST, Fraser WR, Stockton WL, Cliver M (1991) Grounding of the *Bahia Paraiso* at Arthut Harbor, Antarctica. I. Distribution and fate of oil spill related hydrocarbons. Environ Sci Technol 25:509–518

Kennicutt MC II, McDonald TJ, Denoux GJ, McDonald SJ (1992) Hydrocarbon contamination on the Antarctic Peninsula. Arthur Harbour – subtidal sediments. Mar Pollut Bull 24:499–506

Kennicutt MC II, McDonald SJ, Sericano JL, Boothe P, Oliver J, Safe S, Presley BJ, Liu H, Wolfe D, Wade TL, Crockett A, Bockus D (1995) Human contamination of the marine environment: Arthur Harbour and McMurdo Sound, Antarctica. Environ Sci Technol 29:1279–1287

Kennicutt MC, Wolff GA, Klein A, Montagna P (2003) Spatial and temporal scales of human disturbance – McMurdo Station, Antarctica – preliminary findings. In: Huiskes AHL, Gieskes WWC, Rozema J, Schorno RML, van der Vies SM, Wolff WJ (eds) Antarctic biology in a global context. Backhuys, Leiden, pp 271–277

Kerminen V-L, Teinilä K, Hillamo R (2000) Chemistry of sea-salt particles in the summer Antarctic atmosphere. Atmos Environ 34:2817–2825

Kettle AJ, Andreae MO (2000) Flux of dimethylsulfide from the oceans: a comparison of updated data sets and flux models. J Geophys Res 105:26793–26808

Keys JRH, Williams K (1981) Origin of crystalline, cold desert salts in the McMurdo region, Antarctica. Geochim Cosmochim Acta 45:2299–2309

Kiene RP (1999) Ocean biogeochemistry: sulphur in the mix. Nature 402:363–365

Kiest KA (1993) A relationship of diet to prey abundance and the foraging behavior of *Trematomus bernacchii*. Polar Biol 13:291–296

Kim EY, Saeki K, Tanabe S, Tanaka H, Tatsukawa R (1996) Specific accumulation of mercury and selenium in seabirds. Environ Pollut 94:261–265

King JC (1994) Recent climate variability in the vicinity of the Antarctic Peninsula. Int J Climatol 14:357–369

King JC, Harangozo SA (1998) Climate change in the western Antarctic Peninsula since 1945: observations and possible causes. Ann Glaciol 27:571–576

King CK, Riddle MJ (2001) Effects of metal contaminants on the development of the common Antarctic sea urchin *Sterechinus neumayeri* and comparisons of sensitivity with tropical and temperate echinoids. Mar Ecol Prog Ser 215:143–154

King JC, Turner J (1997) Antarctic meteorology and climatology. Cambridge University Press, Cambridge

King EA, Wagstaff J (1982) Extraterrestrial microparticles from Antarctic ice cores and the search for cometary dust. Antarct J US 17:61–62

Klages N (1989) Food and feeding ecology of emperor penguins in the eastern Weddell Sea. Polar Biol 9:385–390

Klusman RW, Jaacks JA (1987) Environmental influences upon mercury, radon and helium concentrations in soil gases at a site near Denver, Colorado. J Geochem Explor 27:259–280

Knox GA (1994) The biology of the Southern Ocean. Cambridge University Press, Cambridge

Knutson TR, Manabe S (1998) Model assessment of decadal variability and trends in the tropical Pacific Ocean. J Climate 11:2273–2296

Knutson TR, Manabe S, Gu D (1997) Simulated ENSO in a global coupled ocean-atmosphere model: multidecadal amplitude modulation and CO_2 sensitivity. J Climate 10:138–161

Kock K-H (1989) Reproduction in fish around Elephant Island. Arch Fishereiwissen 39:171–210

Kock K-H (1992) Antarctic fish and fisheries. Cambridge University Press, Cambridge

Koeman JH, Peters WHM, Koudstaal-Hol CHM, Tjioe PS, de Goeij JJM (1973) Mercury-selenium correlations in marine mammals. Nature 245:385–386

Koide M, Goldberg ED, Herron MM, Langway CC (1979) Deposition history of artificial radionuclides in the Ross Ice Shelf, Antarctica. Earth Planet Sci Lett 44:205–223

Kol E, Flint EA (1968) Algae in green ice from the Balleny Islands. Antarctica. NZ J Bot 6:249–261

Koop T, Bertram AK, Molina LT, Molina MJ (1999) Phase transitions in aqueous NH_4HSO_4 solutions. J Phys Chem 103:9042–9048

Kulmala M, Wagner P (eds) (1996) Nucleation and atmospheric aerosols. Elsevier, New York

Kumar KS, Kannan K, Corsolini S, Evans T, Giesy JP, Nakanishi J, Matsunaga S (2002) Polychlorinated dibenzo-*p*-dioxins, dibenzofurans and polychlorinated biphenyls in polar bear, penguin and south polar skua. Environ Pollut 119:151–161

Kunito T, Watanabe I, Yasunaga G, Fujise Y, Tanabe S (2002) Using trace elements in skin to discriminate the populations of minke whales in southern hemisphere. Mar Environ Res 53:175–197

Kurtz DA (ed) (1990) Long range transport of pesticides. Lewis, Chelsea

Kushner DJ (ed) (1978) Microbial life in extreme environments. Academic Press, London

Kutzbach JE (1992) Modelling large climatic changes of the past. In: Trenberth KE (ed) Climate system modelling. Cambridge University Press, Cambridge, pp 669–688

Kyle PR, Meeker K, Finnegan D (1990) Emission rates of sulfur dioxide, trace gases and metals from Mount Erebus, Antarctica. Geophys Res Lett 17:2125–2128

Lambert G, Ardouin B, Sanak J (1990) Atmospheric transport of trace elements toward Antarctica. Tellus 42B:76–83

Langone L, Frignani M, Ravaioli M, Bianchi C (2000) Particle fluxes and biogeochemical processes in an area influenced by seasonal retreat of the ice margin (northwestern Ross Sea, Antarctica). J Mar Syst 27:221–234

Langway CC, Clausen HB, Hammer CU (1988) An inter-hemispheric volcanic time-marker in ice cores from Greenland and Antarctica. Ann Glaciol 10:102–108

Larsen RJ, Sanderson CG (1990) Annual report of the surface air sampling program. USDOE, Germantown, Rep EML-524

Larsson P, Järnmark C, Södergren A (1992) PCBs and chlorinated pesticides in the atmosphere and aquatic organisms of Ross Island, Antarctica. Mar Pollut Bull 25:281–287

Lauenstein GG, Robertson S, O'Connor TP (1990) Comparison of trace metal data in mussels and oysters from a mussel watch programme of the 1970s with those from a 1980s programme. Mar Pollut Bull 21:440–447

Law RJ, Allechin CR, Jones BR, Jepson PD, Baker JR, Spurrie CJH (1997) Metals and organochlorines in tissues of a Blainville's beaked whale (*Mesoplodon densirostris*) and a killer whale (*Orcinus orca*) stranded in the United Kingdom. Mar Pollut Bull 34:208–212

Laws RM (ed) (1984) Antarctic ecology, vol 1. Academic Press, London

Laws RM (1985) The ecology of the Southern Ocean. Am Sci 73:26–40

Lee JG, Ahner BA, Morel FMM (1996) Export of cadmium and phytochelatin by the marine diatom *Thalassiosira weissflogii*. Environ Sci Technol 30:1814–1821

Lee DS, Köhler I, Grobler E, Rohrer F, Sauen R, Gallardo-Klenner L, Olivier JJG, Dentener FJ, Bouwman AF (1997) Estimates of global NO_x emissions and their uncertainties. Atmos Environ 31:1735–1749

Lefauconnier B, Hagen JO, Pinglot JF, Pourchet M (1994) Mass-balance estimates on the glacier complex Kongsvegen and Sveabreen, Spitsbergen, Svalbard, using radioactive layers. J Glaciol 40:368–376

Legendre L, Ackley SF, Dieckmann GS, Gullicksen B, Horner R, Hoshiai T, Melnikov I, Reeburgh WS, Spindler M, Sullivan CW (1992) Ecology of sea ice biota. 2. Global significance. Polar Biol 12:429–444

Legrand M, Mayewski P (1997) Glaciochemistry of polar ice cores. A review. Rev Geophys 35:219–243

LeMasurier WE, Rex DC (1982) Eruptive potential of volcanoes in Marie Byrd Land. Antarct J US 17:34–36

LeMasurier WE, Rex DC (1991) The Marie Byrd Land volcanic province and its relation to the Cenozoic West Antarctica rift system. In: Tingey RJ (ed) The geology of Antarctica. Clarendon Press, London, pp 249–284

LeMasurier WE, Thomson JW (eds) (1990) Volcanoes of the Antarctic Plate and Southern Oceans. American Geophysical Union, Washington, DC, Antarct Res Ser 48

LeMasurier WE, Harwood DM, Rex DC (1994) Geology of the Mount Murphy Volcano: an 8 Ma history of interaction between a rift volcano and the West Antarctic ice sheet. Geol Soc Am Bull 106:265–280

Lenihan HS (1992) Benthic marine pollution around McMurdo Station, Antarctica: a summary of findings. Mar Pollut Bull 25:9–12

Lenihan HS, Oliver JS (1995) Anthropogenic and natural disturbances to marine benthic communities in Antarctica. Ecol Appl 5:311–326

Lenihan HS, Oliver JS, Oakden JM, Stephenson MD (1990) Intense and localized benthic marine pollution around McMurdo Station, Antarctica. Mar Pollut Bull 21:422–430

Lenihan HS, Kiest KA, Conlan KE, Slattery PN, Konar BH, Oliver JS (1995) Patterns of survival and behavior in Antarctic benthic invertebrates exposed to contaminated sediments: field and laboratory bioassay experiments. J Exp Mar Biol Ecol 192:233–235

Leonzio C, Massi A (1989) Metal biomonitoring in bird eggs: a critical experiment. Bull Environ Contam Toxicol 43:402–406

Levine JS (ed) (1991) Global biomass burning: atmospheric, climatic and biospheric implications. MIT Press, Cambridge, MA

Levine JS (ed) (1996) Biomass burning and global change, vols 1 and 2. MIT Press, Cambridge, MA

Levy H (1990) Regional and global transport and distribution of trace species released at the earth's surface. In: Kurtz DA (ed) Long range transport of pesticides. Lewis, Chelsea, MI, pp 83–94

Li Y-F (1999) Global technical hexachlorocyclohexane usage and its contamination consequences in the environment from 1948 to 1997. Sci Total Environ 232:121–158

Li Y-F, Macdonald RW, Jantunen LMM, Harner T, Bidleman TF, Strachan WMJ (2002) The transport of β-hexachlorocyclohexane to the western Arctic Ocean: a contrast to α-HCH. Sci Total Environ 291:229–246

Libes SM (1992) An introduction to marine biogeochemistry. Wiley, New York

Ligowski R (2000) Benthic feeding by krill, *Euphausia superba* Dana, in coastal waters off West Antarctica and in Admiralty Bay, South Shetland Islands. Polar Biol 23:619–625

Lindberg SE, Brooks S, Lin C-J, Scott KJ, Landis MS, Stevens RK, Goodsite M, Richter A (2002) Dynamic oxidation of gaseous mercury in the Arctic troposphere at polar sunrise. Environ Sci Technol 36:1245–1256

Lindsay DC (1977) Lichens of cold deserts. In: Seaward MRD (ed) Lichen ecology. Academic Press, London, pp 183–209

Ling HU (1996) Snow algae of the Windmill Islands region, Antarctica. Hydrobiologia 336:99–106

Ling HU, Seppelt RD (1998) Snow algae of the Windmill Islands, continental Antarctica. 3. *Chloromonas polyptera* (Volvocales, Chlorophyta). Polar Biol 20:320–324

Linskens HF, Bargagli R, Cresti M, Bargagli R (1993) Entrapment of long-distance transported pollen grains by various moss species in coastal Victoria Land, Antarctica. Polar Biol 13:81–87

Lizotte MP, Arrigo KR (eds) (1998) Antarctic sea ice: biological processes, interactions and variability. American Geophysical Union, Washington, DC, Antarct Res Ser 73

Llano GA (1962) The terrestrial life of the Antarctic. Sci Am 207:213–230

Llano GA (ed) (1977) Adaptations within Antarctic ecosystems. Smithsonian Institution, Washington, DC

Lock JW, Thompson DR, Furness RW, Bartle JA (1992) Metal concentrations in seabirds of the New Zealand region. Environ Pollut 75:289–300

Loeb NG (2002) Albedo. In: Goudie AS, Cuff DJ (eds) Encyclopedia of global change. Environmental change and human society, vol 1. Oxford University Press, Oxford, pp 29–31

Loewe F (1972) The land of storms. Weather 27:110–121

Lohan MC, Statham PJ, Peck L (2001) Trace metals in the Antarctic soft-shelled clam *Laternula elliptica*: implications for metal pollution from Antarctic research stations. Polar Biol 24:808–817

Longton RE (1988) The biology of polar bryophytes. Cambridge University Press, Cambridge

Loring DH, Rantala RTT (1988) An intercalibration exercise for trace metals in marine sediments. Mar Chem 24:13–28

Lorius C, Jouzel J, Ritz C, Merlivat l, Barkov NI, Korotkevich YS, Kotlyakov (1985) A 150,000-year climatic record from Antarctic ice. Nature 316:591–596

Löscher BM, de Baar HJW, de Jong JTM, Veth C, Dehairs F (1997) The distribution of Fe in the Atlantic Circumpolar Current. Deep-Sea Res II 44:143–187

Löscher BM, de Jong JTM, de Baar HJW (1998) The distribution and preferential biological uptake of cadmium at 6° W in the Southern Ocean. Mar Chem 62:259–286

Louanchi F, Hoppema M (2000) Interannual variations of the Antarctic Ocean CO_2 uptake from 1986 to 1994. Mar Chem 72:103–114

Lough JM, Barnes DJ, Taylor RB (1996) The potential of massive corals for the study of high-resolution climate variation in the past millennium. In: Jones PD, Bradley RS, Jouzel J (eds) Climatic variations and forcing mechanisms of the last 2000 years. Springer, Berlin Heidelberg New York, pp 355–369

Loutre MF, Berger A (2000) No glacial-interglacial cycle in the ice volume simulated under a constant astronomical forcing and a variable CO_2. J Geophys Res 27:783–786

Lu JY, Schroeder WH, Barrie LA, Steffen A, Welch HE, Martin K, Lockhart L, Hunt RV, Boila G, Richter A (2001) Magnification of atmospheric mercury deposition to polar regions in springtime: the link to tropospheric depletion chemistry. Geophys Res Lett 28:3219–3222

Lucchitta BK, Rosanova CE (1998) Retreat of northern margins of George VI and Wilkins ice shelves, Antarctic Peninsula. Ann Glaciol 27:41–46

Luckas B, Vetter W, Fischer P, Heidemann G, Plötz J (1990) Characteristic chlorinated hydrocarbon patterns in the bubbler of seals from different marine regions. Chemosphere 21:13–19

Lugar RM, Harless RL, Dupuy AE, McDaniel DD (1996) Results of monitoring for polychlorinated dibenzo-p-dioxins and dibenzofurans in ambient air at McMurdo Station, Antarctica. Environ Sci Technol 30:555–561

Luke BG, Johnstone GW, Woehler EJ (1989) Organochlorine pesticides, PCBs and mercury in Antarctic and sub-Antarctic seabirds. Chemosphere 19:2007–2021

Lyons WB, Howard-Williams C, Hawes I (eds) (1997) Ecosystem processes in Antarctic ice-free landscapes. Balkema, Rotterdam

Lyons WB, Welch KA, Bonzogno J-C (1999) Mercury in aquatic systems in Antarctica. Geophys Res Lett 26:2235–2238

Macdonald RW, Barrie LA, Bidleman TF, Diamond ML, Gregor DJ, Semkin RG, Strachan WMJ, Li YF, Wania F, Alaee M, Alexeeva LB, Backus SM, Bailey R, Bewers JM, Gobeil C, Halsall CJ, Harner T, Hoff JT, Jantunen LMM, Lockhart WL, Mackay D, Muir DCG, Pudykiewicz J, Reimer KJ, Smith JN, Stern GA, Schroeder WH, Wagemann R, Yunker MB (2000) Contaminants in the Canadian Arctic: 5 years of progress in understanding sources, occurrence and pathways. Sci Total Environ 254:93–234

Macelroy RD (1974) Some comments on the evolution of extremofiles. Biosystems 6:74–75

Mackay D (1991) Multimedia environmental models. The fugacity approach. Lewis, Chelsea, MI

Mackay D, Shiu WY, Ma KC (1992) Illustrated handbook of physical-chemical properties and environmental fate of organic chemicals, vols 1–3. Lewis, Chelsea, MI

Mackensen A, Douglas RG (1989) Down-core distribution of live and dead deep-water borderland. Deep-Sea Res 36:879–900

Maenhaut W, Zoller WH, Duce RA, Hoffman GL (1979) Concentration and size distribution of particulate trace elements in the south polar atmosphere. J Geophys Res 84:2421–2431

Maier-Reimer E, Mikolajewicz U, Winguth A (1996) Future ocean uptake of CO_2. Interaction between ocean circulation and biology. Climate Dynam 12:711–721

Manabe S, Stouffer RJ (1994) Multiple century response of a coupled ocean-atmosphere model to an increase of atmospheric carbon dioxide. J Climate 7:5–23

Manheim BS (1992) The failure of the National Science Foundation to protect Antarctica. Mar Pollut Bull 25:253–254

Marchant DR, Denton DE (1996) Miocene and Pliocene paleoclimate of the Dry Valleys region, southern Victoria Land: a geomorphological approach. Mar Micropaleontol 27:269–271

Marchant S, Higgins PJ (1990) Handbook of Australian, New Zealand and Antarctic Birds, vol 1. Oxford University Press, Melbourne

Marchant DR, Denton DE, Bockheim JG, Wilson SC, Kerr AR (1994) Quaternary ice-level changes of upper Taylor Glacier, Antarctica: implications for paleoclimate and ice sheet dynamics. Boreas 23:29–43

Markert B, Friese K (eds) (2000) Trace elements. Their distribution and effects in the environment. Elsevier, Amsterdam

Marr JS (1962) The natural history and geography of the Antarctic krill *Euphausia superba*. Discovery Rep 32:33–464

Marschall H-P (1988) The overwintering strategy of Antarctic krill under the pack-ice of the Weddell Sea. Polar Biol 9:129–135

Marsh GP (1874) The Earth as modified by human action. Scribner, New York

Marshall DJ, Pugh PJA (1996) Origin of inland acari of continental Antarctica, with particular reference to Dronning Maud Land. Zool J Linn Soc 118:101–118

Martin JH, Fitzwater SE (1988) Iron deficiency limits phytoplankton growth in the north-east Pacific sub-Arctic. Nature 331:341–343

Martin JH, Flegal AR (1975) High copper concentrations in squid livers in association with elevated levels of silver, cadmium, and zinc. Mar Biol 30:51–55

Martin JH, Gordon RM (1988) Northeast Pacific iron distribution in relation to phytoplankton productivity. Deep-Sea Res I 35:177–196

Martin JH, Gordon RM, Fitzwater SE (1990) Iron in Antarctic waters. Nature 345:156–158

Martin JH, Coale KH, Johnson KS, Fitzwater SE, Gordon RM, Tanner SJ, Hunter CN, Elrod VA, Nowicki JL, Coley TL, Barber RT, Lindley S, Watson AJ, van Scoy K, Law CS (1994) Testing the iron hypothesis in ecosystems of the equatorial Pacific Ocean. Nature 371:123–129

Martins JV, Artaxo P, Liousse C, Reid JS, Hobbs PV, Kaufman YJ (1998) Effects of black carbon content, particle size, and mixing on light absorption by aerosols from biomass burning in Brazil. J Geophys Res Atmos 103:32041–32050

Masclet P, Hoyau V, Jaffrezo JL, Cachier H (2000) Polycyclic aromatic hydrocarbon deposition on the ice sheet of Greenland, part I. Superficial snow. Atmos Environ 34:3195–3207

Mason RP, Fitzgerald WF (1997) Sources, sinks and biogeochemical cycling of mercury in the ocean. In: Ebinghaus R (ed) Regional and global cycles of mercury: sources, fluxes and mass balances. NATO series. Kluwer, Amsterdam, pp 249–272

Mason RP, Sullivan KA (1999) The distribution and speciation of mercury in the South and equatorial Atlantic. Deep-Sea Res II 46:937–956

Mason RP, Fitzgerald WF, Morel FMM (1994) The biogeochemical cycling of elemental mercury: anthropogenic influences. Geochim Cosmochim Acta 58:3191–3198

Massom RA (1988) Biological significance of open water within sea ice covers of the polar regions. Endeavour 12:21–27

Masuda N, Nishimura M, Torii T (1982) Pathway and distribution of trace elements in Lake Vanda, Antarctica. Nature 298:154–156

Masuda N, Nakaya S, Burton HR, Torii T (1988) Trace element distributions in some saline lakes of the Vestfold Hills, Antarctica. Hydrobiologia 165:103–114

Mataloni G, Tesolin G (1997) A preliminary survey of cryobiontic algal communities from Cierva Point (Antarctic Peninsula). Antarct Sci 9:250–257

Matsumoto A, Hinkley TK (2001) Trace metal suites in Antarctic pre-industrial ice are consistent with emissions from quiescent degassing of volcanoes worldwide. Earth Planet Sci Lett 186:33–43

Mauri M, Orlando E, Nigro M, Regoli F (1990) Heavy metals in the Antarctic molluscs *Adamussium colbecki*. Mar Ecol Prog Ser 67:27–33

Mayewski PA, Goodwin ID (1997) International Trans Antarctic Scientific Expedition (ITASE) "200 years of past Antarctic climate and environmental change. Science and Implementation Plan". Int Workshop ITASE, 2–3 August 1996, Cambridge, Rep 97, pp 1–48

Maykut GA (1985) The ice environment. In: Horner RA (ed) Sea ice biota. CRC Press, Boca Raton, pp 21–82

Mazzera D, Hayes T, Lowenthal D, Zielinska B (1999) Quantification of polycyclic aromatic hydrocarbons in soil at McMurdo Station, Antarctica. Sci Total Environ 229:65–71

Mazzera DM, Lowenthal DH, Chow JC, Watson JG (2001a) Sources of PM_{10} and sulfate aerosol at McMurdo station, Antarctica. Chemosphere 45:347–356

Mazzera DM, Lowenthal DH, Chow JC, Watson JG, Grubisíc V (2001b) PM_{10} measurements at McMurdo Station, Antarctica. Atmos Environ 35:1891–1902

Mazzuccotelli A, Cosma B, Soggia F (1989) Trace metal distribution in Antarctic sediments (Terra Nova Bay – Ross Sea) by inductively coupled plasma atomic emission spectroscopy. Ann Chim (Rome) 79:617–628

McClintock JB (1989) Toxicity of shallow-water Antarctic echinoderms. Polar Biol 9:461–465

McClurg TP (1984) Trace metals and chlorinated hydrocarbons in Ross seals from Antarctica. Mar Pollut Bull 15:384–389

McCormick MP, Thomason LW, Trepte CR (1995) Atmospheric effects of the Mt. Pinatubo eruption. Nature 373:399–404

McCraw JD (1967) Soils of Taylor Dry Valleys, Victoria Land, Antarctica, with notes on soils from other localities in Victoria Land. NZ J Geol Geophys 10:498–539

McDonald SJ, Kennicutt MC, Brooks JM (1992) Evidence of Polycyclic Aromatic Hydrocarbon (PAH) exposure in fish from the Antarctic Peninsula. Mar Pollut Bull 25:313–317

McDonald SJ, Kennicutt MC, Sericano J, Wade TL, Liu H, Safe SH (1994) Correlation between bioassay-derived P4501A1 induction activity and chemical analysis of clam (*Laternula elliptica*) extracts from McMurdo Sound, Antarctica. Chemosphere 28:2237–2248

McKay DS, Gibson EK Jr, Thomas-Keptra KL, Vali H, Romanek CS, Clemett SJ, Chillier XDF, Maechling CR, Zare RN (1996) Search for past life on Mars: possible relict biogenic activity in Martian meteorite ALH84011. Science 273:924–930

McSween HY (1997) Evidence for life in a martian meteorite? GSA Today 7:1–7

McSween HY (1999) Meteorites and their parent planets. Cambridge University Press, Cambridge

Meador JP, Varanasi U, Robish PA, Chan S-L (1993) Toxic metals in pilot whales (*Globicephala melaena*) from strandings in 1986 and 1990 on Cape Cod, Massachusetts. Can J Fish Aquat Sci 50:2698–2706

Measures CI, Vink S (1999) Seasonal variations in the distribution of Fe and Al in the surface waters of the Arabian Sea. Deep-Sea Res II 46:1597–1622

Medlin LK, Lange M, Baumann MEM (1994) Genetic differentiation among three colony-forming species of *Phaeocystis*: further evidence for the phylogeny of the Prymnesiophyta. Phycologia 33:199–212

Meeker KA, Chuan RL, Kyle PR, Palais JM (1991) Emission of elemental gold particles from Mount Erebus, Ross Island, Antarctica. Geophys Res Lett 18:1405–1408

Meleshko VM, Kattsov BV, Sporyshev PV, Vavulin SV, Govorkova VA (2000) Feedback processes in climate system: cloud radiation and water vapour feedbacks interaction. Russian Meteorol Hydrol 2:22–45

Melick D, Broady PA, Rowan KS (1991) Morphological and physiological characteristics of a non-heterocystous strain of *Mastigocladus laminosus* Cohn from fumarolic soils on Mount Erebus, Antarctica. Polar Biol 11:81–89

Melnikov S, Carroll J, Gorshkov A, Vlasov S, Dahle S (2003) Snow and ice concentrations of selected persistent pollutants in the Ob-Yenisey River watershed. Sci Total Environ 306:27–37

Mercantini R, Marsella R, Cervellati MC (1989) Keratinophilic fungi isolated from Antarctic soil. Mycopathologia 106:47–52

Meyer WB (2002) History of global change. In: Goudie AS, Cuff DJ (eds) Encyclopedia of global change. Environmental change and human society. Cambridge University Press, Cambridge, pp 515–520

Millar DHM (1981) Radio-echo layering in polar ice sheets and past volcanic activity. Nature 292:441–443

Miller DGM, Hampton I (1989) Biology and ecology of the Antarctic krill (*Euphausia superba* Dana: a review. SCAR, Scott Polar Research Institute, Cambridge, BIOMASS, vol 9

Miller HC, Mills GN, Bembo DG, Macdonald JA, Evans CW (1999) Induction of cytochrome P4501A (CYP1A) in *Trematomus bernacchii* as an indicator of environmental pollution in Antarctica: assessment by quantitative RT-PCR. Aquat Toxicol 44:183–193

Minganti V, Capelli R, Fiorentino F, de Pellegrini R, Vacchi M (1995) Variations of mercury and selenium concentrations in *Adamussium colbecki* and *Pagothenia bernacchii* from Terra Nova Bay (Antarctica) during a five years period. Int J Environ Anal Chem 61:239–248

Minutoli R, Fossi MC, Guglielmo L (2002) Evaluation of acetylcholinesterase activity in several zooplanktonic crustaceans. Mar Environ Res 54:799–804

Miramand P, Guary JC (1980) High concentrations of some heavy metals in tissues of the Mediterranean octopus. Bull Environ Contam Toxicol 24:783–788

Mittner P, Ceccato D, del Maschio S (1994) Multielemental characterization of aerosol at Terra Nova Bay. Preliminary results on the coarse fraction during the 1990–1991 austral summer. In. Colacino M, Giovanelli G, Stefanutti I (eds) Italian research on Antarctic atmosphere. Italian Physics Society, Bologna, pp 133–143

Molina MJ, Rowland FS (1974) Stratospheric ozone depletion by halocarbons: chemistry and transport. Nature 249:810

Monaci F, Borrel A, Leonzio C, Marsili L, Calzada N (1998) Trace elements in striped dolphins (*Stenella coeruleoalba*) from the western Mediterranean. Environ Pollut 99:61–68

Monirith I, Ueno D, Takahashi S, Nakata H, Sudaryanto A, Subramanian A, Subramanian K, Ismail A, Muchtar M, Zheng J, Richardson BJ, Prudente M, Hue ND, Tana TS, Tkalin AV, Tanabe S (2003) Asia-Pacific mussel watch: monitoring contamination of persistent organochlorine compounds in coastal waters of Asian countries. Mar Pollut Bull 46:281–300

Monod J-L, Arnaud PM, Arnoux A (1992) The level of pollution of Kerguelen Islands biota by organochlorine compounds during the seventies. Mar Pollut Bull 24:626–629

Montone RC, Taniguchi S, Weber RR (2001a) Polychlorinated biphenyls in marine sediments of Admiralty Bay, King George Island, Antarctica. Mar Pollut Bull 42:611–614

Montone RC, Taniguchi S, Sericano JL, Weber RR, Lara WH (2001b) Determination of polychlorinated biphenyls in Antarctic macroalgae "*Desmaretia* sp". Sci Total Environ 277:181–186

Montone RC, Taniguchi S, Weber RR (2003) PCBS in the atmosphere of King George Island. Antarctica. Sci Total Environ 308:167–173

Moore JC, Narita H, Maeno N (1991) A continuous 770 year record of volcanic activity from East Antarctica. J Geophys Res 96:17353–17359

Moore JK, Abbot MR, Richman JG, Smith WO, Cowles TJ, Coale KH, Gardner WD, Barber RT (1999) Sea WIFS satellite ocean color data at the US Southern Ocean JGOFS line along 170° W. Geophys Res Lett 26:1465–1468

Moran SB, Moore RM, Westerlund S (1992) Dissolved aluminium in the Weddell Sea. Deep-Sea Res 39:537–547

Morel FMM, Hudson RJM, Price NM (1991) Limitation of productivity by trace metals in the sea. Limnol Oceanogr 36:1742–1755

Morel FMM, Reinfelder JR, Roberts SB, Chamberlain CP, Lee JG, Yee D (1994) Zinc and carbon co-limitation of marine phytoplankton. Nature 369:740–742

Morgan VI, van Ommen TD (1997) Seasonality in late Holocene climate from ice core records. Holocene 7:351–354

Morgan VI, Goodwin ID, Etheridge DM, Wookey CW (1991) Evidence from Antarctic ice cores for recent increases in snow accumulation. Nature 354:58–60

Moriarty F (1983) Ecotoxicology. The study of pollutants in ecosystems. Academic Press, London

Morris E, King JC, Turner J, Peel D, Doake C (1997) Antarctic climate change – an assessment by the British Antarctic Survey. UK Global Environment Research Office, Swindon, Globe, no 36

Mosley-Thompson E (1992) Paleoenvironmental conditions in Antarctica since AD 1500. Ice core evidence. In: Bradley RS, Jones PD (eds) Climate since AD 1500. Routledge, London, pp 572–591

Mosley-Thompson E (1996) Rapid Holocene climate changes recorded in an East Antarctica ice core. In: Jones PD, Bradley RS, Jouzel J (eds) Climatic variations and forcing mechanisms of the last 2000 years. Springer, Berlin Heidelberg New York, pp 263–279

Mowbray DL (1986) Pesticide control in the South Pacific. Ambio 15:22–29

Mroz EJ, Alei M, Cappis JH, Guthals PR, Mason AS, Rokop DJ (1989) Antarctic atmospheric tracer experiments. J Geophys Res 94:8577–8583

Muir DCG, Segstro MD, Welbourn PM, Toom D, Eisenreich SI, MacDonald CR, Whelpdale DM (1993) Patterns of accumulation of airborne organochlorine contaminants in lichens from the upper Great Lakes region of Ontario. Environ Sci Technol 27:1201–1210

Murozumi M, Chow TJ, Patterson C (1969) Chemical concentrations of pollutant lead aerosols, terrestrial dusts and sea salts in Greenland and Antarctic snow strata. Geochim Cosmochim Acta 33:1247–1294

Murphy EJ, King JC (1997) Icy message from Antarctica. Nature 389:20–21

Murray J (1876) On the distribution of volcanic debris over the floor of the ocean. Proc R Soc Edinb 9:247–261

Nagao I, Matsumoto K, Tanaka H (1999) Sunrise ozone destruction found in the subtropical marine boundary layer. Geophys Res Lett 26:3377–3380

Naidu AS, Blanchard A, Kelley JJ, Goering JJ, Hameed MJ, Baskaran M (1997) Heavy metals in Chukchi Sea sediments as compared to selected circum-Arctic shelves. Mar Pollut Bull 35:260–269

NAS (1980) The international Mussel Watch. National Academy of Sciences, Washington, DC

Nash TH III, Gries C (1995) The use of lichens in atmospheric deposition studies with emphasis on the Arctic. Sci Total Environ 160/161:729–736

Neale PJ, Davis RF, Cullen JJ (1998) Interactive effects of ozone depletion and vertical mixing on photosynthesis of Antarctic phytoplankton. Nature 392:585–589

Nealson RH (1997) The limits of life on Earth and searching for life on Mars. J Geophys Res 102:23675–23686

Neelin JD, Battisti DS, Hirst AC, Jin F-F, Wakata Y, Yumagata T, Zebiak SE (1998) ENSO theory. J Geophys Res 103:14261–14290

Neff WD (1992) On the influence of stratospheric stability on lower tropospheric circulations over the South Pole. In: Proc 3rd Conf Polar Meteorology and Oceanography. American Meteorological Society, Boston, pp 115–120

Nelson JS (1984) Fishes of the world. Wiley, New York

Nelson DM, Smith WO, Gordon LI, Huber BA (1987) Spring distribution of density, nutrients, and phytoplankton biomass in the ice edge zone of the Weddell-Scotia Sea. J Geophys Res 92:7181–7190

Nelson DM, Brzezinski MA, Sigman DE, Franck VM (2001) A seasonal progression of Si limitation in the Pacific sector of the Southern Ocean. Deep-Sea Res II 48:3973–3995

Nemoto T, Okiyama M, Iwasaki N, Kikuchi T (1988) Squid as predators on krill (*Euphausia superba*) and prey for sperm whales in the Southern Ocean. In: Sahrhage D (ed) Antarctic Ocean and resources variability. Springer, Berlin Heidelberg New York, pp 292–296

Nerem RS (1999) Measuring very low frequency sea level variations using satellite altimeter data. Global Planet Change 20:157–171

Ngabe B, Bidleman TF (1992) Occurrence and vapor particle partitioning of heavy organic compounds in ambient air in Brazzaville, Congo. Environ Pollut 76:147–156

Nicol S, Endo Y (1999) Krill fisheries: development, management and ecosystem implications. Aquat Living Resources 12:105–120

Nicol S, Stolp M, Nordstrom O (1992) Change in the gross biochemistry and mineral content accompanying the moult cycle in the Antarctic krill *Euphausia superba*. Mar Biol 113:201–209

Nienow JA, Friedmann EI (1993) Terrestrial lithophytic (rock) communities. In: Friedmann EI (ed) Antarctic microbiology. Wiley-Liss, New York, pp 343–412

Nigro M, Orlando E, Regoli F (1992) Ultrastructural localization of metal binding sites in the kidney of the Antarctic scallop *Adamussium colbecki*. Mar Biol 113:637–643

Nigro M, Regoli F, Rocchi R, Orlando E (1997) Heavy metals in Antarctic molluscs. In: Battaglia B, Valencia J, Walton DWH (eds) Antarctic communities: species, structure and survival. Cambridge University Press, Cambridge, pp 408–412

Nijampurkar VN, Rao DK (1993) Polar fallout radionuclides ^{32}Si, ^{7}Be and ^{210}Pb and past accumulation rate of ice at Indian Station, Dakshin Gangotri, East Antarctica. J Environ Radioactivity 21:10–117

Nimis PL, Scheidegger C, Wolseley PA (eds) (2002) Monitoring with lichens – monitoring lichens. Kluwer, Amsterdam

Nolting RF, de Baar HJW (1994) Behaviour of nickel, copper, zinc and cadmium in the upper 300 m of the transect in the Southern Ocean (57° -62° S, 49° W). Mar Chem 45:225–242

Nolting RF, de Baar HJW, van Bennekom AJ, Masson A (1991) Cadmium, copper and iron in the Scotia Sea, Weddell Sea, and Weddell/Scotia Confluence (Antarctica). Mar Chem 35:219–243

Nonnis Marzano F, Fiori F, Jia G, Chiantore M (2000) Anthropogenic radionuclides bioaccumulation in Antarctic marine fauna and its ecological relevance. Polar Biol 23:753–758

Norheim G (1987) Levels and interactions of heavy metals in sea birds from Svalbard and the Antarctic. Environ Pollut 47:83–94

Norheim G, Sømme L, Holt G (1982) Mercury and persistent chlorinated hydrocarbons in Antarctic birds from Bouvetøya and Dronning Maud Land. Environ Pollut 28:233–240

Novelli PC, Lang PM, Masarie KA, Hurst DF, Myers R, Elkins JW (1999) Molecular hydrogen in the troposphere: global distribution and budget. J Geophys Res 104:30427–30444

Nowlin WDJ, Klinck JM (1986) The physics of the Antarctic Circumpolar Current. Rev Geophys 24:469–491

Nriagu JO (1989) A global assessment of natural sources of atmospheric trace metals. Nature 338:47–49

Nriagu JO, Pacyna JM (1988) Quantitative assessment of worldwide contamination of air, water and soils by trace metals. Nature 333:134–139

NSF (1991) Final supplemental environmental impact statement for the United States Antarctic Program. Division Polar Programs, National Science Foundation, Washington, DC

Nygård T, Lie E, Røv, Steinnes E (2001) Metal dynamics in an Antarctic food chain. Mar Pollut Bull 42:598–602

Nylander W (1866) Les lichens du Jardin du Luxembourg. Bull Soc Bot R 13:364–372

Ockenden W, Lohmann R, Shears JR, Jones KC (2001) The significance of PCBs in the atmosphere of the Southern Hemisphere. Environ Sci Pollut Res 8:189–194

Oechel WC, Vourlitis GL (1996) Climate change in northern latitudes: alterations in ecosystem structure and function and effect on carbon sequestration. In: Oechel WC, Callaghan T, Gilmanov T, Holten JI, Maxwell B, Molau U, Sveinbjörnsson B (eds) Global change and Arctic terrestrial ecosystems. Springer, Berlin Heidelberg New York, pp 381–401

Oerlemans J (1989) A projection of future sea level. Climatic Change 15:151–174

O'Farrell SP, McGregor JL, Rotstayn LD, Budd WF, Zweck C, Warner R (1997) Impact of transient increases in atmospheric CO_2 on the accumulation and mass balance of the Antarctic ice sheet. Ann Glaciol 25:137–144

Ohlendorf HM, Kilness AW, Simmons JL, Stroud RK, Hoffman DJ, Moorwe JF (1988) Selenium toxicosis in wild aquatic birds. J Toxicol Environ Health 24:67–92

Ohmura A, Wild M, Bengtsson L (1996) A possible change in mass balance of Greenland and Antarctic ice sheets in the coming century. J Climate 9:2124–2135

Olech M, Kwiatek WM, Dutkiewicz EM (1998) Lead pollution in the Antarctic regions. X-ray Spectrom 27:232–235

Olivero EB, Gasparini Z., Rinaldi CA, Scasso R (1991) First record of dinosaurs in Antarctica (Upper Cretaceous, James Ross Island): paleogeographical implications. In: Thomson MRA, Crame JA, Thomson JW (eds) Geological evolution of Antarctica. Cambridge University Press, Cambridge, pp 617–622

Olmastroni S, Corsolini S, Pezzo F, Focardi S (2000) The first five years of the Italian-Australian joint program on the Adélie penguin: an overview. Ital J Zool 67:141–145

Olsgard F (1999) Effects of copper contamination on recolonisation of subtidal marine soft sediment – an experimental field study. Mar Pollut Bull 38:65–109

Oltmans SJ, Levy H II (1994) Surface ozone measurements from a global network. Atmos Environ 28:9–24

Oriondo M (2000) Patagonian dust in Antarctica. Quat Int 68–71:83–86

Orren MJ, Monteiro PMS (1985) Trace element geochemistry in the Southern Ocean. In: Siegfried WR, Condy PR, Laws RM (eds) Antarctic nutrient cycles and food webs. Springer, Berlin Heidelberg New York, pp 30–37

O'Shea TJ, Brownell RL (1994) Organochlorine and metal contaminants in baleen whales: a review and evaluation of conservation implications. Sci Total Environ 14:179–200

Oswald GKA, Robin G de Q (1973) Lakes beneath the Antarctic ice sheet. Nature 245:251–254

Øvstedal DO, Smith RIL (2001) Lichens of Antarctica and South Georgia. Cambridge University Press, Cambridge

Paasivirta J (2000) The handbook of environmental chemistry, vol 3, part K. New types of persistent halogenated compounds. Springer, Berlin Heidelberg New York

Pacyna JM, Pacyna EG (2001) An assessment of global and regional emissions of trace metals to the atmosphere from anthropogenic sources worldwide. Environ Rev 9:269–298

Palais JM, Mosher BW, Lowenthal D (1994) Elemental tracers of volcanic emissions from Mount Erebus in Antarctic snow samples. In: Kyle PR (ed) Volcanological and environmental studies of Mount Erebus, Antarctica. American Geophysical Union, Washington, DC, vol 66, pp 103–112

Palmer Locarnini SJ, Presley BJ (1995) Trace element concentrations in Antarctic krill, Euphausia superba. Polar Biol 15:283–288

Palmisano AC, Garrison DL (1993) Microorganisms in Antarctic sea ice. In: Friedmann EI (ed) Antarctic microbiology. Wiley-Liss, New York, pp 167–218

Parish TR (1988) Surface winds over the Antarctic continent. A review. Rev Geophys 26:169–180

Parish TR, Bromwich DH (1987) The surface windfield over the Antarctic ice sheets. Nature 328:51–54

Park B-K, Chang S-K, Yoon HI, Chung H (1998) Recent retreat of ice cliffs, King George Island. South Shetland Islands, Antarctic Peninsula. Ann Glaciol 27:633–635

Parker BC (1978) Environmental impact in Antarctica. Virginia Polytechnic Institute and State University, Blacksburg

Parker BC, Wharton RA (1985) Physiological ecology of blue-green algal mats (modern stromatolites) in Antarctic oasis lakes. Arch Hydrobiol Suppl 71 Algal Stud 38/39:331–348

Parker DE, Folland CK, Jackson M (1995) Marine surface temperature: observed variations and data requirements. Climatic Change 31:559–600

Parkinson CL (1992) Southern Ocean sea ice distribution and extents. Trans R Philos Soc Lond 338:243–250

Parmelee DF (1992) Antarctic birds. University of Minnesota Press, Minneapolis

Parungo F, Ackerman E, Caldwell W, Weickmann HK (1979) Individual particle analysis of Antarctic aerosol. Tellus 31:521–529

Paterson WSB (1994) The physics of glaciers. Pergamon Press, Oxford

Patterson CC (1974) Lead in seawater. Science 183:553–554

Pauwels J, Kramer GN, Grobecker K-H (2001) Preparation and production control of the certified reference material of Antarctic sediment. In: Caroli S, Cescon P, Walton DWH (eds) Environmental contamination in Antarctica. Elsevier, Amsterdam, pp 293–303

Pearson TH, Rosenberg R (1978) Macrobenthic succession in relation to organic enrichment and pollution of the marine environment. Oceanogr Mar Biol Ann Rev 16:229–311

Peel DA (1975) The study of the global atmospheric pollution in Antarctica. Polar Record 17:639–643

Peel DA (1992) Ice core evidence from the Antarctic Peninsula region. In: Bradley RS, Jones PD (eds) Climate since AD 1500. Routledge, London, pp 549–571

Pereira EB, Setzer AW, Cavalcanti IFA (1988) ^{222}Rn in the Antarctic Peninsula during 1986. Radiation Prot Dosimet 24:85–88

Perissinotto R, Gurney L, Pakhomov EA (2000) Contribution of heterotrophic material to diet and energy budget of Antarctic krill, *Euphausia superba*. Mar Biol 136:129–135

Peterle TJ (1969) DDT in Antarctic snow. Nature 224:620

Peters AJ, Gregor DJ, Teixeira CF, Jones NP, Spencer C (1995) The recent depositional trend of polycyclic aromatic hydrocarbons and elemental carbon to the Agassiz Ice Cap, Ellesmere Island, Canada. Sci Total Environ 160/161:167–179

Petit JR, Briat M, Royer A (1981) Ice age aerosol content from East Antarctic ice core samples and past wind strength. Nature 293:391–393

Petit JR, Jouzel J, Raynaud D, Barkov NI, Barnola JM, Basile I, Benders M, Chappellaz J, Davis M, Delaygue G, Delmotte M, Kotlyakov VM, Legrand M, Lipenkov VY, Lorius C, Pepin L, Ritz C, Saltzman E, Stievenard M (1999) Climate and atmospheric history of the past 420,000 years from the Vostok ice core, Antarctica. Nature 399:429–436

Petit JR, Raynaud D, Lorius C, Jouzel J, Delaygue G, Barkov NI, Kotlyakov VM (2000) Historical isotopic temperature record from the Vostok ice core. Oak Ridge National Laboratory, US Department of Energy, Oak Ridge

Petri G, Zauke G-P (1993) Trace metals in crustaceans in the Antarctic Ocean. Ambio 22:529–536

Pezzo F, Olmastroni S, Corsolini S, Focardi S (2001) Factors affecting the breeding success of the south polar skua *Catharacta maccormicki* at Edmonson Point, Victoria Land, Antarctica. Polar Biol 24:389–393

Philander SGH (1990) El Niño, La Niña and the Southern Oscillation. Academic Press, New York

Phillpot HR (1968) A study of the synoptic climatology of the Antarctic. International Antarctic Meteorological Research Centre, Melbourne, Tech Rep no 12

Phillpot HR (1985) Physical geography – climate. In: Bonner WN, Walton DWH (eds) Key environments: Antarctica. Pergamon Press, Oxford, pp 23–38

Picciotto E, Wilgain S (1963) Fission products in Antarctic snow, a reference level for measuring accumulation. J Geophys Res 68:5965–5972

Pickard J (ed) (1986) Antarctic oasis. Academic Press, London

Pilegaard K (1987) Biological monitoring of airborne deposition within and around the Ilimaussaq intrusion, Southwest Greenland. Bioscience 24:1–28

Pirrone N, Keeler GJ, Niragu JO (1996) Regional differences in worldwide emissions of mercury to the atmosphere. Atmos Environ 30:2981–2987

Pirrone N, Hedgecock IM, Forlano L (2000) The role of the ambient aerosol in the atmospheric processing of semi-volatile contaminants: a parameterised numerical model (GASPAR). J Geophys Res 105D:9773–9790

Pirrone N, Costa P, Pacyna JM, Ferrara R (2001) Atmospheric mercury emissions from anthropogenic and natural sources in the Mediterranean region. Atmos Environ 35:2997–3006

Planchon FAM, Boutron CF, Barbante C, Wolff EW, Cozzi G, Gaspari V, Ferrari CP, Cescon P (2001) Ultrasensitive determination of heavy metals at the sub-picogram per gram level in ultraclean Antarctic snow samples by inductively coupled plasma sector field mass spectrometry. Anal Chim Acta 450:193–205

Planchon FAM, Boutron CF, Barbante C, Cozzi G, Gaspari V, Wolff EW, Ferrari CP, Cescon P (2002a) Changes in heavy metals in Antarctic snow from Coats Land since the mid-19th to the late-20th century. Earth Planet Sci Lett 200:207–222

Planchon FAM, Boutron CF, Barbante C, Cozzi G, Gaspari V, Wolff EW, Ferrari CP, Cescon P (2002b) Short-term variations in the occurrence of heavy metals in the Antarctic snow from Coats Land since the 1920s. Sci Total Environ 300:129–142

Planchon FAM, van de Velde K, Rosman KJR, Wolff EW, Ferrari CP, Boutron CF (2003) One hundred fifty-year record of lead isotopes in Antarctic snow from Coats Land. Goechim Cosmochim Acta 67:693–708

Platt HM (1978) Assessment of the macrobenthos in the Antarctic environment following recent pollution abatement. Mar Pollut Bull 9:149–153

Platt HM (1980) Exploitation and pollution in Antarctica: a case history. Prog Underwater Sci 5:188–200

Platt HM, Mackie PR (1979) Analysis of aliphatic and aromatic hydrocarbons in Antarctic marine sediment layers. Nature 280:576–578

Platt HM, Mackie PR (1980) Distribution and fate of aliphatic and aromatic hydrocarbons in Antarctic fauna and the environment. Helgoländer Meeresunters 33:236–245

Plötz J, Ekau W, Reijnders PJH (1991) Diet of Weddell seals Leptonychotes weddellii at Vestkapp, Eastern Weddell Sea (Antarctica), in relation to local food supply. Mar Mammal Sci 7:136–144

Poblet A, Andrade S, Scagliola M, Vodopivez C, Curtosi A, Pucci A, Marcovecchio J (1997) The use of epilithic Antarctic lichens (Usnea aurantiacoatra and U. antarctica) to determine deposition patterns of heavy metals in the Shetland Islands, Antarctica. Sci Total Environ 207:187–194

Podzimek J (1999) Aerosol particles and droplet scavenging by snow crystals. J Aerosol Sci 30:573–574

Poisson A, Metzl N, Danet X, Louanchi F, Brunet C, Schauer B, Bres B, Ruiz-Pino D (1994) Air-sea CO_2 fluxes in the Southern Ocean between 25° E and 85° E. In: Johannessen OM, Muench RD, Overland JE (eds) The polar oceans and their role in shaping the global environment. American Geophysical Union, Washington, DC, vol 85, pp 273–284

Pomeroy JW, Jones Hg (1996) Wind-blown snow: sublimation, transport and changes to polar snow. In: Wolff EW, Bales RC (eds) Chemical exchange between the atmosphere and the polar snow. Springer, Berlin Heidelberg New York

Pongratz R, Heumann KG (1999) Production of methylated mercury, lead, and cadmium by marine bacteria as a significant natural source for atmospheric heavy metals in polar regions. Chemosphere 39:89–102

Ponzano E, Dondero F, Bouquegneau J-M, Sack R, Hunziker P, Viarengo A (2001) Purification and biochemical characterization of a cadmium metallothionein from the digestive gland of the Antarctic scallop Adamusssium colbecki (Smith 1902). Polar Biol 24:147–153

Pourchet M, Bartaya SK, Maignan M, Jouzel J, Pinglot JF, Aristarain AJ, Furdada G, Kotlyakov VM, Mosley-Thompson E, Preiss N, Young NW (1997) Distribution and fallout of ^{137}Cs and other radionuclides over Antarctica. J Glaciol 43:435–445

Pourchet M, Magnad O, Frezzotti M, Ekaykin A, Winther J-G (2003) Radionuclides deposition over Antarctica. J Environ Radioactivity 68:137–158

Price NM, Morel FMM (1990) Cadmium and cobalt substitution for zinc in a marine diatom. Nature 344:658–660

Priddle J (1985) Terrestrial habitats – inland waters. In: Bonner WN, Walton DWH (eds) Key environments – Antarctica. Pergamon Press, New York, pp 118–132

Priddle J, Brandini F, Lipski M, Thorley MR (1994) Pattern and variability of phytoplankton biomass in the Antarctic Peninsula region: an assessment of the BIOMASS cruises. In: El-Sayed SZ (ed) Southern Ocean ecology. The BIOMASS perspective. Cambridge University Press, Cambridge, pp 49–61

Pride DE, Cox CA, Moody SV, Conelea RR, Rosen MA (1990) Investigation of mineralization in the South Shetland Islands, Gerlache Strait, and Anvers Island, northern Antarctic Peninsula. In Splettstoesser JF, Dreschoff GAM (eds) Mineral resources potential of Antarctica. American Geophysical Union, Washington, DC, vol 51, pp 69–94

Prinn RG (ed) (1994) Global atmospheric – biospheric chemistry. Plenum Press, New York

Prinn RG, Weiss RF, Fraser PJ, Simmonds PG, Cunnold DM, Alyea FN, O'Doherty S, Salameh P, Miller BR, Huang J, Wang RHJ, Hartley DE, Harth C, Steele LP, Sturrock G, Midgley PM, McCulloch A (2000) A history of chemically and radiatively important gases in air deduced from ALE/GAGE/AGAGE. J Geophys Res 105:17751–17792

Priscu JC (ed) (1998) Ecosystem dynamics in a polar desert. The McMurdo Dry Valleys, Antarctica. American Geophysical Union, Washington, DC, Antarct Res Ser 72

Priscu JC, Fritsen CH, Adams EE, Giovannoni SJ, Paerl HW, McKay CP, Doran PT, Gordon DA, Lanoil BD, Pinckney JL (1998) Perennial Antarctic lake ice: an oasis for life in polar desert. Science 280:2095–2098

Priscu JC, Adams EE, Lyons WB, Voytek MA, Mogk DW, Brown RL, McKay CP, Takaes CD, Welch KA, Wolf CF, Kirhtein JD, Avci R (1999) Geomicrobiology of subglacial ice above Lake Vostok, Antarctica. Science 286:2141–2144

Prospero JM (1996) The atmospheric transport of particles to the ocean. In: Ittekkot V, Schäfer P, Depetris PJ (eds) Particle flux in the ocean. Wiley, New York, pp 19–53

Pruppacher HR, Klett JD (1980) Microphysics of clouds and precipitation. Reidel, Dordrecht

Pruppacher HR, Klett JD (1997) Microphysics of clouds and precipitation. Kluwer, Dordrecht

Pruppacher HR, Semonin RG, Slinn WGN (eds) (1983) Precipitation scavenging, dry deposition, and resuspension. Elsevier, Amsterdam

Psenner R, Sattler B (1998) Life at the freezing point. Science 280:2073–2074

Radok U (1973) On the energetics of surface winds of the Antarctic ice cap. Energy fluxes over polar surfaces. WMO, Geneva, WMO Tech Note 129, pp 69–100

Rainbow PS (1989) Copper, cadmium and zinc concentrations in oceanic amphipod and euphausiid crustaceans, as source of heavy metals to pelagic seabirds. Mar Biol 103:513–518

Rankin AM, Wolff EW (2000) Ammonium and potassium in snow around an emperor penguin colony. Antarct Sci 12:154–159

Ratcliffe DA (1970) Changes attributable to pesticides in egg breakage frequency and eggshell thickness in some British birds. J Appl Ecol 7:67–115

Raven JA (1990) Predictions of Mn and Fe use efficiencies of phototrophic growth as function of light availability for growth and of C assimilation pathway. New Phytol 116:1–18

Ravich MG, Fedorov LV, Tarutin OA (1982) Precambrian iron deposits of the Prince Charles Mountains. In: Craddock C (ed) Antarctic geoscience. University of Wisconsin Press, Madison, pp 853–858

Regan CT (1913) The Antarctic fishes of the Scottish National Antarctic Expedition. Trans R Soc Edinb 49:229–292

Regoli F, Principato G (1995) Glutathione, glutathione-dependent and antioxidant enzymes in mussels, *Mytilus galloprovincialis*, exposed to metals under field and laboratory conditions: implications for the use of biochemical biomarkers. Aquat Toxicol 31:143–164

Regoli F, Principato GB, Bertoli E, Nigro M, Orlando E (1997) Biochemical characterization of the antioxidant system in the scallop *Adamussium colbecki*, a sentinel organism for monitoring the Antarctic environment. Polar Biol 17:251–258

Regoli F, Nigro M, Bompadre S, Winston GW (2000) Total oxidant scavenging capacity (TOSC) of microsomal and cytosolic fractions from Antarctic, Arctic, and Mediterranean scallops: differentiation between three potent oxidants. Aquat Toxicol 49:13–25

Reid K (2001) Growth of Antarctic krill *Euphausia superba* at South Georgia. Mar Biol 138:57–67

Reijnder PJH (1986) Reproductive failure in common seals feeding on fish from polluted coastal waters. Nature 324:456–457

Remmert H, Olson JS, Golley FB, Billings WD, Goldammer JG (eds) (1990) Fire in the tropical biota: ecosystem processes and global challenges. Springer, Berlin Heidelberg New York

Ricci CA (1991) Rocks and geological history. In: Baroni C (ed) Antarctica – the Earth's white heart. Enrico Rainero, Florence, pp 66–72

Ricci CA (ed) (1997) The Antarctic region: geological evolution and processes. Terra Antartica, Siena

Ricci CA, Baroni C, Brancolini G, Palmeri R, Salvini F, Talarico F (2001) La storia geologica. In: Baroni C (ed) Antartide. Terra di Scienza e Riserva Naturale. Terra Antartica, Siena, pp 88–112

Richardson MD, Hedgpeth JW (1977) Antarctic soft-bottom, macrobenthic adaptations to a cold, stable, highly productive glacially affected environment. In: Llano GA (ed) Adaptations within Antarctic ecosystems. Gulf Publ, Houston, pp 181–196

Ridley HN (1930) The dispersal of plants throughout the world. Reeve, Ashford, Kent

Ridout PS, Rainbow PS, Roe HSJ, Jones HR (1989) Concentrations of V, Cr, Mn, Fe, Ni, Co, Cu, Zn, As and Cd in mesopelagic crustaceans from the North East Atlantic Ocean. Mar Biol 100:465–471

Riget F, Johansen P, Asmund (1997) Baseline levels and natural variability of elements in three seaweed species from West Greenland. Mar Pollut Bull 34:171–176

Riley JP, Chester R (eds) (1983) Chemical oceanography. Academic Press, London

Risebrough RW (1977) Transfer of organochlorine pollutants to Antarctica. In: Llnao GA (ed) Adaptations within Antarctic ecosystems. Gulf Publ, Houston, pp 1023–1210

Risebrough RW, Carmignani GM (1972) Chlorinated hydrocarbons in Antarctic birds. In: Parker BC (ed) Conservation problems in Antarctica. Allen Press, Lawrence, Kansas, pp 63–80

Risebrough RW, Walker HW, Schmidt TT, de Lappe BW, Connors CW (1976) Transfer of chlorinated biphenyls to Antarctica. Nature 264:738–739

Risebrough RW, De Lappe BW, Younghans-Haug C (1990) PCB and PCT contamination in Winter Quarters Bay, Antarctica. Mar Pollut Bull 21:523–529

Robertson JE, Watson AJ (1995) A summer-time sink for atmospheric carbon dioxide in the Southern Ocean between 88° W and 80° E. Deep-Sea Res 42:1081–1091

Robertson G, Williams R, Green K, Robertson L (1994) Diet composition of emperor penguin chicks *Aptenodytes forsteri* at two Mawson coast colonies, Antarctica. Ibis 136:19–31

Robertson A, Overpeck J, Rind D, Mosley-Thompson E, Zielinski G, Lean J, Koch D, Penner J, Tegen I, Healy R (2001) Hypothesized climate forcing time series for the last 500 years. J Geophys Res 106:14783–14803

Robin G (1983) The climatic record in the polar ice sheets. Cambridge University Press, Cambridge

Robin G (1988) The Antarctic ice sheet, its history and responses to sea level and climatic changes over the past 100 million years. Palaeogeogr Palaeoclimatol Palaeoecol 67:31–50

Robock A (2000) Volcanic eruptions and climate. Rev Geophys 38:191–219

Robock A, Free MP (1996) The volcanic record in ice cores for the past 2000 years. In: Jones PD, Bradely RS, Jouzel J (eds) Climatic variations and forcing mechanisms of the last 2000 years. Springer, Berlin Heidelberg New York

Rodhouse PG (1990) Cephalopod fauna of the South Scotia Sea at South Georgia: potential for commercial exploitation and possible consequences. In: Kerry KR, Hempel G (eds) Antarctic ecosystems: ecological change and conservation. Springer, Berlin Heidelberg New York, pp 289–298

Roos P, Holm E, Persson RBR, Aarkrog A, Nielsen SP (1994) Deposition of ^{210}Pb, ^{137}Cs, $^{239+240}$Pu, ^{238}Pu, and ^{241}Am in the Antarctic Peninsula area. J Environ Radioactivity 24:235–251

Rosenfeld D (1999) TRMM observed first direct evidence of smoke from forest fires inhibiting rainfall. Geophys Res Lett 26:3105–3108

Rosman KJR (2001) Natural isotopic variations in lead in polar snow and ice as indicators of source regions. In: Caroli S, Cescon P, Walton DWH (eds) Environmental contamination in Antarctica: a challenge to analytical chemistry. Elsevier, Amsterdam, pp 87–106

Rosman KJR, Chisholm W, Hong S, Candelone J-P, Boutron CF (1997) Lead from Carthaginian and Roman Spanish mines isotopically identified in Greenland ice dated from 600 B.C. to 300 A.D. Environ Sci Technol 31:3413–3416

Rosman KJR, Ly C, van de Velde K, Boutron CF (2000) A two century record of lead isotopes in high altitude Alpine snow and ice. Earth Planet Sci Lett 176:413–424

Ross RM, Hofmann EE, Quetin LB (eds) (1996) Foundations for ecological research west of the Antarctic Peninsula. American Geophysical Union, Washington, DC, Antarct Res Ser 70

Ross JI, Hobbs PV, Holden B (1998) Radiative characteristics of regional hazes dominated by smoke from biomass burning in Brazil: closure tests and direct radiative forcing. J Geophys Res 103:31925–31941

Rothschild LJ, Mancinelli RL (2001) Life in extreme environment. Nature 409:1092–1101

Rott H, Rack W, Nagler T, Skvarka P (1998) Climatically induced retreat and collapse of northern Larsen Ice Shelf, Antarctic Peninsula. Ann Glaciol 27:86–92

Rowley PD, Ford AD, Williams PL, Pride DE (1983) Metallogenic provinces of Antarctica. In: Oliver RL, James PR, Jago JB (eds) Antarctic earth science. Cambridge University Press, Cambridge, pp 414–419

Rowley PD, Williams PL, Pride DE (1991) Metallic and non-metallic mineral resources of Antarctica. In: Tinley RJ (ed) The geology of Antarctica. Clarendon Press, Oxford, pp 617–651

Roy CR, Gies HP, Tomlinson DW, Lugg DL (1994) Effects of ozone depletion on the ultraviolet radiation environment at the Australian stations in Antarctica. In: Weiler CS, Penhale PA (eds) Ultraviolet radiation in Antarctica: measurements and biological effects. American Geophysical Union, Washington, DC, vol 62, pp 1–15

Ruberto L, Vazquez SC, Mac Cormack WP (2003) Effectiveness of the natural bacterial flora, biostimulation and bioaugmentation on the bioremediation of a hydrocarbon contaminated Antarctic soil. Int Biodeterior Biodegrad 52:115–125

Rühling Å, Tyler G (1968) An ecological approach to the lead problem. Bot Notiser 121:321–342

Saager PM, de Baar HJW, Howland RJ (1992) Cd, Zn, Ni and Cu in the Indian Ocean. Deep-Sea Res 39:9–35

Saarhage D (ed) (1998) Antarctic Ocean and resources variability. Springer, Berlin Heidelberg New York

Sakshaug E, Holm-Hansen O (1984) Factors governing pelagic production in polar oceans. In: Holm-Hansen O, Bolis OL, Gilles R (eds) Marine phytoplankton and productivity. Springer, Berlin Heidelberg New York, pp 1–18

Salbu B, Steinnes E (eds) (1995) Trace elements in natural waters. CRC Press, Boca Raton, FL

Salbu B, Krekling T, Oughton DH (1998) Characterisation of radioactive particles in the environment. Analyst 123:843–849

Sansom J (1989) Antarctic surface temperature time series. J Climate 2:1164–1172

Santovito G, Irato P, Piccinni E, Albergoni V (2000) Relationship between metallothionein and metal contents in red-blooded and white-blooded Antarctic teleosts. Polar Biol 23:383–391

Sañudo-Wilhelmy SA, Olsen KA, Scelfo JM, Foster TD, Flegal AR (2002) Trace metal distributions off the Antarctic Peninsula in the Weddell Sea. Mar Chem 77:157–170

Sarmiento JL, Le Quere C (1996) Oceanic carbon dioxide uptake in a model of century scale global warming. Science 274:1346–1350

Sarmiento JL, Hughes TMC, Stouffer RJ, Manabe S (1998) Simulated response of the ocean carbon cycle to anthropogenic climate warming. Nature 395:245–249

Sassen K (1992) Evidence for liquid-phase cirrus cloud formation from volcanic aerosol: climatic implications. Science 257:516–519

Satheesh SK, Ramanathan V (2000) Large differences in tropical aerosol forcing at the top of the atmosphere and Earth's surface. Nature 405:60–63

Satheesh SK, Ramanathan V, Li-Jnes X, Lobert JM, Podgorny IA, Prospero JM, Holben BN, Loev NG (1999) A model for the natural and anthropogenic aerosols over the tropical Indian Ocean derived from INDOEX data. J Geophys Res 104:27421–27440

Savage ML, Stearns CR, Weidner GA (1988) The Southern Oscillation signal in Antarctica. In: Proc 2nd Conf Polar Meteorology and Oceanography. American Meteorological Society, Madison, pp 141–144

Savarino J, Boutron Cf, Jaffrezo J-L (1994) Short-term variations of Pb, Cd, Zn and Cu in recent Greenland snow. Atmos Environ 28A:1731–1737

SCAR (1989) The role of Antarctica in global change. Scientific priorities for the International Geosphere-Biosphere Programme (IGBP). ICSU Press/SCAR, Cambridge

SCAR (1993) The role of the Antarctic in global change. An international plan for a regional research programme. SCAR, Cambridge

SCAR/COMNAP (1996) Monitoring of environmental impacts from science and operations in Antarctica. Rep SCAR/COMNAP Worksh Environmental Monitoring in Antarctica, 17–20 October 1995, Oslo, Norway, and 25–29 March 1996, College Station, Texas. SCAR, Cambridge

SCAR/COMNAP (2000) Antarctic environmental monitoring handbook. COMNAP Secretariat, Hobart

Scarponi G, Barbante C, Cescon P (1994) Differential pulse anodic stripping voltammetry for ultratrace determination of cadmium and lead in Antarctic snow. Analysis 22:47–50

Scarponi G, Capodaglio G, Barbante C, Toscano G, Cecchini M, Gambaro A, Cescon P (2000) Concentration changes in cadmium and lead in Antarctic coastal seawater (Ross Sea) during the austral summer and their relationship with the evolution of biological activity. In: Faranda FM, Guglielmo L, Ianora A (eds) Ross Sea ecology. Springer, Berlin Heidelberg New York, pp 585–594

Scheringer M, Wegmann F, Fenner K, Hungerbuhler K (2000) Investigation of the cold condensation of persistent organic pollutants with a global multimedia fate model. Environ Sci Technol 34:1842–1850

Schneider R, Steinhagen-Schneider G, Drescher HE (1985) Organochlorines and heavy metals in seals and birds from Weddell Sea. In: Siegfried WR, Condy PR, Laws RM (eds) Antarctic nutrient cycles and food webs. Springer, Berlin Heidelberg New York, pp 652–655

Schneppenheim R (1981) Results of the biological investigations in the Weddell Sea during a site survey 1979/80. Polarforschung 51:91–99

Schroeder WH, Analuf KG, Barrei LA, Lu JY, Steffen A, Schneeberger DR, Berg T (1998) Arctic springtime depletion of mercury. Nature 394:331–332

Schultz MG, Jacob DJ, Wang YH, Logan JA, Atlas EL, Blake DR, Blake NJ, Bradshaw JD, Browell EV, Fenn MA, Flocke F, Gregory GL, Heikes BG, Sachse GW, Sandholm ST, Shetter RE, Singh HB, Talbot RW (1999) On the origin of tropospheric ozone and NOx over the tropical South Pacific. J Geophys Res 104:5829–5843

Schulz-Baldes M (1992) Baseline study on Cd, Cu and Pb concentrations in Atlantic neuston organisms. Mar Biol 112:211–222

Schürmann G, Markert B (1997) Ecotoxicology. Ecological fundamentals, chemical exposure and biological effects. Wiley, New York

Schuster RM (1979) On the persistence and dispersal of transantarctic Hepaticae. Can J Bot 57:2179–2225

Schwartz SE, Slinn WGN (eds) (1992) Precipitation, scavenging and atmospheric surface exchange, vol 3. Hemisphere, New York

Schwerdtfeger W (1970) The climate of the Antarctic. In: Orvig S (ed) World survey of climatology, vol 14. Elsevier, Amsterdam, pp 253–355

Schwerdtfeger W (1984) Weather and climate of Antarctica. Elsevier, Amsterdam

Scudiero R, Capasso C, Carginale V, Riggio M, Capasso A, Ciaramella M, Filosa S, Parisi E (1997) PCR amplification and cloning of metallothionein complementary DNAs in temperate and Antarctic sea urchin characterized by a large difference in egg metallothionein content. Cell Molec Life Sci 53:472–477

Seaward MRD, Bylinska EA, Goyal R (1981) Heavy metal content of *Umbilicaria* species from the Sudety region of SW Poland. Oikos 36:107–113

Seckmeyer G, Mayer B, Bernhard G, McKenzie RL, Johnston PV, Kotkamp M, Booth CR, Lucas T, Mestechkina T, Roy CR, Gies HP, Tomlinson D (1995) Geographical differences in the UV measured by intercompared spectroradiometers. Geophys Res Lett 22:1889–1892

Seinfeld JH, Pandis SN (1998) Atmospheric chemistry and physics from air pollution to climate change. Wiley, New York

Sen Gupta R, Sarkar A, Kureishey TW (1996) PCBs and organochlorine pesticides in krill, birds and water from Antarctica. Deep-Sea Res II 43:119–126

Serreze MC, Walsh FSI, Chapin T, Osterkamp T, Dyurgerov M, Romanovsky V, Oechel WC, Morison J, Zhang T, Barry RG (2000) Observational evidence of recent change in the northern high-latitude environment. Climatic Change 46:159–207

Severinghaus JP, Grachev A, Battle M (2001) Thermal fractionation of air in polar firn by seasonal temperature gradients. Geochem Geophys Geosyst G^3 2:2000GC000146

Shaw GE (1988) Antarctic aerosols: a review. Rev Geophys 26:89–112

Shen GT, Boyle EA, Lea DW (1987) Cadmium in corals as a tracer of historical upwelling and industrial fallout. Nature 328:794–796

Sheppard DS, Patterson JE, McAdam MK (1991) Mercury content of Antarctic ice and snow: further results. Atmos Environ 25A:1657–1660

Sheppard DS, Le Guern F, Christenson BW (1994) Composition and the mass flux of the Mount Erebus volcanic plume. In: Kyle PR (ed) Volcanological and environmental studies of Mount Erebus, Antarctica. American Geophysical Union, Washington, DC, vol 66, pp 83–96

Sheppard DS, Campbell IB, Claridge GGC (2000) Metal contamination of soils at Scott Base, Antarctica. Appl Geochem 15:5130–5530

Sherrell RM, Boyle EA, Falkner KK, Harris NR (2000) Temporal variability of Cd, Pb, and Pb isotope deposition in central Greenland snow. Geochem Geophys Geosyst G^3 1:1999GC000007

Sheuhammer AM (1988) Chronic dietary toxicity of methylmercury in the zebra finch, *Phoepila guttata*. Bull Environ Contam Toxicol 40:123–130

Shine KP, Forster PM (1999) The effects of human activities on radiative forcing of climate change: a review of recent development. Global Planet Change 20:205–225

Shotyk W, Weiss D, Appleby PG, Cheburkin AK, Frei R, Gloor M, Kramers JD, Reese S, van der Knaap WO (1998) History of atmospheric lead deposition since 12,370 ^{014}C yr BP from a peat bog, Jura Mountains, Switzerland. Science 281:1635–1640

Siegert MJ (2001) Ice sheets and late quaternary environmental changes. Wiley, Chichester

Siegert MJ, Kwok R, Mayer C, Hubbard B (2000) Water exchange between the subglacial Lake Vostok and the overlying ice sheet. Nature 403:643–646

Siegert MJ, Ellis-Evans JC, Tranter M, Mayer C, Petit J-R, Salamatin A, Priscu JC (2001) Physical, chemical and biological processes in Lake Vostok and other Antarctic subglacial lakes. Nature 414:603–609

Sigmon DE, Nelson DM, Brzezinski MA (2002) The Si cycle in the Pacific sector of the Southern Ocean: seasonal diatom production in the surface layer and export to the deep sea. Deep-Sea Res II 49:1747–1763

Simmonds I (1992) Modelling the reaction of Antarctica to climate changes at its periphery. Impact of climate change on Antarctica-Australia. Australian Government Publ Serv, Canberra, pp 16–23

Simmonds I (1998) The climate of the Antarctic region. In: Hobbs JE, Lindesay JA, Bridgemann HA (eds) Climates of the southern continents. Wiley, New York, pp 137–159

Simmonds PG, Derwent RG, O'Doherty S, Ryall DB, Steele LP, Langenfelds RL, Salameh P, Wang HJ, Dimmer CH, Hudson LE (2000) Continuous high-frequency observations of hydrogen at the Mace Head baseline atmospheric monitoring station over the 1994–1998 period. J Geophys Res 105:12105–12121

Simmons JM, Vestal JR, Wharton RA (1993) Environmental regulators of microbial activity in continental Antarctic lakes. In: Friedmann EI (ed) Antarctic microbiology. Wiley-Liss, New York, pp 491–541

Simoneit BRT, Elias VO (2000) Organic tracers from biomass burning in atmospheric particulate matter over the ocean. Mar Chem 69:301–312

Simoneit BRT, Schauer JJ, Nolte CG, Oros DR, Elias VO, Fraser MP, Rogge WF, Cass GR (1999) Levoglucosan, a tracer for cellulose in biomass burning and atmospheric particles. Atmos Environ 33:173–182

Simpson-Housley P (1992) Antarctica, exploration, perception and metaphor. Routledge, London

Sinclair MR (1981) Record-high temperatures in the Antarctic – a synoptic case study. Month Weather Rev 109:2234–2242

Siniff DB, DeMaster DP, Hofman RJ, Eberhart LL (1977) An analysis of the dynamic of a Weddell seal population. Ecol Monogr 47:319–335

Skotnicki ML, Bargagli R, Ninham JA (2002) Genetic diversity in the moss *Pohlia nutans* on geothermal ground of Mount Rittmann, Victoria Land, Antarctica. Polar Biol 25:771–777

Slemr F (1996) Trends in atmospheric mercury concentrations over the Atlantic Ocean and the Wank Summit, and the resulting constraints on the budget of atmospheric mercury. In: Baeyens W, Ebinghaus R, Vasiliev O (eds) Global and regional mercury cycles: sources, fluxes and mass balance. Kluwer, Dordrecht, pp 33–84

Smith RIL (1984) Terrestrial plant biology of the sub-Antarctic and Antarctic. In: Laws RM (ed) Antarctic ecology, vol 1. Academic Press, London, pp 61–162

Smith RIL (1990) Signy Island as a paradigm of biological and environmental change in Antarctic terrestrial ecosystems. In: Kerry KR, Hempel G (eds) Antarctic ecosystems. Ecological changes and conservation. Springer, Berlin Heidelberg New York, pp 32–50

Smith RIL (1991) Exotic sporomorpha as indicators of potential immigrants colonists in Antarctica. Grana 30:313–324

Smith RIL (1993) Dry coastal ecosystems of Antarctica. In: Maarel E (ed) Ecosystems of the world, vol 2A. Dry coastal ecosystems. Elsevier, Amsterdam, pp 52–71

Smith RIL (1996) Terrestrial and freshwater biotic components of the western Antarctic Peninsula. In: Ross RM, Hofmann EE, Quetin LB (eds) Foundations for ecological research west of the Antarctic Peninsula. American Geophysical Union, Washington, DC, vol 70, pp 15–59

Smith FB, Clark MJ (1986) Radionuclide deposition from the Chernobyl cloud. Nature 322:690–691

Smith WO, Garrison DL (1990) Marine ecosystem research at the Weddell Sea ice edge: the AMIEREZ program. Oceanography 3:22–29

Smith WO, Gordon LI (1997) Hyperproductivity of the Ross Sea (Antarctica) polynya during austral spring. Geophys Res Lett 24:233–236

Smith WO, Nelson DM (1985) Phytoplankton bloom produced by a receding ice edge in the Ross Sea: spatial coherence with the density field. Science 227:163–166

Smith WO, Nelson DM (1986) The importance of ice edge phytoplankton production in the Southern Ocean. Bioscience 36:251–257

Smith HR, Stearns CR (1993) Antarctic pressure and temperature anomalies surrounding the minimum in the Southern Oscillation index. J Geophys Res 98:13071–13083

Smith GA, Nichols PD, White DC (1986) Fatty acid composition and microbial activity of benthic marine sediments from McMurdo Sound, Antarctica. FEMS Microbiol Ecol 38:219–231

Smith RC, Prézelin BB, Baker KS, Bidigare RR, Boucher NP, Coley T, Karentz D, MacIntyre S, Matlick HA, Menzies D, Ondrusek M, Wan Z, Waters KJ (1992) Ozone depletion: ultraviolet radiation and phytoplankton biology in Antarctic waters. Science 255:952–959

Smith IN, Budd WF, Reid P (1998) Model estimates of Antarctic accumulation rates and relationship to temperature changes. Ann Glaciol 27:246–250

Smith AM, Vaughan DG, Doake CSM, Johnson AC (1999) Surface lowering of the ice ramp at Rothera Point, Antarctic Peninsula, in response to regional climate change. Ann Glaciol 27:113–118

Smith RC, Ainley D, Baker K, Domack E, Emslie S, Fraser B, Kennett J, Leventer A, Mosley-Thompson E, Stammerjohn S, Verent M (1999) Marine ecosystem sensitivity to climate change. Bioscience 49:393–404

Smith WO, Marra J, Hiscock MR, Barber RT (2000a) The seasonal cycle of phytoplankton biomass and primary productivity in the Ross Sea, Antarctica. Deep-Sea Res II 47:3119–3140

Smith WO, Anderson RF, Moore JK, Codispoti LA, Morrison JM (2000b) The US Southern Ocean Joint Global Ocean Flux Study: an introduction to AESOPS. Deep-Sea Res II 47:3073–3093

Snape I, Riddle MJ, Stark JS, Cole CM, King CK, Duquesne S, Gore DB (2001) Management and remediation of contaminated sites at Casey Station, Antarctica. Polar Record 37:199–214

Soevik T, Braekhan OR (1979) Fluoride in Antarctic krill (*Euphausia superba*) and the Atlantic krill (*Megayctiphanes norvegica*). J Fish Res Board Can 36:1414–1416

Solomon S (1990) Progress towards a quantitative understanding of Antarctic ozone depletion. Nature 347:347–354

Souchez RA, Lorrain RD (1991) Ice composition and glacier dynamics. Springer, Berlin Heidelberg New York

Sparmacher H, Fülber K, Bonka H (1993) Below-cloud scavenging of aerosol particles: particle-bound radionuclides. Experimental. Atmos Environ A 27:605–618

Spezie G, Manzella GMR (eds) (1999) Oceanography of the Ross Sea, Antarctica. Springer, Berlin Heidelberg New York

Spigel RH, Priscu JC (1998) Physical limnology of the McMurdo Dry Valleys lakes. In: Priscu JC (ed) Ecosystem dynamics in a polar desert. The McMurdo Dry Valleys, Antarctica. Am Geophys Union 72:153–187

Spiro PA, Jacob DJ, Logan JA (1992) Global inventory of sulphur emissions with 1 × 1 resolution. J Geophys Res 97:6023–6036

Splettstoesser JF (1992) Antarctic global warming. Nature 355:503

Splettstoesser JF, Dreschoff GAM (eds) (1990) Mineral resources potential of Antarctica. American Geophysical Union, Washington, DC, Antarct Res Ser 51

Sprovieri F, Pirrone N, Hedgecock IM, Landis MS, Stevens RK (2002) Intensive atmospheric mercury measurements at Terra Nova Bay in Antarctica during November and December 2000. J Geophys Res 107:4722–4429

Squibb KS, Cousin RJ (1977) Synthesis of metallothionein in a polysomal cell-free system. Biochem Biophys Res Comm 75:806–812

Staden WLJ, Menzie CM, Reichel WL (1966) DDT residues in Adélie penguins and a crabeater seal from Antarctica. Nature 210:670–673

Staley JT, Gosink JJ (1999) Poles apart: biodiversity and biogeography of sea ice bacteria. Annu Rev Microbiol 53:189–215

Stammerjohn SE, Smith RC (1996) Spatial and temporal variability of western Antarctic Peninsula sea ice coverage. In: Ross RM, Hofmann EE, Quetin LB (eds) Foundations for ecological research west of the Antarctic Peninsula. American Geophysical Union, Washington, DC, vol 70, pp 81–104

Stark PS (1994) Climatic warming in the central Antarctic Peninsula area. Weather 49:215–220

Stark JS (2000) The distribution and abundance of soft-sediment macrobenthos around Casey Station, East Antarctica. Polar Biol 23:840–850

Stark JS, Riddle MJ (2003) Human impacts in marine benthic communities at Casey Station: description, determination and demonstration of impacts. In: Huiskes AHl, Gieskes WWC, Rozema J, Schorno RML, van der Vies SM, Wolff WJ (2003) Antarctic biology in a global context. Backhuys, Leiden, pp 278–284

Stark JS, Snape I, Riddle MJ (2003) The effects of petroleum hydrocarbon and heavy metal contamination of marine sediments on recruitment on Antarctic soft-bottom assemblages: a field experimental investigation. J Exp Mar Biol Ecol 283:21–50

Starmans A, Gutt J, Arntz WE (1999) Mega-epibenthic communities in Arctic and Antarctic shelf areas. Mar Biol 135:269–280

Stearns CR, Weidner GA (1993) Sensible and latent heat flux estimates in Antarctica. In: Bromwich DH, Stearns CR (eds) Antarctic meteorology and climatology: studies based on automatic weather stations. American Geophysical Union, Washington, DC, vol 61, pp 109–138

Stearns CR, Keller ML, Weidner GA, Sievers M (1993) Monthly mean climatic data for Antarctic automatic weather stations. In: Bromwich DH, Stearns CR (eds) Antarctic meteorology and climatology: studies based on automatic weather stations. American Geophysical Union, Washington, DC, vol 61, pp 1–21

Stenberg M, Eriksson C, Heintzenberg J (1998) Trace substances in snow and firn from the vicinity of two small research stations in Antarctica. Ambio 27:451–455

Stenni B, Caprioli R, Cimino L, Cremisini C, Flora O, Gragnani R, Longinelli A, Maggi V, Torcini S (1999) 200 years of isotope and chemical records in a firn core from Hercules Névé, northern Victoria Land, Antarctica. Ann Glaciol 29:106–112

Stephens BB, Keeling RF (2000) The influence of Antarctic Sea ice on glacial-interglacial CO_2 variations. Nature 404:171–174

Stewart FM, Phillips RA, Catry P, Furness RW (1997) Influence of species, age and diet on mercury concentrations in Shetland seabirds. Mar Ecol Prog Ser 151:237–244

Stockton WL (1984) The biology and ecology of the epifaunal scallop *Adamussium colbecki* on the west side of the McMurdo Sound. Antarct Mar Biol 78:171–178

Stoeppler M, Brandt K (1979) Comparative studies on trace metal levels in marine biota. II. Trace metal in krill, krill products and fish from the Antarctic Scotia Sea. Z Lebensm Unters Forsch 169:95–98

Stoll MHC, de Baar HJW, Hoppema M, Fahrbach E (1999) New early winter fCO_2 data reveal continuous uptake of CO_2 by the Weddell Sea. Tellus 51:679–687

Stonehouse B (1989) Polar ecology. Blackie, Glasgow

Storey BC (1995) The role of mantle plumes in continental breakup: case studies from Gondwanaland. Nature 377:301–308

Strand A, Hov Ø (1996) A model strategy for the simulation of chlorinated hydrocarbon distributions in the global environment. Water Air Soil Pollut 86:283–316

Stumm W, Morgan JJ (1996) Aquatic chemistry, 3rd edn. Wiley, New York

Stump E (1995) The Ross Orogen of the Transantarctic Mountains. Cambridge University Press, Cambridge

Sturges WT (ed) (1991) Pollution of the Arctic atmosphere. Elsevier Science, London

Sturges WT, McIntyre HP, Penkett SA, Chappellaz J, Barnola J-M, Mulvaney R, Atlas E, Stroud V (2001) Methyl bromide, other brominated methanes, and methyl iodide in polar firn air. J Geophys Res 106:1595–1606

Subramanian AN, Tanabe S, Hidaka H, Tatsukawa R (1983) DDTs and PCB isomers and congeners in Antarctic fish. Arch Environ Contam Toxicol 12:621–624

Subramanian AN, Tanabe S, Hidaka H, Tatsukawa R (1986) Bioaccumulation of organochlorines (PCBs and *p,p'*-DDE) in Antarctic Adélie penguins *Pygoscelis adeliae* collected during a breeding season. Environ Pollut 40:173–189

Subramanian AN, Tanabe S, Tatsukawa R (1988) Estimating some biological parameters of Baird's beaked whales using PCBs and DDE as tracers. Mar Pollut Bull 19:284–287

Sugden DE (1982) Arctic and Antarctic: a modern geographical synthesis. Blackwell, Oxford

Sugden DE (1996) The east Antarctic Ice Sheet: unstable ice or unstable ideas? Trans Inst Br Geogr 21:443–456

Sugden DE, Clapperton CM (1977) The maximum ice extent on island groups in the Scotia Sea, Antarctica. Quat Res 7:268–282

Sugden DE, John BS (1976) Glaciers and landscape. Arnold, London

Sullivan CW, Ainley DG (1987) Antarctic marine ecosystem research at the ice-edge zone, 1986. Antarct J US 22:167–169

Sullivan CW, McClain CR, Comiso JC, Smith WO (1988) Phytoplankton standing crops within and Antarctic ice edge assessed by satellite remote sensing. J Geophys Res 93:12487–12498

Sullivan CW, Cota GF, Krempin DW, Smith WO (1990) Distribution and activity of bacterioplankton in the marginal ice zone of the Weddell-Scotia Sea during austral spring. Mar Ecol Prog Ser 63:239–252

Sun D-Z, Trenberth KE (1998) Coordinated heat removal from the tropical Pacific during the 1986–1987 El Niño. Geophys Res Lett 25:2659–2662

Sunda WG (1994) Trace metal/phytoplankton interactions in the sea. In: Bidoglio G, Stumm W (eds) Chemistry of aquatic systems: local and global perspectives. Kluwer, Boston, pp 213–247

Sunda WG, Huntsman S (1995) Cobalt and zinc interreplacement in marine phytoplankton: biological and geochemical implications. Limnol Oceanogr 40:1401–1417

Suttie ED, Wolff EW (1993) The local deposition of heavy metal emissions from point sources in Antarctica. Atmos Environ 27A:1833–1841

Svensson A, Biscaye PE, Grousset FE (2000) Characterisation of the late glacial continental dust in the Greenland Ice Core project. J Geophys Res 105:4637–4656

Swanson AL, Blake NJ, Dibb JE, Albert MRR, Blake DR, Rowland FS (2002) Photochemically induced production of CH_3Br, CH_3I, C_2H_5I, ethene, and propene within surface snow at Summit, Greenland. Atmos Environ 36:2671–2682

Sweeney C, Hansell DA, Carlson CA, Codispoti LA, Gordon LI, Marra J, Millero FJ, Smith WO, Takahashi T (2000) Biogeochemical regimes, net community production and carbon export in the Ross Sea, Antarctica. Deep-Sea Res I 47:3369–3394

Symonds RB, Rose WI, Bluth GJS, Gerlach TM (1994) Volcanic gas studies: methods, results and applications. In: Carrol R, Holloway JR (eds) Volatiles in magma. Rev Mineral 30:1–66

Szefer P, Czarnowski W, Pempkowiak J, Holm E (1993) Mercury and major essential elements in seals, penguins, and other representative fauna of the Antarctic. Arch Environ Contam Toxicol 25:422–427

Szefer P, Szefer K, Pempkowiak J, Skwarzec B, Bojanowski R, Holm E (1994) Distribution and coassociations of selected metals in seals of the Antarctic. Environ Pollut 83:341–349

Takeda S (1998) Influence of iron availability on nutrient consumption ratio of diatoms in oceanic waters. Nature 393:774–777

Talbot RW, Dibb JE, Scheuer EM, Blake DR, Blake NJ, Gregory GL, Sachse GW, Bradshaw JD, Sandholm ST, Singh HB (1999) Influence of biomass combustion emissions on the distribution of acidic trace gases over southern Pacific basin during austral springtime. J Geophys Res 104:5623–5634

Tamburrini M, Romano M, Giardina B, di Prisco G (1999) The myoglobin of emperor penguin (*Aptenodytes forsteri*): amino acid sequence and functional adaptation to extreme conditions. Comp Biochem Physiol 122:235–240

Tanabe S, Tatsukawa R (1986) Distribution, behaviour and load of PCBs in the Oceans. In: Waid JS (ed) PCBs and the environment. CRC Press, Boca Raton, pp 143–162

Tanabe S, Tatsukawa R, Kawano M, Hidaka H (1982) Global distribution and atmospheric transport of chlorinated hydrocarbons: HCH (BHC) isomers and DDT compounds in the western Pacific, eastern Indian and Antarctic Oceans. J Oceanogr Soc Jpn 38:137–148

Tanabe S, Hidaka H, Tatsukawa R (1983a) PCBs and chlorinated biphenyls in Antarctic atmosphere and hydrosphere. Chemosphere 12:277–288

Tanabe S, Mori T, Tatsukawa R (1983b) Global pollution of marine mammals by PCBs, DDTs and HCHs (BHC). Chemosphere 12:1269–1275

Tanabe S, Subramanian AN, Hidaka H, Tatsukawa R (1986) Transfer rates and pattern of PCB isomers and congeners and p,p'-DDE from mother to egg in Adélie penguin (*Pygoscelis adeliae*). Chemosphere 15:343–351

Tanabe S, Loganathan BG, Subramanian AN, Tatsukawa R (1987) Organochlorine residues in short-fined pilot whale: possible use as tracers of biological parameters. Mar Pollut Bull 18:561–563

Tedrow JCF, Ugolini FC (1966) Antarctic soils. In: Tedrow JCF (ed) Antarctic soils and soil-forming processes. Antarct Res Ser 8:161–177

Tegen I, Lacis AA, Fung I (1996) The influence on climate forcing of mineral aerosols from disturbed soils. Nature 380:419–422

Temme C, Einax JW, Ebinghaus R, Schroeder WH (2003) Measurements of atmospheric mercury species at a coastal site in the Antarctic and over the south Atlantic Ocean during polar summer. Environ Sci Technol 37:22–31

Testa JW, Oehlert G, Ainley DG, Bengston JL, Siniff DB, Laws RM, Rounsevell D (1991) Temporal variability in Antarctic marine ecosystems: periodic fluctuations in the phocid seals. Can J Zool 48:631–639

Thiel E, Schmidt EA (1961) Spherules from the Antarctic ice cap. J Geophys Res 66:307–310

Thomas WH, Duval B (1995) Sierra Nevada, California, USA, snow algae: snow albedo changes, algal-bacterial interrelationships, and ultraviolet radiation effects. Arctic Alpine Res 27:389–399

Thompson LG (1996) Climatic changes for the last 2000 years inferred from ice-core evidence in tropical ice cores. In: Jones PD, Bradley RS, Jouzel J (eds) Climatic variations and forcing mechanisms for the last 2000 years. Springer, Berlin Heidelberg New York, pp 281–295

Thompson LG (2000) Ice core evidence for climate change in the Tropics: implications for our future. Quat Sci Rev 19:19–35

Thompson DR, Furness RW (1989) Comparison of the levels of total and organic mercury in seabird feathers. Mar Pollut Bull 20:577–579

Thompson LG, Mosley-Thompson E, Bolzan JF, Koci BR (1985) A 1500-year record of tropical precipitation in ice cores from the Quelcaya Ice Cap, Perù. Science 229:971–973

Thompson DR, Stewart FM, Furness RW (1990) Using seabirds to monitor mercury in marine environments: the validity of conversion ratio for tissue comparisons. Mar Pollut Bull 21:339–342

Thompson DR, Furness RW, Walsh PM (1992) Historical changes in mercury concentrations in the marine environment of the north and north-east Atlantic Ocean as indicated by seabird feathers. J Appl Ecol 29:79–84

Thompson DR, Furness RW, Lewis SA (1993) Temporal and spatial variation in mercury concentrations in some albatrosses and petrels from the sub-Antarctic. Polar Biol 13:239–244

Thompson AM, Pickering KE, McNamara DP, Schoeberl MR, Hudson RD, Kim JH, Browell EV, Kirchhoff VWJH, Nganga D (1996) Where did tropospheric ozone over southern Africa and the tropical Atlantic come from in October 1992? Insights from TOMS, GTE TRACE A and SAFARI 1992. J Geophys Res 101:24251–24278

Thompson BW, Riddle MJ, Stark JS (2003) Cost-efficient methods for marine pollution monitoring at Casey Station, east Antarctica: the choice of sieve mesh-size and taxonomic resolution. Mar Pollut Bull 46:232–243

Thomson MRA, Crame JA, Thomson JW (eds) (1991) Geological evolution of Antarctica. Cambridge University Press, Cambridge

Timmermans KR, van Leeuwe MA, de Jong JTM, McKay RM, Nolting RF, Witte HJ, van Ooyen J, Swagerman MJW, Kloosterhuis H, de Baar HJW (1998) Iron stress in the Pacific region of the Southern Ocean: evidence from enrichment bioassays. Mar Ecol Prog Ser 166:27–41

Tingey RJ (1991) The geology of Antarctica. Clarendon Press, Oxford

Tittlemier SA, Simon M, Jarman WM, Elliott JE, Norstrom RJ (1999) Identification of a novel $C_{10}H_6N_2Br_4Cl_2$ heterocyclic compound in seabird eggs. A bioaccumulating marine natural product? Environ Sci Technol 33:26–33

Titus JG, Narayanan V (1996) The risk of sea level rise. Climatic Change 33:151–212

Trathan PN, Everson I, Miller DGM, Watkins JL, Murphy EJ (1995) Krill biomass in the Atlantic. Nature 373:201–202

Tréguer PJ, Jacques G (1986) L'océan antarctique. Recherche 178:746–755

Trenberth KE (ed) (1992) Climate system modeling. Cambridge University Press, Cambridge

Trenberth KE, Hoar TJ (1996) The 1990–1995 El Niño-Southern Oscillation event: longest on record. Geophys Res Lett 23:57–60

Trenberth KE, Caron JM, Stepaniak DP (2001) The atmospheric energy budget and implications for surface fluxes and ocean heat transport. Climate Dynam 17:259–276

Triulzi C, Mangia A, Casoli A, Albertazzi S, Nonnis-Marzano A (1990) Artificial and natural radionuclides, alkaline and earth-alkaline elements in some environmental abiotic samples of Antarctica. Ann Chim 79:723–733

Trivelpiece WZ, Aimley DG, Fraser WR, Trivelpiece SG (1990) Reply to letter of Eppley and Rubega. Nature 245:211

Truswell EM (1991) Antarctica: a history of terrestrial vegetation. In: Tingey RJ (ed) The geology of Antarctica. Clarendon Press, Oxford

Tubertini O, Bettoli G, Cantelli L, Tositti L, Valcher S, Triulzi C, Nonnis Marzano F, Mori A, Vaghi M, Sbrignadello G, Degetto S, Faggin M (1995) Italian Antarctic Research Program: environmental radioactivity survey around the Italian base (1987–1991) Terra Nova Bay-Ross Sea Region. J Environ Radioactivity 28:35–41

Tuncel G, Aras NK, Zoller WH (1989) Temporal variations and sources of elements in the South Pole atmosphere. 1. Nonenriched and moderately enriched elements. J Geophys Res 94:13025–13038

Twiss RJ, Moores EM (1992) Structural geology. Freeman, New York

Ugolini FC (1963) Soil investigations in the lower Wright Valley. Natl Acad Sci Natl Res Council 1287:55–61

Ugolini FC (1970) Antarctic soils and their ecology. In: Holdgate MW (ed) Antarctic ecology, vol 2. Academic Press, London, pp 673–692

Ugolini FC, Anderson DM (1973) Ionic migration and weathering in frozen Antarctic soils. Soil Sci 115:461–470

UNEP (1996) Question of Antarctica. State of the environment in Antarctica. United Nations Environment Programme, Chemicals, Geneva

UNEP (2002a) Regionally based assessment of persistent toxic substances – Antarctica regional report. United Nations Environment Programme, Chemicals, Geneva

UNEP (2002b) Global mercury assessment. United Nations Environment Programme, Chemicals, Geneva

Vacchi M, La Mesa M, Castelli A (1994) Diet of two coastal nototheniid fish from Terra Nova Bay, Ross Sea. Antarct Sci 6:61–65

Vacchi M, La Mesa M, Greco S (2000a) The coastal fish fauna of Terra Nova Bay. In: Faranda F, Guglielmo L, Ianora A (eds) Ross Sea ecology. Springer, Berlin Heidelberg New York, pp 457–468

Vacchi M, Cattaneo-Vietti R, Chiantore M, Dalu M (2000b) Predator-prey relationship between the nototheniid fish *Trematomus bernacchii* and the Antarctic scallop *Adamussium colbecki*. Antarct Sci 12:64–68

Vali G (1992) Memory effect in the nucleation of ice on mercuric iodide. In: Wagner E (ed) Nucleation and atmospheric aerosol. Deepak, Hampton, pp 259–262

Vallack HW, Bakker DJ, Brandt I, Broström-Lundén E, Brouwer A, Bull KR, Gough C, Guardans R, Holoubek I, Jansson B, Koch R, Kuylenstierna J, Lecloux A, Mackay D,

McCutcheon P, Mocarelli P, Taalman RDF (1998) Controlling persistent organic pollutants – what next? Environ Toxicol Pharmacol 6:143–175

Vallee BL, Auld DS (1990) Zinc coordination, function and structure of zinc enzymes and other proteins. Biochemistry 29:5647–5659

Vallelonga P, Candelone J-P, Van der Velde K, Curran MAJ, Morgan VI, Rosman KJR (2003) Lead, Ba and Bi in Antarctic Law Dome ice corresponding to the 1815 AD Tambora eruption: an assessment of emission sources using Pb isotopes. Earth Planet Sci Lett 211:329–341

Vandal GM, Fitzgerald WF, Boutron CF, Candelone J-P (1993) Variations in mercury deposition to Antarctica over the past 34,000 years. Nature 362:621–623

Vandal GM, Mason RP, McKnight D, Fitzgerald W (1998) Mercury speciation and distribution in a polar desert lake (Lake Hoare, Antarctica) and two glacial meltwater streams. Sci Total Environ 213:229–237

Van de Velde K, Boutron C, Ferrari C, Bellomi T, Barbante C, Rudnev S, Bolshov M (1998) Seasonal variations of heavy metals in the 1960s Alpine ice: sources versus meteorological factors. Earth Planet Sci Lett 164:521–533

Van den Brink NW (1997) Direct transport of volatile organochlorine pollutants to polar regions: the effect on the contamination pattern of Antarctic seabirds. Sci Total Environ 198:43–50

Van den Brink NW, van Franeker JA, De Ruiter-Dukman EM (1998) Fluctuating concentrations of organochlorine pollutants during a breeding season in two Antarctic seabirds: Adélie penguin and southern fulmar. Environ Toxicol Chem 17:702–709

Van Dorland R, Dentener FJ, Lelieveld J (1997) Radiative forcing due to tropospheric ozone and sulphate aerosols. J Geophys Res 102:28079–28100

Van Zanten BO (1983) Possibilities of long-distance dispersal in bryophytes with special reference to the Southern Hemisphere. Naturwissen V Hamburg 7:49–64

Vaughan DG, Doake CSM (1996) Recent atmospheric warming and retreat of ice shelves on the Antarctic Peninsula. Nature 379:328–331

Vaughan DG, Bamber JL, Giovinetto M, Russell J, Cooper APR (1999) Reassessment of net surface mass balance in Antarctica. J Climate 12:933–946

Venkatesan MI (1988) Organic geochemistry of marine sediments in Antarctic region: marine lipids in McMurdo Sound. Org Geochem 12:13–27

Venkatesan MI, Kaplan IR (1987) The lipid geochemistry of Antarctic marine sediments: Bransfield Strait. Mar Chem 21:347–375

Venkatesan MI, Kennicutt MC II (1996) Pollutants in Antarctica: hydrocarbons, metals and synthetic chemicals. In: Kennicutt MC II, Sayers JCA, Walton D, Wratt G (eds) Monitoring of environmental impacts from science and operations in Antarctica. SCAR-COMNAP, Oslo, pp 15–16

Venkatesan MI, Mirsadeghi FH (1992) Coprostanol as sewage tracer in McMurdo Sound, Antarctica. Mar Pollut Bull 25:328–333

Venkatesan MI, Ruth E, Kaplan IR (1986) Coprostanols in Antarctic marine sediments: a biomarker for marine mammals and not human pollution. Mar Pollut Bull 17:554–557

Vernadsky VI (1945) The biosphere and the noosphere. Am Sci 33:1–12

Vetter W, Krock B, Luckas B (1997) Congener specific determination of compounds of technical toxaphene (CTTs) in different Antarctic seal species. Chromatographia 44:65–73

Vetter W, Alder L, Kallenborn R, Schlabach M (2000) Determination of Q1, an unknown organochlorine contaminant, in human milk, Antarctic air, and further environmental samples. Environ Pollut 110:401–409

Veysseyre A, Moutard K, Ferrari C, van de Velde K, Barbante C, Cozzi G, Capodaglio G, Boutron C (2001) Heavy metals in fresh snow collected at different altitudes in the

Chamonix and Maurienne valleys, French Alps: initial results. Atmos Environ 35:415–425

Viarengo A, Canesi L, Pertica M, Poli G, Moore MN, Orunesu M (1990) Heavy metal effects on lipid peroxidation in the tissues of *Mytilus galloprovincialis* Lam. Comp Biochem Physiol 97C:37–42

Viarengo A, Canesi L, Mazzuccotelli A, Ponzano E (1993) Cu, Zn and Cd content in different tissues of the Antarctic scallop *Adamusssium colbecki*: role of metallothionein in heavy metal homeostasis and detoxication. Mar Ecol Prog Ser 95:163–168

Villalba R, Cook ER, D'Arrigo RD, Jacoby GC, Jones PD, Salinger MJ, Palmer J (1997) Sea-level pressure variability around Antarctica since AD 1750 inferred from sub-Antarctic tree-ring records. Climate Dynam 13:375–390

Vincent WF (1988) Microbial ecosystems of Antarctica. Cambridge University Press, Cambridge

Vincent WF (1997) Polar desert ecosystems in a changing climate: a north-south perspective. In: Lyons WB, Howard-Williams C, Hawes I (eds) Ecosystems processes in Antarctic ice-free landscapes. Balkema, Rotterdam, pp 3–14

Vincent WF, Ellis-Evans JC (eds) (1989) High latitude limnology. Hydrobiologia 172:5–9

Vincent WF, Howard-Williams C (1986) Antarctic stream ecosystems: physiological ecology of a blue-green algal epilithon. Freshwater Biol 16:219–233

Vincent WF, Roy S (1993) Solar ultraviolet-B radiation and aquatic primary production: damage, protection and recovery. Environ Rev 1:1–12

Vincent WF, Vincent CL (1982) Factors controlling phytoplankton production in Lake Vanda (77° S). Can J Fish Aquat Sci 39:1602–1609

Vincent WF, Howard-Williams C, Broady PA (1993) Microbial communities and processes in Antarctic flowing waters. In: Friedmann EI (ed) Antarctic microbiology. Wiley-Liss, New York, pp 543–569

Vink S, Measures CI (2001) The role of dust deposition in determining surface water distributions of Al and Fe in the South West Atlantic. Deep-Sea Res II 48:2787–2809

Vishniac HS (1985) *Cryptococcus friedmannii*, a new species of yeast from the Antarctic. Mycologia 77:149–153

Vodopivez C, Curtosi A (1998) Trace metals in some invertebrates, fishes and birds from Potter Cove. Ber Polarforsch 299:296–303

Volkening J, Baumann H, Heumann KG (1988) Atmospheric distribution of particulate lead over the Atlantic Ocean from Europe to Antarctica. Atmos Environ 22:1169–1174

Voloshchuk VM, Sedunov YS (eds) (1973) Hydrodynamics and thermodynamics of aerosols. Wiley, New York

Vorobyova EA, Soina VS, Mulukin AL (1996) Microorganisms and enzyme activity in permafrost after removal of long-term cold stress. Adv Space Res 18:103–108

Voytek MA (1990) Addressing the biological effects of decreased ozone on the Antarctic environment. Ambio 19:52–61

Waddington ED, Cunningham J, Harder SL (1996) The effects of snow ventilation on chemical concentrations. In: Wolff EW, Bales RC (eds) Processes of chemical exchange between the atmosphere and polar snow. Springer, Berlin Heidelberg New York, pp 403–451

Wadhams P (1991) Atmosphere-ice-ocean interactions in the Antarctic. In: Harris CM, Stonehouse B (eds) Antarctica and global climatic changes. Belhaven Press, London, pp 65–81

Wadhams P, Davis NR (1997) Climate-related research in the UK and Antarctic sea ice. UK Global Environment Research Office, vol 36, pp 11–13

Wagemann R, Trebacz E, Boila G, Lockhart WL (1998) Methymercury and total mercury in tissues of Arctic marine mammals. Sci Total Environ 218:19–31

Wagenbach D (1989) The environmental records in alpine glaciers. In: Oeschger H, Langway CC (eds) The environmental record in glaciers and ice sheets. Wiley, Chichester, pp 69–83

Wagenbach D (1996) Coastal Antarctica: atmospheric chemical composition and atmospheric transport. In: Wolff EW, Bales RC (eds) Chemical exchange between the atmosphere and polar snow. NATO ASI series, Global Environmental Change, vol 43. Springer, Berlin Heidelberg New York, pp 173–199

Wagenbach D, Ducroz F, Mulvaney R, Keck L, Minikin A, Legrand M, Hall JS, Wolff EW (1998) Sea-salt aerosol in coastal Antarctic regions. J Geophys Res D 103:10961–10974

Wagner G, Beer J, Laj C, Kissel C, Masarik J, Muscheler R, Synal HA (2000) Chlorine-36 evidence for the Mono Lake event in the Summit Grip ice core. Earth Planet Sci Lett 181:1–6

Walton DWH (1990) Colonization of terrestrial habitats: organisms, opportunities and occurrence. In: Kerry KR, Hempel G (eds) Antarctic ecosystems. Ecological change and conservation. Springer, Berlin Heidelberg New York, pp 51–60

Walton DWH, Vincent WF, Timperley MH, Hawes I, Howard-Williams C (1997) Synthesis: polar deserts as indicators of change. In: Lyons WB, Howard-Williams C, Hawes I (eds) Ecosystem processes in Antarctic ice-free landscapes. Balkema, Rotterdam, pp 275–279

Wang XL, Ropelewski CF (1995) An assessment of ENSO-scale secular variability. J Climate 8:1584–1599

Wang W-X, Dei RCH, Xu Y (2001) Cadmium uptake and trophic transfer in coastal plankton under contrasting nitrogen regimes. Mar Ecol Prog Ser 211:293–298

Wangberg I, Munthe J, Pirrone N, Iverfeldt Å, Bahlman E, Costa P, Ebinghaus R, Geng X, Ferrara R, Gårdfeldt K, Kock H, Lanzillotta E, Mamane Y, Mas F, Melamed F, Osnat Y, Prestbo E, Sommar J, Spain G, Sprovieri F, Tuncel G (2001) Atmospheric mercury distribution in northern Europe and the Mediterranean region. Atmos Environ 35:3019–3025

Wania F, Mackay D (1993) Global fractionation and cold condensation of volatile organochlorine compounds in polar regions. Ambio 22:10–18

Wania F, Mackay D (1999) Global chemical fate of a-hexa-chloro-cyclohexane. 2. Use of a global distribution model for mass balancing, source apportionment, and trend prediction. Environ Toxicol Chem 18:1400–1407

Wanwimolruk S, Zhang H, Covillek PF, Savillek J, Davisk LS (1999) In vitro hepatic metabolism of CYP3A-mediate drug quinine, in Adélie penguins. Comp Biochem Physiol 124:301–307

Warneck P (1988) Chemistry of the natural atmosphere. Academic Press, London

Warner RC, Budd WF (1998) Modelling the long-term response of the Antarctic ice sheet to global warming. Ann Glaciol 27:161–168

Warnke DA, Marzo B, Hodell DA (1996) Major deglaciation of East Antarctica during the early Late Pleistocene? Not likely from a marine perspective. Mar Micropaleontol 27:237–251

Warren SG (1982) Optical properties of snow. Rev Geophys Space Phys 20:67–89

Warren CR (1993) Rapid recent fluctuations of the calving San Rafael Glacier, Chilean Patagonia: climatic or non-climatic? Geogr Ann 75:11–125

Warren SG, Clarke AD (1990) Soot in the atmosphere and snow surface of Antarctica. J Geophys Res 95:1811–1816

Warrick R, Oerlemans J (1990) Sea level rise. In: Houghton JT, Jenkins GJ, Ephraums JJ (eds) Climate change. The IPCC Scientific Assessment. Cambridge University Press, Cambridge, pp 257–281

Watanabe I, Kunito T, Tanabe S, Amano M, Koyama Y, Miyazaki N, Petrov EA, Tatsukawa R (2002) Accumulation of heavy metals in Caspian seals (*Phoca caspica*). Arch Environ Contam Toxicol 43:109–120

Watanuki Y, Kato A, Naito Y, Robertson G, Robinson S (1997) Diving and foraging behaviour of Adélie penguins in areas with and without fast sea-ice. Polar Biol 17:296–304

Waterhouse EJ (1997) Implementing the protocol on ice-free land: the New Zealand experience at Lake Vanda. In: Lyons W, Howard-Williams C, Hawes I (eds) Ecosystems and processes in Antarctic ice-free landscapes. Balkema, Rotterdam, pp 265–274

Waterhouse EJ, Hemmings A, Harris C, Fitzsimmon S, Knox G, Clarkson T (2001) Ross Sea Region. A state of the environment report for the Ross Sea Region, Antarctica. New Zealand Antarctic Institute, Christchurch

Watson R (1999) Common themes for ecologists in global issues. J Appl Ecol 36:1–10

Watson AJ, Bakker DCE, Ridgwell AJ, Boyd PW, Law CS (2000) Effect of iron supply on Southern Ocean CO_2 uptake and implications for glacial atmospheric CO_2. Nature 407:730–733

Watterson IG, Dix MR, Colman RA (1999) A comparison of present and doubled CO_2 climates and feedbacks simulated by three general models. J Geophys Res 104:1943–1956

Watzin MC, Roscigno PR (1997) The effects of zinc contamination on recruitment and early survival of benthic invertebrates in an estuary. Mar Pollut Bull 34:443–455

Webb PN, Harwood DM, McKelvey BC, Mercer JN, Stott LD (1984) Cenozoic marine sedimentation and ice-volume variation on the East Antarctic craton. Geology 12:287–291

Weber RR, Bicego MC (1990) Petroleum aromatic hydrocarbons in surface waters around Elephant Island, Antarctic Peninsula. Mar Pollut Bull 21:448–449

Weber K, Goerke H (1996) Organochlorine compounds in fish off the Antarctic Peninsula. Chemosphere 33:377–392

Weber RR, Montone RC (1990) Distribution of organochlorines in the atmosphere of the South Atlantic and Antarctic Oceans. In: Kurtz DA (ed) Long-range transport of pesticides. Lewis, Chelsea, MI, pp 185–197

Webster J, Hawes I, Downes MT, Timperley M, Howard-Williams C (1996) Evidence for regional climate change in the recent evolution of a high latitude, pro-glacial lake. Antarct Sci 8:49–59

Webster J, Webster K, Nelson P, Waterhouse E (2003) The behaviour of residual contaminants at a former station site, Antarctica. Environ Pollut 123:163–179

Weiss HV, Herron MM, Langway CC (1978) Natural enrichment of element in snow. Nature 274:352–353

Weiss D, Shotyk W, Kempf O (1999) Archives of atmospheric lead pollution. Naturwissenschaften 86:262–275

Wendler G, Ishikawa, Kodama Y (1988) On the heat budget of an icy slope of Adélie Land, eastern Antarctica. J Appl Meteorol 27:52–65

Westerlund S, Öhman P (1991a) Iron in the water column of the Weddell Sea. Mar Chem 35:199–217

Westerlund S, Öhman P (1991b) Cadmium, copper, cobalt, nickel, lead, and zinc in the water column of the Weddell Sea, Antarctica. Geochim Cosmochim Acta 55:2127–2146

Wharton RA, McKay CP, Clow GD, Andersen DT (1993) Perennial ice covers and their influence on Antarctic lake ecosystems. In: Green WJ, Friedmann EI (eds) Physical and biogeochemical processes in Antarctic lakes. American Geophysical Union, Washington, DC, vol 59, pp 53–70

White MG (1984) Marine benthos. In: Laws RM (ed) Antarctic ecology, vol 2. Academic Press, London, pp 421–461

Whitworth T, Orsi AH, Kim SJ, Nowlin WD (1998) Water masses and mixing near the Antarctic slope front. In: Jacobs SS, Weiss RF (eds) Ocean, ice and atmosphere interaction at the continental margin. American Geophysical Union, Washington, DC, vol 75, pp 1–27

WHO (1992) Cadmium, environmental health criteria. World Health Organization, Geneva

Wienecke BC, Lawless R, Rodary D, Bost C-A, Thomson R, Pauly T, Robertson G, Kerry KR, LeMaho Y (2000) Adélie penguin foraging behaviour and krill abundance along the Wilkes and Adélie Land coasts, Antarctica. Deep-Sea Res II 47:2573–2587

Wiersma GB, Slaughter C, Hilgert J, McKee A, Halpern C (1986) Reconnaissance of Noatak National Preserve and biosphere reserve as a potential site for inclusion in the integrated global background monitoring network. Natl Tech Inform Serv, US Department of Commerce, Springfield

Willan RCR, Macdonald DIM, Drewry DJ (1990) The mineral resource potential of Antarctica: geological realities. In: Cook G (ed) The future of Antarctica. Exploitation versus preservation. Manchester University Press, Manchester, pp 25–43

William RS, Hall DK (1993) Glaciers. In: Gurney RJ, Foster JL, Parkinson CL (eds) Atlas of satellite observations related to global change. Cambridge University Press, Cambridge, pp 401–422

Williams AJ (1984) The status and conservation of seabirds on some islands in the African sector of the Southern Ocean. In: Croxall JP, Evans PGH, Schreiber RW (eds) Status and conservation of the world's seabirds. International Council for Bird Preservation, Cambridge, Tech Publ no 2, pp 627–635

Williams PM, Michel R, Weiss H (1974) Tritium and mercury results from Eltanin cruise 51. Antarct J US 9:221–222

Wilson AT (1964) Evidence for chemical diffusion of a climate change in the McMurdo Dry Valleys 1200 years ago. Nature 201:176–177

Wilson PR, Ainley DG, Nur N, Jacobs SS, Barton KJ, Ballard G, Comiso JC (2001) Adélie penguin population change in the pacific sector of Antarctica: relation to sea-ice extent and the Antarctic Circumpolar Current. Mar Ecol Prog Ser 213:301–309

Wingham DJ, Ridout AJ, Scharroo R, Arthern RJ, Schun CK (1998) Antarctic elevation change from 1992 to 1996. Science 282:456–458

Winston GW (1991) Oxidants and antioxidants in aquatic animals. Comp Biochem Physiol 100C:173–176

Wirth WW, Gressitt JL (1967) Diptera: Chironomidae (midges). In: Gressit JL (ed) Entomology of Antarctica. Smithsonian Institution, Washington, DC, vol 10, pp 197–203

WMO (1999) Scientific assessment of ozone depletion: 1998. WMO, Geneva, Global Ozone Research and Monitoring Project Rep no 44

Wolff EW, Bales RC (eds) (1996) Processes of chemical exchange between the atmosphere and polar snow. Springer, Berlin Heidelberg New York

Wolff EW, Cachier H (1998) Concentrations and seasonal cycle of black carbon in aerosol at a coastal Antarctic station. J Geophys Res D 103:11033–11041

Wolff EW, Suttie ED (1994) Antarctic snow record of southern hemisphere lead pollution. Geophys Res Lett 21:781–784

Wolff EW, Legrand MR, Wagenbach D (1998) Coastal Antarctic aerosol and snowfall chemistry. J Geophys Res D 103:10927–10934

Wolff EW, Suttie ED, Peel DA (1999) Antarctic snow record of cadmium, copper, and zinc content during the twentieth century. Atmos Environ 33:1535–1541

Wong APS, Bindoff NL, Church JA (1999) Large-scale freshening of intermediate waters in the Pacific and Indian Oceans. Nature 400:440–443

Wood WF, Marsh KV, Buddmeier RW, Smith C (1990) Marine biota as detection agents for low-level radionuclide contamination in Antarctica and the Southern Hemisphere Oceans. In: Kerry KR, Hempel G (eds) Antarctic ecosystems. Ecological change and conservation. Springer, Berlin Heidelberg New York, pp 372–378

Woodruff SD, Slutz RJ, Jenne RJ, Steurer PM (1987) A comprehensive ocean-atmosphere dataset. Bull Am Meteorol Soc 68:1239–1250

Worland MR (1996) The relationship between body water content and cold tolerance in the Arctic collembolan *Onychiurus arcticus* (Collembola: Onychiuridae). Eur J Entomol 93:341–348

Wu X, Budd WF, Lytle VI, Massom RA (1999) The effect of snow on Antarctic sea ice simulations in a coupled atmosphere sea ice model. Clim Dynam 15:127–143

Wynn-Williams DD (1993) Microbial processes and initial stabilization of fellfield soil. In: Miles J, Walton DWH (eds) Primary succession on land. Blackwell, Oxford, pp 17–32

Yamamoto Y, Honda K, Hidaka H, Tatsukawa R (1987) Tissue distribution of heavy metals in Weddell seals (*Leptonychotes weddellii*). Mar Pollut Bull 18:164–169

Yamamoto Y, Honda K, Endo Y, Tatsukawa R (1990) Sex- and maturity-related heavy metal accumulation in the Antarctic krill *Euphausia superba*. Proc NIPR Symp Polar Biol 3:57–63

Yamazaki K (1992) Moisture budget in the Antarctic atmosphere. Proc NIPR Symp Polar Meteorol Glaciol 6:36–45

Yeats PA, Campbell JA (1983) Nickel, copper, cadmium and zinc in the northwest Atlantic Ocean. Mar Chem 12:43–58

Yeats PA, Westerlund S, Flegal AR (1995) Cadmium, copper and nickel distributions at four stations in the eastern central and south Atlantic. Mar Chem 49:283–293

Yoshimura Y, Kohshima S, Ohtani S (1997) A community of snow algae on a Himalayan glacier: change of algal biomass and community structure with altitude. Arctic Alpine Res 29:126–137

Yuan X, Cane MA, Martinson DG (1996) Cycling around the South Pole. Nature 380:673–674

Yurukova L, Ganeva A (1999) Bioaccumulative and floristic characteristics of mosses near St. Kliment Ohridski Antarctic Base Station of Bulgaria. J Balkan Ecol 2:65–71

Zhang MJ, Li ZQ, Xiao CD, Qin DH, Yang HA, Kang JC, Li J (2002) A continuous 250-year record of volcanic activity from Princess Elizabeth Land, East Antarctica. Antarct Sci 14:55–60

Zillman JW (1967) The surface radiation balance in high southern latitudes. Polar meteorology. WMO, Geneva, WMO Tech Note no 87, pp 142–171

Zipan W, Norman FI (1993) Foods of the south polar skua *Catharacta maccormicki* in the eastern Lansermann Hills, Princess Elizabeth Land, East Antarctica. Polar Biol 13:255–262

Zolensky M (1998) The flux of meteorites to Antarctica. In: Grady MM, Hutchison R, McCall GJH, Rothery DA (eds) Meteorites: flux with time and impact effects. Geol Soc Lond Spec Publ 140:93–104

Zoller WH, Gladney ES, Duce RA (1974) Atmospheric concentrations and sources of trace metals at the South Pole. Science 183:198–200

Zreda-Gostynska G, Kyle PR, Finnegan D, Prestbo KM (1997) Volcanic gas emissions from Mount Erebus and their impact on the Antarctic environment. J Geophys Res 102:15039–15055

Zwally HJ (1991) Breakup of Antarctic ice. Nature 350:274

Zwally HJ, Giovinetto MB (1995) Accumulation in Antarctica and Greenland derived from passive-microwave data: a comparison with contoured compilations. Ann Glaciol 21:123–130

Zwally HJ, Gloersen P (1977) Passive microwave images of the polar regions and research applications. Polar Rec 18:431–450

Geographical Index

Subject Index

Taxonomic Index

Ecological Studies

Volumes published since 2003

Printed in the United Kingdom
by Lightning Source UK Ltd.
134515UK00001B/391/A